Process Development Division 2013

Core Programming Topic at the 2013 AIChE Spring Meeting & 9th Global Congress on Process Safety

San Antonio, Texas, USA
28 April – 2 May 2013

ISBN: 978-1-62748-463-3

Printed from e-media with permission by:

Curran Associates, Inc.
57 Morehouse Lane
Red Hook, NY 12571

Some format issues inherent in the e-media version may also appear in this print version.

Printed by Curran Associates, Inc. (2013)

AIChE
3 Park Avenue
New York, NY 10016-5991

Phone: (203) 702-7660
Fax: (203) 775-5177

www.aiche.org

Additional copies of this publication are available from:

Curran Associates, Inc.
57 Morehouse Lane
Red Hook, NY 12571 USA
Phone: 845-758-0400
Fax: 845-758-2634
Email: curran@proceedings.com
Web: www.proceedings.com

TABLE OF CONTENTS

Pilot Plants: Yesterday, Today and Tomorrow

Richard Palluzi
ExxonMobil Research and Engineering Co
1545 Route 22 East
Annandale, NJ 08801
908-730-2323
richard.p.palluzi@exxonmobil.com

I appreciate that the Process Development Division felt that I had the experience and knowledge to talk to you about such a broad and interesting topic; I fear, however, that it was mostly because most of my contemporaries who were around 40 years ago have expired leaving me as the last one standing!

Yesterday

I will forgo the usual attempt to prove that the Egyptians, Greeks, or Romans had the first pilot plant and jump ahead to around 1970. A mere fifty years ago, pilot plants were a lot different from what we see today.

In 1970 pilot plants occupied numerous small buildings scattered around a company's operating organization or perhaps occupied their own research center. Then most pilot plant personnel were happy to have finally gotten their own sites and fondly (or not so fondly) reminisced about the days when the units were only found in corners of buildings, concrete pads near process units and the occasional ill constructed small shed. The units were larger than used today. Many still used ¼" and ½" piping as their basis although most organizations had already transitioned to ¼" and ½" tubing and compression fittings. This progression was ending the old arguments as to whether the process was well enough defined to used welded pipe and flanges (thereby minimizing leakage problems) or was still too uncertain not to use easily modified threaded pipe and fittings (at the cost of continual leaks). The availability of small valves with compression connections had led this change. Most pilot plants were built in open bays, behind barricades or in blast cells or occasionally on laboratory benches or floors. Hood units were small and usually short lived. Most of these were designed for low pressure and involved at least some batch operation. Larger scale outdoor units comprised of several decks were common although less so than in the decades before.

Controls were primarily manual except on larger units. When available these were grouped in large, dedicated instrument cabinets and central control rooms. Control displays were un-ergonomic, scattered and difficult to interpret unless familiar with the unit. The number of control loops was carefully controlled to the minimum due to the high cost of reliable controllers, $8,500 in today's dollars and the size of a large shoebox. A few forward looking companies were starting to experiment with a new type of controller, a software one, based on the new "minicomputers". The size of a large instrument cabinet, prone to sudden hardware and software failures and requiring a highly temperature and humidity controlled environment these expensive units held the promise of both data gathering and control in one package. Their ease of modification matched the control and data gathering needs of pilot plants, promising that we could reap the advantages process plants had already been experiencing in their much more sophisticated and expensive systems. They required specialty programing which was both tedious and difficult. Few pilot plant personnel had the required skills so most had to be done by specialists, totally unfamiliar with pilot plant operations, a learning experience for both sides and a preview of the increasing use of specialists overtaking the industry. Operator interfaces were teletypes and line printers which were starting to eat forests of trees. Data still needed to be transcribed to other places, control information was limited and often obscure but the promise was there to those forward thinking enough to look past the current limitations.

Pilot plants were generally continuously staffed and multiple operators per unit not uncommon. Data gathering involved the use of a good clipboard, numerous run sheets (photocopied now instead of mimeographed!) and a good eye for pressure gauges, thermometers and rotameters. Budget discussions centered on the need for large analog gauges for accuracy, the potential for cost cutting by using manual devices instead of automatic controls and endless arguments over the minimum size of the reactor. Operating procedures were hand written and often varied between shifts. Each operator brought their own bag of tricks to the unit; some helpful some less so.

Multiple runs were common and data scatter, errors, and anomalies prevalent. Was the operator recording the highest pressure during the hour, their guess as to the average or the instantaneous value at the time of reading? Were they always consistent in the way they made the reading? What in heaven's name value did that chicken scratch represent? Strip chart recorders abounded. Operator's drawers overflowed with ink bottles, cleaning brushes and fine wires to unclog pen tips. The results of the data were graphed, analyzed, pontificated upon, mulled over and then used to design process units. Surprises, requiring more research or investigation of new problems were, if not common, at least far from unknown. More data required more units, more operating staff and more research staff and, of course, more money and time.

Safety was always important. Safety reviews were mandatory but often highly organizationally specific. What you were concerned about was often not what I was concerned about. How you handled the concern was often very different from how I did. Organized risk assessment on a common basis was unknown. Safety systems were fairly reliable as almost all were hardwired and were forced to be independent from the control system systems due to impedance limitations of many analog systems.

Individual instruments were the only things that was available save for the occasional multi-point thermocouple selector switch. Readouts were analog and often in "percent of full scale" requiring one to memorize the calibrated range and convert mentally. Wiring was complex, complicated, ran to every separate device and expensive; modifications time consuming and tedious. Taped notes indicating the current ranges abounded on instrument panels.

Common diagnostic tools included analog volt meters, bubble solution, screwdrivers, wrenches and – occasionally – a hammer. Most common problems were fixed by the operating staff with little reference to other groups. Systematic troubleshooting was rare; one would start playing with something and see what happened. This resulted in varying explanations, solutions and occasional really incredible analysis which led to highly individualistic units and very personal preferences for future changes since many times the root cause escaped notice. A pilot plant engineer was expected to realize that Dr. Smith always wanted dual isolation valves and a snubber on every pressure gauge because of a problem in 1952 while Operator Jones demanded that only a specific type of globe valves every be used anywhere on his unit based on a 1960 experience. We were expected to commit these preferences to memory.

Analytical equipment was primarily off line making sample taking a continual challenge. Most sampling systems were totally manual and relied upon the operator's skill for accuracy and repeatability. Transient analysis based on this sampling always included anomalies due to timing errors, operator errors or the occasional relabeled sample based on someone's "memory". A few units had process analyzers installed but their expense made them the exception rather than the rule. Most did fairly simple analysis; few were believed without off line confirmation. The need for a certain sample size often dictated the size of the unit as often as the confidence in the minimum size of the reactor or the process needs.

The engineering staff could be identified by the multiple rulers in their offices (for extending nomographs well beyond their published ranges). They increasingly used the new "scientific" calculators instead of slide rules, a development greatly regretted by the older engineering staff. Design information was only rarely documented, never published and frequently contradictory. Instead it resided in the collective memory of the specific organization. Information gathering meant sitting down and listening to a raft of war stories and trying to abstract the key learnings with a few well-placed questions and multiple beverages. It often required trying to decide if one believed Jones or Smith more as each suggested a different approach for totally contradictory reasons.

Pilot plants were considered overly expensive to construct and much too time consuming to build and operate, but necessary.

I think you'll agree the environment I describe is quite different from today.

Today

Pilot plants today have shrunk in size. Except for the occasional demonstration unit for a new or very novel technology, most "large" pilot plants are the size of the older "small" units. Units which fit in laboratory hoods, once rare, are common - often more prevalent than their larger cousins. Large, exterior units are few and far between. Units designed to handle multiple runs simultaneously while sharing common systems are often the norm replacing the older versions which essentially replicated entire units. Research sites have shrunk, consolidated, and even closed as organizations find ways to get the results with less equipment, space and staff.

More importantly the rise of a new generation of computer modeling based on the ability to harness the power of computational fluid dynamics, 3-D modeling and highly predictive computing tools has effectively replaced the pilot plant as the primary process design tool. Instead the pilot plant has become the way to validate and improve the model that is used for the design. This has dramatically changed the scale of the units, greatly reduced the number of data runs required and greatly expanded the ability to extend research results with less grassroots facilities and equipment. Units are modified again and again as program needs change. Most are designed whenever possible to accommodate multiple programs extending their life. A 10 year old pilot plant in 1970 was a rare and novel unit; a ten year old pilot plant today excites little interest.

Pilot plants today are almost entirely automated using inexpensive stand-alone personal computers (PC's) or programmable logic controllers (PLC's). The advent of cheap stand-alone microprocessor controllers was a Godsend to pilot plant designers. Yet in less than a decade these faded away for all but the simplest units with the advent of low cost PC's and readily available data acquisition and control software. Today, data acquisition is usually entirely automatic and transmitted to its final location for analysis, typically another software program. Almost all variables are automatically controlled which has led to much better data, fewer lost runs and more stable operations. It has also led to a significant reduction in operating staff.

Today, few units are continuously attended. Most operate unattended for long periods, often over nights and weekends with little or no operator attention. Most formerly bustling research sites appear to be inhabited by numerous running units with sparse operating personnel. Trying to find a specific operator once meant walking out to their unit. If they were on site you would find them there, in the break room or in the bathroom. Today, locating them without a cell phone can require a walking tour of the site!

Experiments today are carefully planned and designed to minimize the quantity which have to be run. This reduces the cost and time to generate the required information and is only made possible by the sophisticated control systems now available. Aborted experiments, bad data, excessive data scatter and similar problems are still not unknown but are much rarer and (being less common today) more of an issue.

Organized pilot plant safety systems are routine today. They draw on methodologies common to operating plants and generally implement mitigative measures that are more common across the industry albeit with modifications appropriate to pilot plant scale. Management of change has come of age with robust controls in place to make sure that the frequent pilot plant modifications are carefully reviewed and safely implemented. Safety systems are more prevalent although requiring more review as the potential failure modes of computers and microprocessor based systems must be taken into account. While recognized at the plant scale, this is an area of continued concern at the pilot plant scale.

Instruments are more complex and more robust. Most are accessible over data highways and can be modified, calibrated or checked remotely anywhere along the wires. Dedicated instrument cabinets have effectively disappeared or at least shrunk dramatically. A CRT or LCD or multiple units are the replacement. Large central control rooms have shrunk or often vanished. Those that remain look more like computer rooms than control rooms.

Diagnostic tools have grown exponentially. Flow, temperature, and pressure calibrators abound; many are relatively inexpensive, highly portable and utterly mandatory for the accuracy modern systems require. The soap solution remains but has been supplemented by helium and combustible gas leak detectors that are faster and easier to use. The wrench remains as integral as ever but gageable compression fittings, vacuum style fittings, sealless pumps and mixers, and bellows seal valves are making leaks less frequent. The hammer has, thankfully, been replaced more with software driven diagnostic programs. The laptop is rapidly overtaking many other tools.

On-line analytical equipment has become common. Few pilot plants today are built without an integral GC, LC or similar analytical device. Many have several. On line sampling remains a key feature as always but has been reduced to the occasional specialty analysis or, more often, the need for an actual product sample except in some difficult analysis or problem systems. Rarely do the analytical needs drive the size of the unit any more. In fact disposal of unused sample volumes are a painful headache on some units.

The engineering staff has, strangely, gotten remarkably younger than me. They work on laptops and rely on cell phones for not just communication but coordination, information and their personal life. Design information is more readily available, at least internally. Specialized training is more readily available both internally and externally. The organizational memory remains as important and difficult to access as ever. Automatic individual cup beverage makers have, however, made accessing it both somewhat easier and cheaper.

Pilot plants today, however changed, are still considered very expensive and very time consuming to construct and operate but necessary.

Tomorrow

Predicting where pilot plants will go tomorrow requires a good crystal ball as well as a willingness to be proved spectacularly wrong. (Remember the person who predicted the paperless society!) While I am not looking forward to the inevitable ridicule over whichever of my predictions prove spectacularly wrong, I will venture my personal opinion of what the future probably holds.

First, I do not think pilot plants will ever go away. Computer modeling will continue to grow and advance and the way that pilot plants will be used to validate and support it will change accordingly. However some real world process validation will always be necessary. And sometimes, well, things just don't turn out the way the first model suggested. Even though the second (or third or ninth) model is much better, getting there without some level of pilot plant activity is, at least in my opinion, unlikely. Modeling will continue to augment and validate pilot plant operations and, in their always symbiotic relationship, pilot plants will continue to validate and augment modeling. In the words of a famous song "you can't have one without the other".

I also think that pilot plants will be forced to operate at ever more demanding conditions. The low hanging process fruits have all been harvested, only the more difficult ones remain. Higher pressures, higher temperatures, more corrosive materials, more toxic environments; all will contribute to the need for the next generation of units to handle even more challenging conditions. This will raise design and operating problems which the next generation will struggle to solve. This will also provide them with a new crop of war stories with which to regale the new hires and pick up some free beverages.

I think pilot plants will continue to get somewhat smaller as newer instrumentation and advances in technology allow a further reduction in size. However I think that this reduction has probably come close to bottoming out. The old ¼" OD tubing is now commonly 1/8". Some will go to 1/16" but I think most will stay about where it is today. I think any further reduction in the size of the equipment will be more than compensated for in the complexity of the units. The next generation of pilot plants will need to produce even more accurate and focused data to support the next generation of modeling. This higher accuracy and focus will involve more complexity which translates into more equipment and space, albeit of different types (analytical versus process, instrumentation versus valving). The use of micro sized high throughput "pilot plants" (actually more a complex piece of experimental equipment than a true pilot plant) will increase but not dominate. These high throughput units will provide much of the screening capability currently performed more slowly and expensively in today's more conventional small pilot plants. The promising leads produced will then be developed further in highly automated pilot plants which will run in a more realistic and scalable range to allow evaluation of synergistic and transient effects. The combination, when applied properly, will produce more high quality leads faster and provide a means to screen these for the next generation of process or product improvements.

As pilot plants will grow more expensive, complex, and sophisticated their support requirements will become correspondingly more complex and costly. Those organizations that recognize this need and address it organizationally, rather than bemoaning the additional expense, will be best positioned for competitive advantage. Organizations that provide the appropriate (albeit expensive) level of support will reap the most benefits. Many of the required maintenance functions will routinely be contracted out for cost or flexibility to specialty suppliers. Just as the "tooth to tail" ratio in the military gets smaller as the lethality of weaponry and their associated complexity increase so the "unit to support" costs of pilot plants will shrink as the data becomes better, more useful and more focused but at a higher cost and effort to keep the pilot plant working properly. The traditional process and mechanical engineering support requirements will remain but be joined by new computing, automation, safety, calibration, troubleshooting, analytical, and electronics support requirements. Specialists will abound and using them effectively as part of an integrated team will mean the difference between success and failure.

Pilot plant automation will continue and expand. Current trends which allow data gathering, changing conditions and remote operational monitoring will grow and expand. Feeding the data into a proprietary algorithm which determines the next best set of operating conditions is already here; this will grow and expand in the future as this area evolves. The next generation of units will require even lower operator presence and maximize use of recipe driven menus where the operator selects the sequence of operations from a master list, secure in the knowledge that the pilot plant will execute each step properly (at least as much as Murphy allows!) Operating staff requirements will probably not change as dramatically as they have in the past as we have probably captured the majority of those savings. However there will be a growing tendency to split operators into personnel skilled in software, computing and analysis and those skilled in loading, cleaning and maintaining the equipment. Specialization, a concern to many yesterday, is inevitable and organizations will have to continue to address the best way to pool these specialized resources together for the best effect.

Standardization within organizations will also increase. The desire to have operators able to handle numerous different units, the desire to have more units able to handle a broader range of operations, the desire to minimize lead time by avoiding the time to design and construct a new unit as well as the savings attendant in standardizing on design and construction will all promote the use of more general purpose, standardized and easily modifiable units. These will vary among organizations depending on their needs; even within an organization there will likely be several different types. However there will be more commonalities than differences then ever in the past. Dr. Smith will need to accept that the pressure transducer will be installed and configured the approved way unless there is a compelling reason otherwise. Some sequences and approaches will begin to be more common across different pilot plants as organizations strive to develop a more standardized approach to common operation. Efforts to develop the "best" approach will tend to make these operating sequences more uniform and sharable.

There will be increasingly greater use of contracted services. The conventional in house design, construction, start up and operation process will be less able to meet efficiency and responsiveness needs in tomorrow's even faster paced research world. Organizations will make more use of contractor options from design, construction and maintenance to analytical, programming and operations. The flexibility contracted services offer to gear up for a sudden short term need or conversely scale down during an industry downturn will prove irresistible much as has happened in the process design industry. This will create problems as in house expertise begins to diminish. Organizations will need to recognize the need for continued in house expertise; While less numerous, it will need to have greater depth and breath. It will need to be more of a strategic resource to find, evaluate, review and select the best contractor as well as do some or all of the unit process design or at least review and approve the contract design. How organizations will produce this in house expertise within smaller organizations will probably involve rethinking appropriate career paths and conventional job assignments.

Increased material toxicity, reduced exposure limits, and continued concerns for the long term health effects of any exposure will continue to promote design and construction options that are more leak free, particularly given that there will be less operator time to find and correct leaks. Sealless pumps and mixers, bellows seal valves, and high integrity fittings are all examples becoming common on pilot plants. Automatic tube welding and mechanical joining systems will continue to supplant (but never eliminate) conventional compression and pipe fittings. Instrumentation will continue to replace numerous individual seals with sealed units and higher integrity seals. Specialty closures, assemblies, and equipment, once considered prohibitively expensive, will become more commonplace while still, of course, considered horribly expensive. Automatic on line leak detection will become more prevalent to ensure safety and data integrity in the absence of an operator.

Pilot plants will begin to catch up to process industry standards on the reliability and implementation of control and safety systems. Increased concern over safety system effectiveness during a control system problem will force pilot plant control systems in new directions. Simplified layer of protection analysis will be more common as will a more detailed analysis of safety system failure modes and appropriate mitigative measures.

There will be increased use of wireless technology for pilot plant operations. Even though the long distance between sensors and control sytems is rare in pilot plants, the (eventual) lower cost, higher flexibility and reduced construction time will make this approach attractive. They will replace common hardwired systems such as in small tank farms or remote operations or even process components across the bay or room. We will see greater use of wireless data and increased use of wireless transmission of data to end users and storage.

Pilot plants will be designed and constructed to produce more data per unit. This may be as simple as multiple train units or as complex as high throughput experimentation. The higher productivity this approach confers will be offset somewhat by the increased complexity of the pilot plants as well as their support and maintenance requirements. The most effective organizations will realize that the increased effectiveness does carry a cost and plan their support organizations accordingly.

Process control based on real time analytical data will become more widespread. Process optimization will become common. More importantly, the ease of integrating this data into the pilot plant's control system will become more uniform and hence easier and less expensive. Third party programs, integrated systems and common bus structures will allow the data to be fed to the process control system in a more straightforward, common and less unique manner. The complexity of the pilot plant control system will grow as these inputs are integrated to the maximum feasible amount. This area will encompass more difficult analysis such as particle size distribution, complex product compositions as well as more routine information such as real time density measurement and composition.

New and novel instrumentation will change how we design units. Fifty years ago no one ever heard of Coriolis meters much less ones that can be used with ¼" tubing. Small scale magnetic flow meters, vortex meters, corrosion probes and numerous other devices were but a pilot plant designer's daydream. The growth in this area will continue. The availability of these devices will allow pilot plant designers to solve some issues which have plagued them for years much as the advent of thermal mass flow meters finally solved the search for ultra-small size control valves to use with differential pressure devices in the 70's. The resultant increase in accuracy and reliability will continue the growth of pilot plants better able to produce valid useful data with every run as opposed to statistical analysis of several runs to address inaccuracy and non-repeatability.

Increased use of multi-functional instruments will be common. Pressure transmitters will simultaneously measure temperatures; Flow meters will also provide pressure or density in a single unit. Most new transmitters will be able to be calibrated through addressing them anywhere in the loop as opposed to finding, removing and replacing the transmitter for off line calibration. While the individual transmitter will be more expensive, it will be smaller, more accurate and more reliable. The decrease in installation costs will more than offset the higher purchase cost. These multi-functional units will also be easier to interface with control and data acquisition systems generating additional savings in programming and set up.

One thing I am sure of is that the pilot plants of tomorrow, will, still be considered expensive and time consuming to construct and operate but necessary. So some things apparently never change.

How accurate will my predictions be? Only time will tell.

As my final prediction I suggest that something new and unexpected will arise. It will greatly change the way we design, construct and/or operate our pilot plants in a way or ways no one even remotely envisions today. After all, that's what research is all about, looking to the future, trying to understand things better to take better advantage of what we know now to make it a better world tomorrow. And, OK to make our companies more money so they can pay our salaries. The future is always about change, planned and predicted as well as new and unexpected! It's what makes working in pilot plants one of the more rewarding careers there is in chemical engineering.

Except of course when you are over budget, late and that new widget really doesn't work like you expected .....

Thank you and good luck.

Process Development: Managing Risks and Expectations

Joseph B. Powell, Ph.D.
Shell Chief Scientist – Chemical Engineering

Abstract

Process development is a key phase which determines whether ideas can be successfully commercialized, or whether they should be abandoned. Spending accelerates dramatically upon progression from research to full development, and with this comes greatly increased expectations from all stakeholders. Risk of failure may not be fully apparent to business sponsors, as many of the failure modes will not be evident until development is attempted. On the other hand, technology opportunities with high earning potential will provide incentives for risk taking and rapid deployment. Effective management of these competing interests requires close communication between the technology organization and business sponsors in assessing and balancing risk vs. opportunity.

Process Development as a discipline earns its value in ability to rapidly identify show stoppers, and in finding ways around identified roadblocks, where possible, before extensive spending and time have been allocated to major programs or pilot tests. Approaches and best practices in managing spending vs. level of risk reduction will be the subject of this seminar, including some case histories and examples.

Risk Management in Process Development

Research and development is inherently risky [1]. The following presents a few rules to maximum efficiency in selecting projects, and executing the development. Development can be a success either upon commercialization, or in quickly identifying why a new technology will not work, either in a technical or business sense, so the organization can move on to more lucrative opportunities.

Economic and Project Viability Evaluation

Risk management begins with an assessment of the value of the opportunity. Economic evaluations can be difficult to perform at the ideation phase due to lack of information on what is feasible. However, some basic questions can be asked, such has if the technology were to work at 100% yields and selectivity, how much would it be worth? How much capital could one afford given the maximum margin one can expect, and is this capital reasonable for the notional conversion costs? Is the business opportunity a fit for the organization, or would it be better developed elsewhere? What is the size of the prize: is it worth devoting organizational resources to the proposed target, or would efforts be better spent on a larger target with multiple potential applications? Does your organization have the skills and competencies to develop the new process or product, or are new competencies required, suggesting new unknowns to be uncovered? Are there significant safety or environmental risks, or social and/or other stakeholder issues that should be considered before initiating a formal research effort?

Identify and examine the readily identifiable show stoppers

A new idea will generally have a central hypothesis that must be tested or evaluated to assess potential. This may entail yields or rates for a proposed reaction or separation, or the performance of an essential step of a newly engineered device or unit operation. Rather than simply investigating what is feasible or the extended parameter space, propose targets that must be approached if the concept is to be viable, and use these targets to make an initial assessment of feasibility. If targets are not indicated to be achievable in initial screening it may be possible to redesign aspects of the concept or process to examine a concept that could deliver against required targets. In this way, preliminary economic evaluations can help set direction, by identifying levels of performance that must be achieved to have commercial value, and technology approaches that may be able to deliver against these targets.

A second paradigm is to strive to test the hypothesis as quickly as possible. There is a trap in R&D of waiting for design and implementation the perfect apparatus or system to begin testing or assessment. Crude scouting experiments or back-of-the-envelope calculations are often more valuable, as they can lead to a change in direction or design, on the basis of preliminary learnings. Strive to start testing and evaluation early, not later.

Prepare a risk matrix for full commercialization

Once the initial hypothesis is tested, "Begin with the end in mind" [2] and work backwards to determine what information and development must take place for full commercial deployment. Often, one can become trapped in doing what one is set up to do or knows how to do, and overlook other developments that must take place in parallel. As an example, research groups with skill and existing apparatus in catalyst synthesis and testing may focus on study of catalyst performance, which often is the central idea behind a new concept in the chemical industry and refining. However, at some point the performance metrics achieved may exceed those required for commercial viability, and other engineering factors such as reactor design or separations, or ability to handle impurities in real feeds, may become the bottleneck in deployment. It is important to proportion resources and, where needed, bring in new skills and apparatus to address the newly defined critical issues defining a program development.

Peer challenge can be quite useful in augmenting one's own background in anticipating other risks and other project needs which which will require additional development. One does not know what one does not know, and exposing the technology to a broader peer group engaged in similar developments is critical in identifying other potential show stoppers, or significant development risks. In addition, these exercises can also help identify known solutions which may already exist, either inside or outside of the organization.

Develop a comprehensive low-cost exploration plan

Most R&D projects fail due to business or technical reasons. Business pull and enthusiasm is usually strong at project onset (which is why the project is initiated), but often wanes periodically with economic ebb and flow [3]. Maintaining low costs can prevent a project from falling victim from periodic cost cutting initiatives, until it is ready for commercialization and can be revenue producing.

Incorporate Sustainability and Efficiency Drivers into Early Decision Making

Many factors in sustainability such as carbon footprint, land use, water, or societal issues have weak direct economic indicators, yet are highly important to many stakeholders. Early consideration of these factors is the best way to achieve sustainability at low cost, by driving decisions at the early stages of process development where more sustainable options can often be selected at low incremental cost [4]. After-the-fact retrofit of energy or resource efficiency including heat integration, is typically much more expensive.

Managing Expectations

It is easy for business sponsors to equate assessment of economic potential with technical viability. The more detailed the methodology used for economic evaluation of earning potential and capital cost, the more "real" the technology appears to become. A second trap is in project execution timelines. The more detailed the timeline and project plan, the more "certainty" the project appears to take on. One cannot however reduce risk simply by increasing the level of detail of the economic assessment or project planning. Evaluation and planning insure that the proper targets are available and that a comprehensive plan with resourcing is in place to address and mitigate risks, but individual risks are not reduced simply by planning to address them.

One of the means to manage expectations is via frequent communication of project development and risk to business sponsors, even if technical detail is not fully comprehended. In addition, the technical team lead can serve as an "intrepreneur" to help assess status of the technology development relative to business value, and routinely evaluate this together with business leads.

As one learns during the course of research and development, there are many highs and lows in assessment of project potential (Figure 2). A key metric is to assess the rate of learning, and use this to decide whether or not to continue the program, or whether the program has been de-risked and developed to the point of being ready for commercial investment. Frequent oscillations in degree optimism / pessimism often indicates a high rate of learning typically in the middle of the process development effort, and is therefore neither a time to abandon or invest, provided economic targets remain viable.

In order to succeed in R&D, one must strive for ambitious targets, or the developments will never be pushed to the potential or at the pace that they may be capable of achieving. This requires an optimist or champion. When finalizing a design for commercialization, however, and prior to spending at high rates on detailed design with an engineering contractor, one must be a pessimist to look for and mitigate deviant process behaviors or upsets.

Managing process development thus entails balancing the fine art of optimization over what can be, with the reality of what is needed to be robust, economic, and rigorously defined for final investment decision and detailed engineering development.

References

1. A. Gerstenfeld, *Effective Management of Research & Development*, Addison- Wesley (1970).
2. S. R. Covey, *The Seven Habits of Highly Effective People,* Free Press (1989).
3. R. Leifer, C. M. McDermott, G. Colarelli O'Connor, L. S. Peters M. P. Rice , R. W. Veryzer, *Radical Innovation: How Mature Companies Can Outsmart Upstarts*, Harvard Business Review Press, 2000.
4. J. Harmsen and J. B. Powell, *Sustainable Development in the Process Industries: Cases and Impact*, John Wiley & Sons, New York, April 2010 (ISBN-10: 0-470-18779-4; ISBN-13: 978-0-470-18779-1).

Figure 1: Hypothetical Example for Rates of Spending for Research, Development, and Demonstration

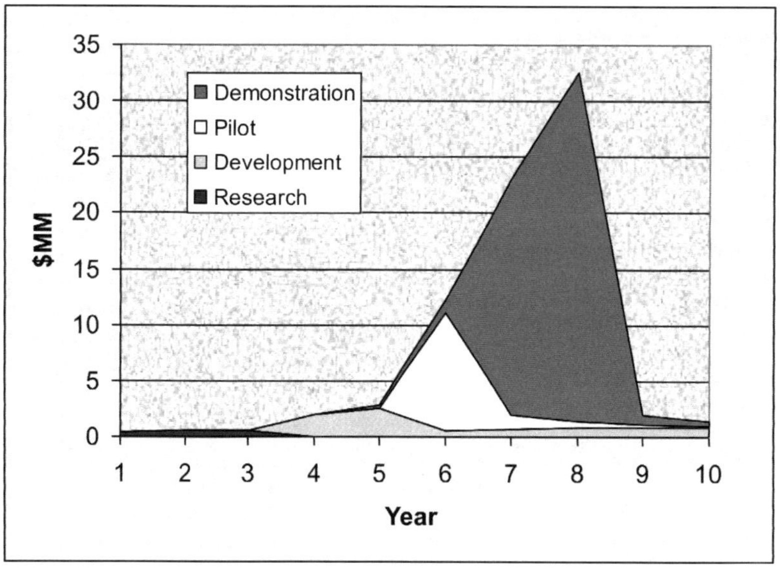

Order of magnitude increase in spending as progress from research, development, demonstration, and commercialization.

Figure 2: Managing Optimism vs. Pessimism

Perspective:

Don't stop program until stop learning. Some "facts" or paradigms are generally wrong.

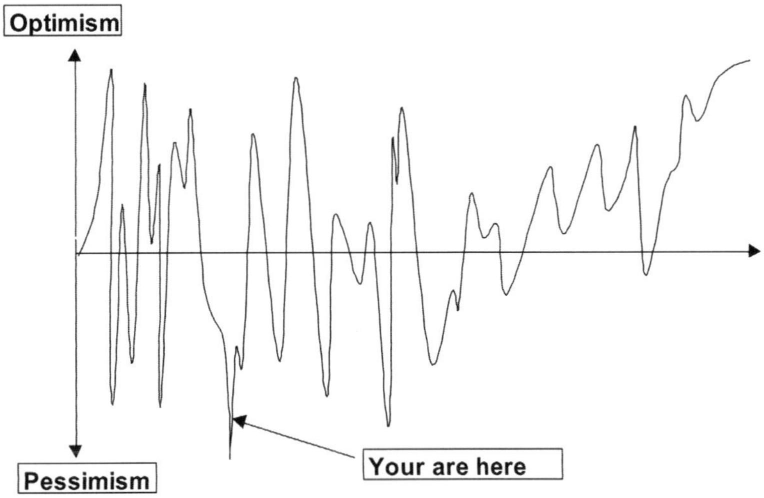

(33b) Degumming Process Design for Biodiesel Plants

Maoqi Feng, Div. of Chemistry and Chemical Engineering, Southwest Research Institute, San Antonio, TX

Phosphorus is a contaminant in biodiesel. Phosphorus originates from phospholipids in biodiesel feedstock. Phospholipids removal is required in biodiesel production. Degumming is the most commonly used method for phospholipids removal. Different degumming processes, such as acid hydrolysis, enzymatic degumming, membrane, ultrafiltration, and supercritical CO_2 processing, etc., can be selected for different feedstocks.

Continuous ultrasonic degumming improves mass transfer and thus higher efficiency. In-situ FT-IR analysis of phospholipids content provides a method for monitoring the phospholipids during degumming. Adsorbent could be used as a post-treatment method for quality assurance. Case study for scaling up the degumming process of soybean oil will be presented.

(33c) The Role of a Maxi Lab On the Scale-up of New Processes

César González Serrano, Matthew Kuhlman and Gregory Hemmer, SABIC, Mt Vernon, IN

The role of a Maxi Lab on the scale-up of new processes

The SABIC Innovative Plastic's High Performance Polymer Business constantly strives to engineer and commercialize new products for the marketplace, which often require that new processes are developed or existing ones are modified.

Initial laboratory trials leave questions opened to be addressed during pilot plant trials. However, availability, cost and feasibility may challenge the answers that can be realistically obtained from the pilot plant. A Maxi Lab provides an intermediate scale between the lab and the pilot plant where solutions can be found in a quick, visual and cost effective manner.

The utilization of the Maxi Lab allows for the early identification of critical steps in the operating procedures and preliminary scale-up parameters. Furthermore, due to the relatively small size of equipment, it allows for higher flexibility in changes of units of operation and multiple configurations of a piece of equipment. Such characteristics will contribute to reduce the number of runs required in the pilot plant and increase their effectiveness.

The Maxi Lab scale allows for production of a few kilograms of material, which grants collaboration with costumers via sample testing. These early samples are a key factor to understand the costumer CTQ's and unveil possible hidden CTQ's that may hinder the process at later stages.

(33d) Challenges in Commercial Scale up of Novel Poly(phenylene ether)-Polysiloxane Block Copolymer Process

Alvaro Carrillo and Scott Fisher, Noryl* Technology, SABIC Innovative Plastics, Selkirk, NY

Poly(phenylene ether) is an engineering thermoplastic known for its excellent hydrolytic stability, dimensional stability and inherent flame retardancy. A block copolymer of poly(phenylene ether)-polysiloxane (PPE-Si), produced by the oxidative coupling polymerization of 2,6 xylenol and polysiloxane, has been developed to deliver significant improvements in mechanical properties and flame retardancy.

Process development for new technology to culminate in production in an existing operating commercial chemical plant requires conscientious planning and risk mitigation efforts. This presentation illustrates the challenges involved in the commercial scale up of PPE-Si copolymer, covering areas such as: lab and pilot plant scale testing, determination of feasibility and process conditions for all unit operations, cross contamination of existing processes, management of transitions, quality specifications and testing methods, EHS and product stewardship considerations.

(34a) Process Development for Savannah River Site High Level Radioactive Waste

Dan P. Lambert[1], Michael E Stone[1], David C. Koopman[2] and Bradley R. Pickenheim[1], (1)Savannah River National Laboratory, Aiken, SC, (2)Process Tech Programs, Savannah River National Laboratory, Aiken, SC

The Defense Waste Processing Facility (DWPF) treats legacy nuclear waste generated at the Savannah River Site (SRS) during production of plutonium and tritium required during the Cold War. Approximately 35 million gallons of nuclear waste is first treated via a complex sequence of controlled chemical reactions and then vitrified into a borosilicate glass form and poured into stainless steel canisters. Converting the nuclear waste into borosilicate glass canisters is a safe, effective way to reduce the volume of the waste and stabilize the radionuclides.

Radioactive processing in DWPF began in 1996.

To date 3.7 million gallons of waste have been processed and 2,562 two-ton canisters of glass (10 million pounds) have been produced.

However, higher facility throughput is needed to shorten the processing life of DWPF.

The current processing flowsheet is unable to handle this higher throughput.

In addition, there are safety (hydrogen generation) and processing issues (pH and anion depletion due to catalytic formate decomposition) that are impacting production.

Savannah River National Laboratory was asked to investigate potential flowsheet improvements.

Flowsheet development was completed in four stages, (1) identification of alternative reductants and acids in simple screening tests, (2) extensive testing of two best alternative flowsheets, (3) selection of the best alternative using a systems engineering evaluation, and (4) lab-scale testing to determine optimal processing conditions and to evaluate downstream processing impacts.

Bench-scale flowsheet testing was performed to develop the new nitric-glycolic acid flowsheet as an alternative to the existing nitric-formic acid flowsheet currently used by the DWPF. Testing established that the new flowsheet can effectively reduce mercury and remove it by steam stripping with minimal catalytic hydrogen generation. All other processing objectives were met, including greatly reducing the Slurry Mix Evaporator (SME) product yield stress and increasing throughput compared to the baseline nitric/formic flowsheet.

The vast majority of the experiments performed to date have been completed using nonradioactive simulants.

A successful demonstration of the flowsheet with radioactive waste was completed in August 2011.

The paper will describe the process flowsheet development and the remaining work required prior to implementing the flowsheet in DWPF.

(34b) NIST Thermodata Engine: Tool for Chemical Product Design

Vladimir Diky[1], Robert D. Chirico[1], Chris Muzny[1], Andrei Kazakov[1], Ilmutdin Abdulagatov[1], Joseph W. Magee[1], Kenneth Kroenlein[1], Jeong Won Kang[2], Rafiqul Gani[3] and Michael Frenkel[1], (1)Applied Chemicals and Materials Division, National Institute of Standards and Technology, Boulder, CO, (2)Department of Chemical and Biological Engineering, Korea University, Seoul, South Korea, (3)CAPEC, Department of Chemical and Biochemical Engineering, Technical University of Denmark, Kgs. Lyngby, Denmark

ThermoData Engine (TDE, NIST Standard Reference Database 103) is the first product that implements the concept of Dynamic Data Evaluation in the fields of thermophysics and thermochemistry. That includes maintaining of the comprehensive and up-to-date database of experimentally measured property data and expert system for data analysis and generation of recommended property values along with uncertainties on demand. Different areas of chemical engineering, including product design, benefit from the services provided by TDE.

Chemical product design can be based on experimental data or property prediction methods based on chemical structure. TDE provides unprecedented collection of evaluated experimental data and uncertainty estimation that distinguishes it from any other product design tool. TDE product design functions include pure-compound design and solvent selection. Pure-compound design is search of a pure compound possessing required properties at specified conditions. A dynamically grown collection, now about 30,000 compounds, is available for search. User can restrict selection by experimental data or allow property prediction for the compounds with no experimental data. Properties supported are normal melting and boiling temperature, heat capacity, density viscosity, vapor pressure, etc.

Solvent Design function serves three tasks: (1) selection of best solvent for a solid solute, (2) search for a selective solvent for separation of a solid binary mixture, and (3) selection of best solvent for extraction. Solvents are selected from a pool containing more than 27,000 compounds. Selection is made by best efficiency (solubility or selectivity, depending on the task) and matching other requirements requested by the user. Efficiency criteria are evaluated based on experimental data for binary mixtures or predictive models (original and NIST-KT UNIFAC). Underlying experimental data can be accessed and inspected in order to prove the reliability of the models used in the selection.

Examples will be shown and benefits of ThermoData Engine for product design demonstrated.

(34c) Acetylation of Wood - New Tricks for 100 Year Old Technology

David C. Attride, Eastman Chemical Company, Kingsport, TN

The acetylation of cellulose was first practiced in the early 1900s. Application of the reaction is more complex in a solid, heterogeneous natural composite such as wood. The acetylation of southern yellow pine (SYP) imparts improved properties that are of significance in the building & construction area. This paper will provide background on how the process enhances and alters certain properties.

THREE KEYS TO SUCCESSFUL NEW PRODUCT and PROCESS DEVELOPMENT

Douglas H. Lenz, PhD

SABIC
SABIC Innovative Plastics
1 Noryl Avenue
Selkirk, NY 12158

Prepared for Presentation at the Spring 2013 AIChE Meeting, San Antonio, Texas, April 28- May 2

ABSTRACT

Managing the development and implementation of new technology or a new process from lab through start-up is an exciting challenge that is certainly the goal of many process engineers. However, very few engineers and chemists actually have the opportunity to manage and successfully implement new technologies that start out as an idea all the way through to manufacturing a profitable product. More common is the challenge to manage the re-development of an existing process or technology with the goal of making a product or process more profitable, safer, more environmentally compliant, etc. but the same approach can be successful.

Three key techniques to the successful management of process development will be discussed, using examples from a coated product / process R&D development program to illustrate the techniques. First, intellectual downstreaming was used, which helped to envision the ideal product and process to produce it. A cross-functional team comprising engineers, operators, management and consultants from the plant, divisional, and corporate levels accomplished this step. Second, a systematic program was managed by focusing initially on how to make the "ideal product" from both a fundamental first principles basis and ideal plant basis compared to the new customer requirements or the existing plant basis. Managing this "gap analysis" then lead to either the development of the R&D plan with decision points or "gates" or the incorporation of the correct and sufficient capital improvements necessary for the existing process to produce the desired material. Third, managing the superb execution of the all resources working in parallel or consecutively was required to minimize the process development or re-development time-frame while pursuing ideas and options that can eventually lead to revolutionize the process. These diverse resources to be managed included teams of chemists and external consultants at the lab-scale, teams of engineers & co-op students at the pilot and plant-scale, state-of-the-art surface science / powder technology techniques & analytical methods, and sophisticated computer models.

These techniques for the development of new and existing technology will be discussed in a case study. This case study describes R&D New Product & Process Development for a novel coated product that could consistently meet key customer quality criteria for a new application. Meeting constantly evolving market projections and new product specifications required equipment for this coating process that was larger than anything previously built and operated. Success was achieved through the utilization of the intellectual downstreaming, a systematic program using gap analysis and gates, and superb execution.

INTRODUCTION

Managing the development and implementation of new product or a new technology or both from lab through start-up is an exciting challenge that is certainly the goal of R&D. However, very few engineers and chemists actually have the opportunity to manage and successfully implement new technologies that start out as an idea all the way through to manufacturing a profitable product. Three key techniques to the successful management of new product design and process development will be discussed. The first key concept is intellectual downstreaming which helps to envision the ideal product and process to produce it. A cross-functional team is best for accomplishing this step. The second concept is seamless engineering which incorporates a systematic program with decision points or gates for managing how to make the "ideal product" from both a fundamental first principles basis and ideal plant basis compared to the new customer requirements or the existing plant basis. An equally important aspect of seamless engineering is managing the superb execution of the all resources working in parallel or consecutively in order to minimize the product design and process development time-frame. The final key consideration is the use of the combination of fundamental science, statistical tools and methods and sophisticated computer simulations and models throughout the design and development process.

Several[i ii iii] process innovation / process development articles have described many of the above elements and their benefits. These techniques for the development of new products and technology will be illustrated in a case study. This case study[iv] describes R&D new product & process development for a novel coated product that could consistently meet key customer quality criteria for a new application. An additional challenge to the project was the constantly evolving market projections and new product specifications that required equipment for this coating process that was larger than anything previously built and operated. Success was achieved through the application of intellectual downstreaming, seamless engineering, and statistical & modeling tools to supplement the science.

DISCUSSION: THREE KEYS FOR A SUCCESSFUL PRODUCT DESIGN\PROCESS DEVELOPMENT PROGRAM

While many factors are important and indeed necessary for the successful development of a new product or process, from lab to commercialization, this paper will highlight three important concepts that may not always be a part of an R&D program. These 3 factors are: 1) Intellectual Downstreaming; 2) Seamless Engineering; and 3) the 3 S's - Science, Statistics & Simulations.

The first key concept can be thought of as INTELLECTUAL DOWNSTREAMING. The critical objective of this R&D factor is to define the ideal product or process that will be the measurable end result of the R&D program and commercialization effort. It is an element of Steven Covey's 7 Habits of Highly Effective People - "Begin with the End in Mind" [v]. Ultimately, the group or the R&D effort should be avoiding starting with the approach that if you don't know where you are going, any route will get you there! Therefore, intellectual downstreaming is like reading a map where you see where you want to go, know where you are starting from and can quickly find the easiest or best route to get there. For a corporate R&D program, it will help to envision the ideal product and process to produce it.

An important tool to get the most value from the intellectual downstreaming process is a FOCUSED TECHNOLOGY REVIEW. These are technical discussions, organized and managed similar to the systematic "HAZOP" style safety reviews, which focus on the state of the art related to the defined ideal product or process. To be most successful, these focused technology sessions include as broad and diverse a group of participants as possible. Ideally, this cross-functional team comprises engineers, operators, management and consultants from the plant, divisional, and corporate levels accomplished this step. A corporate or high level director, who will not be part of the working team, should facilitate this activity so that the resulting set of ideas and action plans is as objective as possible and independent of any one person's or groups bias.

FMC frequently conducted focused technology reviews facilitated by the corporate technology director. This meeting format assures that the plans and actions for technology improvement and/or development are understood and supported by corporate management. It is always important to get management support at the highest levels possible as early as possible for a successful project or program. These technology reviews bring together plant, divisional, and corporate technical personnel and often specific outside experts or consultants for broad, thorough, often state-of-the art technical discussions. Often, each area of the process' technology is reviewed in detail and as part of the entire system using an approach similar to a HAZOP (Hazard and Operability Review) session.

The outcome will therefore be a consensus action plan. An excellent tool known well by project managers to both communicate this plan and help the group manage the plan is a GANTT CHART[vi]. A Gantt chart is a compilation of all the project task timeline including the critical path or paths. In this way, each project member understands the importance of their individual task and how it impacts the other members and the project goals and objectives. So it can be a very valuable tool for this product improvement and process optimization program by helping communicate project progress and/or barriers and resource needs. An example is shown below with the critical paths highlighted in red.

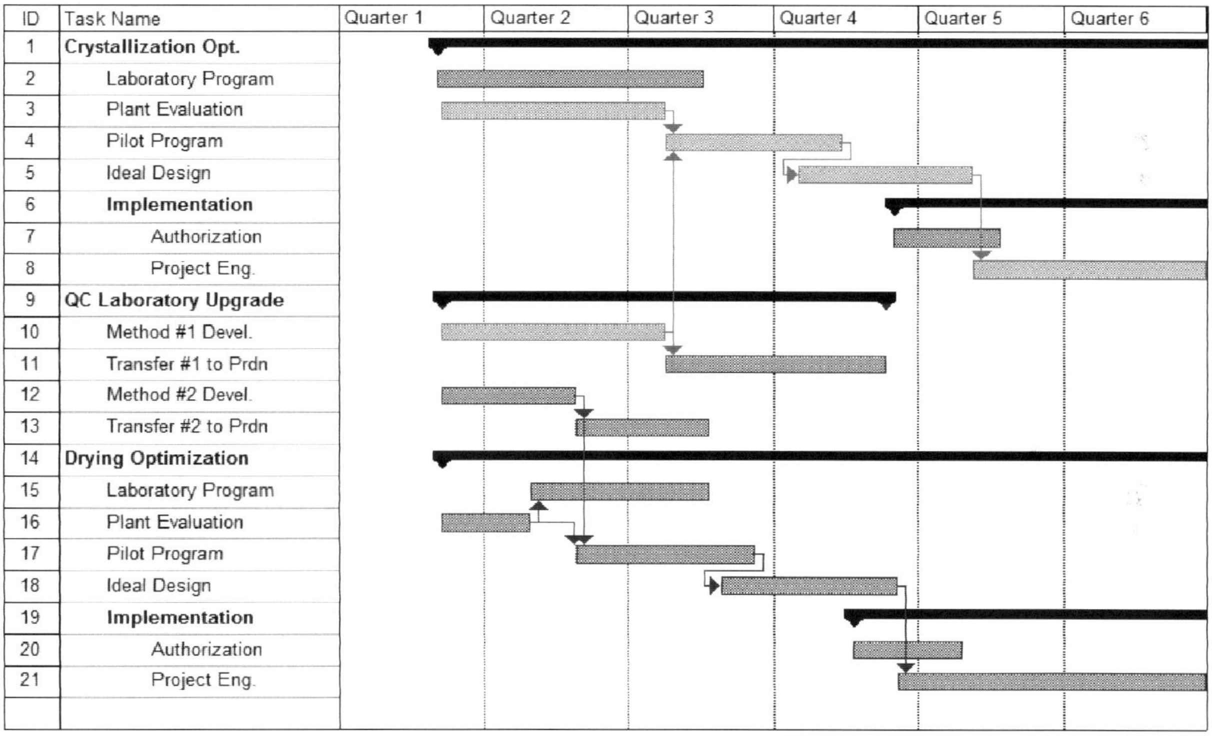

The second key concept is often referred to as SEAMLESS ENGINEERING. The "seamless engineering" concept was not developed at FMC, but with hindsight tends to be a part of many successful technical projects. Seamless engineering is more than just concurrent engineering or a group of individual projects. Seamless engineering can be thought of as a systematic program managed by focusing initially on how to make the "ideal product" from both a fundamental first principles basis and ideal plant basis compared to the new customer requirements or the existing plant basis. Managing this "gap analysis" then leads to either the development of the R&D plan with decision points or "gates" or the incorporation of the correct and sufficient capital improvements necessary for the existing process to produce the desired material. As demonstrated by the Gantt chart example above, it will be a combination of multiple, parallel or partial-parallel tasks like concurrent engineering.

The critical difference is that seamless engineering is an integrated program versus a series of individual projects handed over to the next team upon completion. The cross-functional, diverse team(s) that participated in the intellectual downstreaming now participate in making the defined "End" or Ideal Product/Process a commercial reality. To ensure success, this engineering effort must have a champion & sponsor leading and coordinating the multiple concurrent tasks. The program champion is involved in all efforts which facilitated communication of discoveries, ideas, problem resolution, project timing. The tools of first principle science, statistical analyses & methods and modeling or simulations (the 3 S's) are an integral part of a successful seamless engineering implementation. All these resources work in parallel or consecutively to minimize the program time frame. A final element of successful execution is superb communication fostered by frequent progress review meetings that often need to be no more than 1 hour on a biweekly or monthly basis or perhaps longer if on a bimonthly basis.

One of the values and benefits of a superbly executed seamless engineering effort is reduced development & implementation time. This result occurs because of the excellent communication and coordination from Production advising R&D during lab program which lead to the R&D concepts rapidly implemented by Production w/ minimal disruptions and all with confidence and support from upper management.

A second benefit is often obtained from reduced costs for recommended capital improvements. Since the technical fix accounting for "Whole" process or solution was developed and embraced by the team during the intellectual downstreaming, these designs are most often the best and most cost effective capital improvements, not necessarily requiring all new capital investment and certainly minimizing costly overdesigned solutions. In essence, it incorporates the tried and true concept of Right First Time (RFT) !!

The third and final key concept to be discussed is the incorporation of the 3 S's of SCIENCE & STATISTICS & SIMULATIONS or MODELING throughout the entire seamless engineering process. Many if not all R&D projects incorporate first principle science as part of the program. But to translate that science successfully and rapidly from the test tube to the customers application via a commercial process should also incorporate the tools of statistics and simulation or modeling.

What kind of statistics tools? Today many of these tools and approaches are found in the Six Sigma programs recognized and used by many corporations. These include designed experiments (DOEs) and statistical analyses on the lab, pilot, and plant levels. At every level, the baseline behavior of the process must be studied and documented to understand the measurement and process variability so the meaningful interpretations are obtained and statistically significant results can be found.

Often due to time or resource constraints, the fundamental science must come from plant level studies or the technology must be scaled directly from lab to plant scale. Now the cost of the experiment is high and the number and types of experiments must be very selective and well planned so as to not disrupt production or affect product quality. It is these situations where the 3 S's ensure a successful and useful result. Plant-scale DOEs are particularly effective as they can quickly determine the significant variables and/or interactions for that specific system from the resultant Scree plots, and estimate the effect of the variables on the process all while minimizing disruption to production.

Again, due to time and resource constraints, a full understanding or scale-up of the technology may not be achievable even with science and statistics. This situation is addressed with process simulation and/or product modeling. Examples of these can be heat & energy balance models, regression or multivariate analyses, fault-tree analysis, and dynamic behavior models.

Why model? Once a good model or simulation is developed and validated with actual data (ideally from multiple scales such as lab, pilot and plant data), then these simulations can improve process understanding, allow for an assessment of wider range of process conditions and atypical &/or new conditions without disrupting production. The end result is a more complete understanding and solution with reduced engineering time and costs!

UNUSUAL PRODUCT DESIGN & DEVELOPMENT CASE STUDY: CREATING A STABLE SOLID OXIDIZER USING AN AQUEOUS COATING AT HIGH TEMPERATURE IN THE "WORLD'S LARGEST" COATER

Extensive laboratory research determined that only a coated peroxygen product (CPP) could consistently meet key customer quality criteria for a new application. The preferred coating was applied as an aqueous solution onto a fluidized bed of this solid reactive oxidizer at high temperatures – a very difficult challenge! However, the commercial scale-up would be too capital intensive for the initially demonstrated Wurster, or bottom, coating technology (Figure 1). Top-coating technology was the method of choice for a commercial scale plant (Figure 2).

FIGURE #1: Wurster Technology

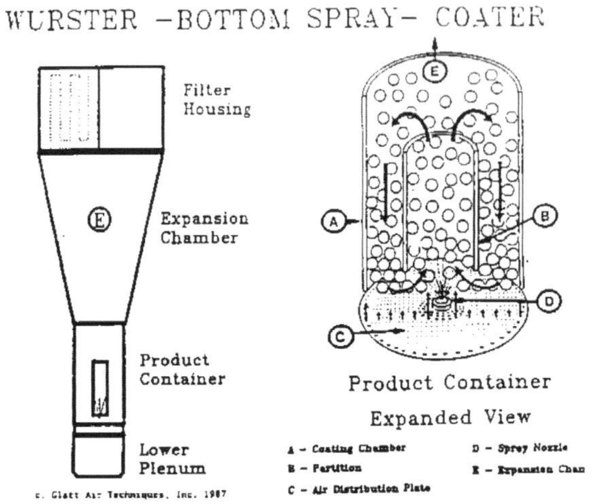

FIGURE #2: Top-Coater Technology

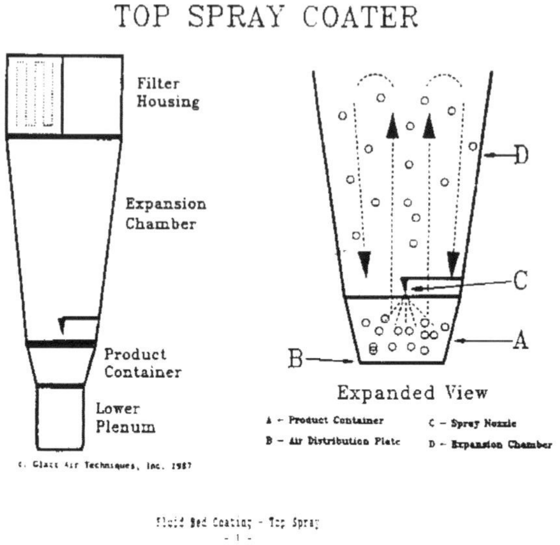

This new top-coater technology required a rapid demonstration and scale-up program for determining its viability as an alternative for coating the reactive material to keep the process / product development program on track. Success was achieved due to several factors including the 3 key factors described above: defining the ideal product; a cross-functional, responsive team took charge of the project up-front; the project had a champion; piloting, scale-up, lab work, and modeling were conducted very often in parallel both within FMC and at several equipment manufacturer's sites; fundamental science and statistical tools were utilized throughout the program; identified problems & issues were communicated early and resolved by the team rapidly.

Early in the program, the team created the ideal product definition for this application of a solid oxidizer (bleach) in a detergent box that could survive long periods at warm, humid conditions (a laundry room). These criteria included measurable specifications for 1) Shelf Life / Stability, 2) Active Oxygen, 3) Bulk Density and 4) Coating Level. Knowing this target upfront was important as the measurable specifications often changed throughout the project due to changing legal and customer requirements.

The cross-functional team comprised R&D chemists, R&D process engineers, project engineers, plant engineers, marketing & sales representatives, and legal representatives from within FMC. But as importantly for success, external consulting and support was provided by tollers and equipment manufacturers as well.

The program had a champion. The champion was responsible for 1) planning & coordinating all project activity; 2) ensuring resource availability; 3) coordinating the data sharing & analyses; 4) communicating program progress as well as issues and 5) understanding all aspects of the technology (very important!).

After an extensive program, top-coater development and scale-up was ultimately successful. Initial feasibility & scale-up trials were successfully conducted at a manufacturer's site. However, the initial semi-work scale-up trials were unsuccessful as an excessively agglomerated, unstable product was produced. The two main reasons for the set-back were investigated and identified. A designed experimental program was conducted to establish proper process conditions for successful scale-up and to understand the causes of the excessive agglomeration observed. A model was subsequently developed, based on a first principles understanding gained from the statistical designed experiments, to correlate key parameters to final bulk density (an agglomeration indicator). Extensive scale-up work was conducted with this technology for FMC's CPP product resulting in the successful installation and operation of semi-works plant. Whereas this semi-works size was considered commercial by the manufacturers, our commercial plant would require an innovative design & operation for an order-of-magnitude larger coater which had never been built at that time.

The top-coating technology development program was initiated after the successful scale-up at the Coating Place of the Wurster technology for coating CPP with FMC's preferred aqueous coating. If the Wurster technology had proven unfeasible, top-coating technology would not have even been considered.

Initial feasibility & scale-up trials were successfully conducted at Glatt Air Techniques, Inc. A basic three variable full factorial (8-run) designed experiment (DOE) was conducted at the lab scale which identified key parameters and conditions and potential problem areas. The real value of incorporating a statiscal-based tool early in the program, such as the DOE rather than conducting Edisonian trial & error testing is that one can quickly determine the significant variables and/or interactions, can estimate the effect of these variables on the process and reduce the development time & costs. Scree plots from this DOE are shown in Figures #3 and #4.

Figure #3: DOE Scree Plot for Coating Effectiveness

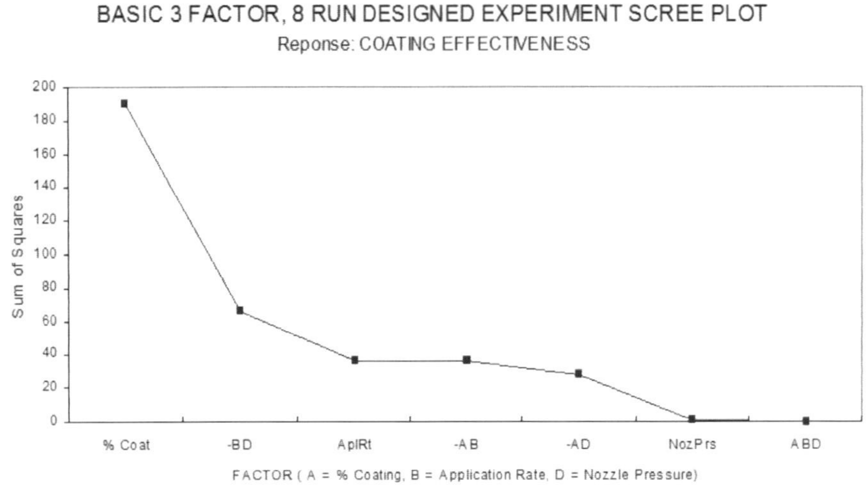

Figure #4: DOE Scree Plot for Coated Product Bulk Density

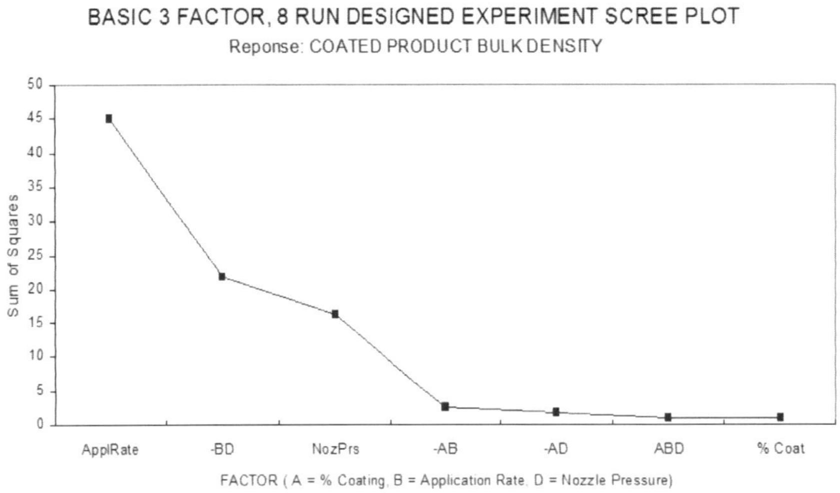

However, Murphy's Law often prevails at points throughout a development program but the use of the 3 KEYS can mitigate the consequences. The initial semi-work scale-up trials also conducted at Glatt were unsuccessful. An excessively agglomerated product was produced, with bulk densities well below the minimum target. The two main reasons for the set-back were; 1) using a nozzle design not suited for our product on this scale; 2) using semi-works scale top coater not properly designed for our product needs with respect to multiple nozzles, heat input, disengagement height, and filter design. The excessive agglomeration occurred because the coating zone was oversaturated due to the combination of process and design conditions. Two key causes of this oversaturation were the nozzle design and the unit design specifications. The likely mechanisms for oversaturation were either more overlapping of the zones than desired, or localized high humidity concentrated in the center of the coater only, or both. Heat input was limited due to design limitations of fluidizing air volume & temperature, disengagement height, and filter design encountered with our product. The coater design needs to take into consideration both the CPP particle size distribution and the coating solution properties. Attempts, listed above, to obtain adequate heat input resulted in filter pluggage with the semi-works unit, further limiting the amount of heat input.

Four major factors play a role in the dynamics of spray coating. These factors are 1) liquid/coating properties and characteristics; 2) substrate properties and characteristics; 3) bed humidity; 4) agglomeration versus coating dynamics.

The liquid/coating properties primarily affect the size and behavior of the droplet during the coating process. Due to properties of FMC's CPP substrate material, only aqueous solvents were considered. The physical properties of this coating solution as a function of temperature were measured and well understood which contributed to the success of this program.

Prediction of droplet size is well known[vii]. For a two-fluid atomizer a well-accepted equation[viii ix x] for predicting the Sauter Mean Diameter, D32, (in microns) is

$$D_{32} = \frac{1920}{U}\left(\frac{\sigma}{\rho_l}\right)^{0.5} + 597\left(\frac{\mu_l}{\sqrt{\sigma\rho_l}}\right)^{0.45}\left(1000\frac{L}{G}\frac{\rho_g}{\rho_l}\right)^{1.5} \qquad 1$$

where U = liquid to air relative velocity (ft/sec); L/G = liquid to gas mass ratio; s = liquid surface tension (dyne/cm); r = density, liquid - l or gas - g (g/cc); m = liquid viscosity (poise). Note that the equation is not dimensionally correct.

As can be seen, droplet size is a strong function of liquid velocity (spray rate & port diameter), atomizing air velocity (air pressure) and a weaker function of the temperature & physical properties of the liquid. For example, a 10 C increase in solution temperature decreases the mean diameter roughly 10% in the 40 C - 60 C range. The mean droplet size of the coating solution as a function of atomizing air pressure, liquid spray rate, and port diameter were all determined.

Particle size, friability, porosity, and solubility can all impact coating dynamics.

Particle size influences surface area. The smaller the particle, the more coating is required to achieve an acceptable film thickness, as determined by performance and customer specification. Larger particle sizes are preferred.

Particle friability can lead to particle attrition. The fine particles produced can accumulate in the layers of film, possibly decreasing its uniformity and effectiveness. The fines may also agglomerate with larger particles, requiring possibly higher percentage coatings to assure some minimum film thickness.

Porosity would only be a factor for solids with high surface porosity and large pore sizes. In this situation, it may be difficult to bridge the pores and additional coating would be required.

If the substrate is soluble in the spray solution, the chance of increased core penetration exists. One remedy for this is the concept of staged coating. Initial coating conditions can be low spray rates at possibly higher inlet air temperatures until the product is protected by a thin film. Then the spray rate can be increased and the temperature decreased to optimize coating quality without core penetration.

The humidity of the bed influences the evaporation rate of the coating solution both as a droplet and as a surface film. A balance is required, as discussed above. The bed humidity is a function of fluidization air volume, temperature, and dew point, plus coating solution concentration and spray rate.

The fluidized bed top spraying machine is designed to perform as either a top coater, an agglomerator / granulator, or a spray dryer. Therefore, it is critical to understand the mechanisms and conditions which cause or influence agglomeration and operate the top coater at conditions to minimize this effect.

Three major mechanisms are responsible for particle agglomeration or bed granulation to occur. Each mechanism involves material bridging and are the following:

1) Agglomeration by partial wetting/recrystallization;
2) Agglomeration by deposition of a film forming binder;
3) Agglomeration by layering

Agglomeration by wetting and recrystallization occurs at wet bed conditions of high humidity, high spray rate, low temperature. Highly porous and friable granules are produced.

Agglomeration by the deposition of a film forming binder is the most common mechanism. It is possible for solutions intended as coatings to become binders if the coating solution becomes viscous enough to inhibit proper spreading and remain tacky. Generally, granules produced are still porous but stronger, less friable.

Agglomeration by layering is the least common and controllable mechanism. It involves formation of a very wet agglomerate of fine particles engulfed by a larger droplet which become tightly packed upon drying. These small agglomerates coalesce and grow with a "snowballing" effect. Again, excessively wet bed conditions are required. Granules produced are typically very dense with minimal porosity.

Three key conditions are required to minimize agglomeration during the coating process. These conditions are small droplet size, low bed humidity, and short droplet travel distance from nozzle port to fluidized particle(s).

The concept of the coating zone is key to understanding the complex interactions involved during the coating process and to reliably scale-up the technology. The coating zone is the area in the immediate vicinity of the nozzle port where the liquid droplets mix with and contact the solid particles. Localized conditions within this zone influence the coating dynamics, not overall bed or process conditions. Once the proper combination of process conditions are determined for a single coating zone, then scale-up is accomplished primarily by multiplying the number of coating zones and scaling the conditions accordingly. The coating zone conceptxi is shown schematically below.

Figure#5: Coating Zone Schematic

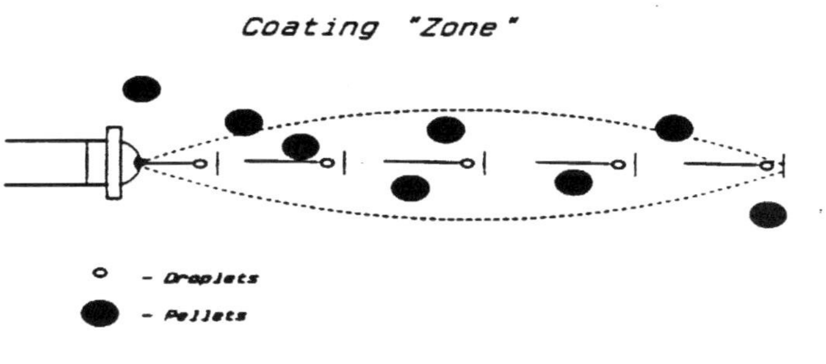

The coating solution spray or application rate is almost always dependent on the "saturation" of this zone and not the saturation of the fluidizing air. The same variables which control bed humidity influence the saturation of the coating zone. If the coating zone is saturated, then agglomeration will occur via one of the above mechanisms. This saturation condition can occur even if the fluidization air is not saturated and still has excess drying capacity. The remedy for this situation is to increase the number of zones by using multiple nozzles wands and/or multiple port nozzles. With this modification, the application rate can be significantly increased and/or the excess drying capacity can be utilized.

At this point, prior to further technology development, extensive technical and commercial discussions were held between FMC and the equipment manufacture personnel. Critical scale-up and design issues were discussed. These scale-up and design issues were:

- FLUIDIZATION AIR VOLUME/VELOCITY - More air is needed for fluidization and water removal.
- FLUIDIZATION AIR TEMPERATURE/HEAT INPUT - More heat is needed for drying. Increased air volume will allow for additional heat or greater flexibility in providing adequate heat.
- EXPANSION CHAMBER HEIGHT/TDH - More disengagement height is required to minimize fines loss and to minimize fines and/or coating solution droplets plugging the filter bags.
- FILTERS: MESH SIZE - Large, course filters are preferred (20-25 micron). Larger mesh sizes will decrease the pressure drop through the filter system allowing for higher air volumes. Larger mesh sizes will also minimize the filter plugging problem.
- NOZZLE DESIGN & CONFIGURATION - The key is to determine the conditions which provide for an effective coating zone at the nozzle, irrespective of application rate. Then if the application rate from one nozzle is less than required, the number of nozzles is increased until the required rate is obtained. A unit with multiple nozzle capabilities was required. Essentially, this was the approach learned in developing and scaling the Wurster technology. Once effective coating zone conditions were established on the lab and pilot scale, this same nozzle design was replicated to provide the desired application rates with the semi-works coater. It is important to recognize that technology scale-up set-backs can provide valuable learning experience and improved design data.

Incorporating the key concept of the 3 S's (Science, Statistics & Simulations) proved invaluable at this point in the program due to the complexity of the fundamental science and the shortened project timeline. Based on the first principles discussed above, both lab and pilot designed experiments were conducted. Besides the using the DOE methodology, another statistically-based method was used on the pilot scale know as self-directed optimization (SDO). The resulting data was further analyzed by simulation and statistical modeling to prepare as complete an understanding as possible prior to making a second semi-works trial which needed to succeed for the program to advance.

This pilot-scale designed experimental program was conducted to establish proper process conditions for successful scale-up relative to the coating zone concept and to understand the causes of the excessive agglomeration observed. The coating zone studies were designed as modified self-directed optimization (SDO) studies.

Effective coating zone conditions were determined using 8 runs by examining three fluidization air rates/velocities, two solution application rates, and two nozzle pressures. The application rate and nozzle pressure combinations yielded two ranges of droplet sizes. The air rate application rate combinations provided five drying capacities. The bed temperature and solution temperature remained constant. This set of conditions allowed for the replication of previous effective and poor coating zone found in the lab DOE while also exploring the limits and evaluating a preliminary bulk density prediction model. The SDO matrix of conditions is shown in Table 1, including run numbers and measured bulk densities.

TABLE #1: Modified Coating Zone SDO Matrix & Bulk Density Results

High Spray Rate (lb/hr/mm2)

Air Volume / Nozzle Pressure	Low (Bar)	High (Bar)	Drying Capacity
High (SCFM)	Run 4: 48 lb/ft3	Run 3: 56 lb/ft3	High (BTU/lb)
Middle (SCFM)	Run 2: 48 lb/ft3	Run 1: 52 lb/ft3	Middle (BTU/lb)
Low (SCFM)	Run 6: 44 lb/ft3	Run 5: 51 lb/ft3	Low (BTU/lb)
Droplet Size	Large (Microns)	Small (Microns)	

Low Spray Rate (lb/hr/mm2)

Air Volume / Nozzle Pressure	Low (Bar)	High (Bar)	Drying Capacity
High (SCFM)			High (BTU/lb)
Middle (SCFM)	Run 8: 50 lb/ft3	Run 7: 56 lb/ft3	Middle (BTU/lb)
Low (SCFM)			Low (BTU/lb)
Droplet Size	Large (Microns)	Small (Microns)	

As can be seen, higher bulk density (less agglomeration) is favored by higher nozzle pressure, lower spray application rate and higher air volumes/velocities. In terms of the coating zone, desired bulk densities are favored by smaller droplets combined with higher drying capacities. The data from all top-coating runs was also regressed and correlated into a bulk density model.

A model was subsequently developed to correlate key parameters to final bulk density (an agglomeration indicator). This bulk density model was developed for two reasons. First, to understand, control and minimize agglomeration by identifying the critical process parameters which lead to low bulk density. Second, to provide a predictive tool for evaluating coater designs and establishing future commercial process conditions.

The bulk density model was derived statistically. Several techniques were used. First, a database was constructed from all the top-coating trial process conditions and results. Using SYSTAT, Pearson correlations were calculated for bulk density with 28 variables and 22 potentially relevant interactions. The resulting correlation matrix primarily identified the variables & interactions to remove from further consideration. The matrix also provided a measure of the interdependence or intercorrelation of all the remaining variables. This information was useful in understanding possible compounding effects in the model. Multiple linear regression (MLG) was performed with the remaining variables and interactions. The goal was to derive as simple a model as possible which maximized the correlation (R, R2) and the F-Ratio (indicating minimized model residuals). Fifty (50) models were derived based on various combinations of the variables. In some cases, the regression was performed on large subsets of randomly chosen variables. In some cases, the regression was performed on smaller subsets of specific variables selected based on coating and agglomeration theory. Both forward and backward step-wise MLG's were done to validate the most promising models.

Ultimately, three models were generated in this manner. Each contains some subset of only eight variables and three of their interactions out of the original fifty. These key variables and interactions are:

1. Solution Application Rate per Port
2. Atomization Nozzle Pressure
3. Fluidization Air Volume
4. Fluidization Air Velocity
5. Fluidization Air Temperature
6. Total SPP charge
7. Solution Droplet Size
8. Heat Input
9. Zone Contacting Time(Interaction)
10. Droplet Evaporation Dynamic (Interaction)
11. Zone Drying Capacity (Interaction)

Model-by-model comparisons were made of the actual measured density versus the calculated density and of each model's residuals or fractional error. For the final models, the fractional errors are predominately less than 5%. The simplest and the best models are summarized below.

MODEL 1: $R = 0.84$ $F = 45.3$

BULK DENSITY = f (Air Flow, Mass Solid, Air Temperature, Droplet Size)

MODEL 25: R = 0.92 F = 87.6

BULK DENSITY = ƒ (Air Flow , Mass & Volume Solid, Spray Rate, Nozzle Pressure, Drying Capacity)

Each model was tested for its predictive capabilities using a different top-coater trial bulk density results. Three models provided very good to excellent agreement with the Vector trial data, with Model #25 being the best. Each model predicted that the commercial designs can provide product with acceptable bulk density. The models assumed that nozzle positioning and height are properly designed.

This model was successfully used to determine scale-up conditions which provided a good coated product. With the information provided by the model, a successful semi-works scale-up trial was conducted using a different Glatt coater. This coater design was a closer match to our product needs with respect to multiple nozzles and additional heat input.

The coating process variables can be divided into six broad categories. Many of the variables have been identified above, but will be reiterated for completeness. The categories, including all relevant variables, are:

1) Coater Design - Wurster or Bottom Spray; Top Spray
2) Atomizing Air Conditions - Pressure; Volume
3) Nozzle Design - Port Size; Ports per Nozzle Head; # of Nozzle Wands; Spray Angle; Nozzle Position or Height (relative to the static bed)
4) Fluidizing Air Conditions - Volume; Superficial Velocity; Dew Point; Temperature
5) Coating Solution Conditions - Application/Spray Rate; Solids Concentration; Application Time; Temperature; Physical Properties (Viscosity, Surface Tension, Specific Gravity); Solvent
6) Solid Characteristics - Particle Size/Surface Area; Friability; Porosity; Solubility (relative to coating solution)

The critical variables for coating in general, either by Wurster or top coating, are atomizing air conditions, fluidization air dew point and temperature, and solution spray rate.

The atomizing air pressure and volume were selected so that the droplet size is small relative to the particle to be coated. The higher the pressure and volume, the smaller the droplets. However, above 3-4 bar (45-60 psi) pressure has a minimal effect on droplet size. Also, if the air velocity is too high at the nozzle tip, the shear forces may disrupt the fluidization pattern in the coating zone or damage friable particles.

The fluidization air temperature and dew point are critical for controlling the solvent evaporation rate and maintaining the desired product or bed temperature. These parameters in turn dictate the degree of saturation in the coating zone. The higher the air temperature, the less seasonal variation on drying capacity and the lower the bed humidity.

The spray rate influences both the droplet size and the bed humidity. The desire is to determine the maximum spray rate possible without saturating the coating zone. The trick is that at lower spray rates, the droplet size is smaller and bed humidity is lower. These are the conditions that favor coating over agglomeration with a top spray machine. For scale-up, the increase in spray rate is often proportional to the drying capacity.

Three additional key process variables for top coating are the fluidizing air volume, superficial velocity, and solution concentration.

The air volume/velocity control the fluidization pattern in the bed and the particle velocity through the coating zone as well as the excess drying capacity. It is important that the volume/velocity be high enough to provide smooth fluidization but not slugging. Slugging would disrupt the dynamics of the coating zone, effecting coating quality, and should be avoided. On scale-up, the bed depth should be maintained so as to prevent slugging. Expanding the bed diameter, while maintaining a minimum air velocity, is one way to accomplish scale-up.

The concentration of the coating solution needs to be optimized. Generally, lower viscosity and less water/solvent are preferred to minimize agglomeration. However, higher concentrations containing less water tend to have higher viscosities.

For design and scale-up of a coating technology system, the following factors need to be well understood:

A. BATCH SIZE - Typically the target is approximately 80% of the bowl capacity for final coated product. Process conditions are not scaled-up according to batch size.

A critical scale-up parameter is the nozzle height relative to the static bed. This distance should remain the same as in the lab/pilot units if the fluidizing velocity remains the same. If too high on start-up, excessive agglomeration or spray-drying can occur since the nozzle will be farther from the bed and spraying into a more disperse substrate pattern.

B. FLUIDIZATION PROPERTIES - Since bowl volume, bowl diameter, and bed depth increase on scale-up, the key is to maintain smooth fluidization and minimize slugging. If smooth fluidization is achieved in the lab/pilot units, scale-up is based on the corresponding air velocity (screen area) or air volume, whichever is lower to be conservative. Scale-up is not based on the increase in batch size. For example, if the scaled-up coater unit has 5x the bowl volume but 3x the bowl screen cross-sectional area, then the air volume is increased 3x to maintain the same face velocity, not 5x to maintain the same air/solids ratio.

Slugging is caused primarily by large air bubbles and is not usually seen in lab/pilot units due to the shallow bed depths and wall effects. Air bubble growth rate increases with increasing bed depth, particle size and density. In general, the SPP substrate properties were not in the range (diameter larger than 1 mm, density at or above 1 g/cc) which is known to cause fluidization problems for top-coating scale-up.

C. COATING ZONE: DROPLET SIZE - Droplet size is established from the spray rate, atomizing air pressure, and port size in the lab/pilot units. The droplet size should be maintained in the larger coater. This is done by simply using multiple nozzles wands with multiple heads while maintaining the same per port spray rate and atomizing pressure.

D. COATING ZONE: SATURATION - Agglomeration due to saturation of the coating zone often occurs well before the fluidization air is saturated. Again, the use of multiple nozzles can take advantage of the excess drying capacity. Key to start-up will be the positioning of the multiple nozzles so that minimum overlapping of the spray pattern occurs. Spray angle may need to be modified to achieve this minimum overlapping. Standard nozzles have a 60 degree spray pattern. However, special nozzles with 45 degree spray patterns exist.

E. SOLUTION SPRAY RATE - Spray rate is generally scaled proportionally to the increase in drying capacity. Drying capacity is determined by the fluidizing air volume and heat exchanger design. Since air volume scale-up is based on maintaining the same velocity, initial spray rate scale-up is based on the same factor. From the above example, the initial spray rate increase would be 3x. However, this factor may require modification if the coating zone is saturated. For example, additional nozzles and/or additional heat exchanger duty may be required to achieve the targeted or design application rates.

F. PRODUCT TEMPERATURE - The same product temperature should be maintained as used in the lab/pilot units. The inlet air temperature is adjusted or balanced to achieve this condition.

G. MASS EFFECTS - The larger batch sizes and bed depths associated with scale-up may lead to increased friability. The friability will be dependent on the strength or robustness of the substrate, and on the batch cycle time (dictated by the spray rate and coating level). Higher spray rates and lower coating levels would be beneficial for more friable material.

A phenomenon called "smushing" can also exist at larger scales. The mass of the bed may tend to compact weakly agglomerated particles, maintaining or increasing the bulk density. Since we want to minimize agglomeration, this effect may be beneficial for our product or may offset excessive agglomeration.

As an approach to reducing the scale-up risk and increasing the probability of commercial unit success for this technology, alternate top-coater vendors were evaluated to assure that the final commercial-size coater would be meet our design needs and be cost effective via competitive bidding. Vector Corporation and Fluid Air, Inc. were identified as potential vendors in addition to Glatt Air Technique. Both companies had different designs than Glatt and so scale-up trials were conducted by both at each manufacturing site. Vector's trials were successful whereas Fluid Air's trials were not for our particular product.

Commercial designs were obtained based on the results of the scale-up trials and additional engineering specifications. The final design was chosen based on its robustness and the manufacturer's experience in building industrial fluid-bed coater/granulators. Both manufacturers' designs addressed our capacity requirements based on current and potential optimized conditions. These potential optimized conditions included larger bed charges, higher solution concentrations, faster application rates, and/or lower coating levels either alone or in some combination.

CONCLUSIONS

1. Three Key Factors for successful product design and process development are intellectual downstreaming; seamless engineering and combined use of the 3 S's.
2. Good quality coated reactive peroxygen product was produced using top coating technology which had been successfully scaled-up from the lab to the semiworks.
3. Acceptable commercial scale top-coater designs for coating CPP were obtained from multiple fluid-bed coater manufacturers.
4. Successful process / product development of this innovative coating technology within the required project timeframe was due to the following key factors: defining the ideal product; a cross-functional, responsive team; a project champion; multiple parallel development efforts; systematic utilization of fundamental science, statistical tools and simulation & modelling; rapid resolution of process issues & problems.

---

[i] Basu, P.K., Mack, R.A, and Vinson, J.M., "Consider a New Approach to Pharmaceutical Process Development", Chem. Eng. Progress **95** (8), August 1999

[ii] Mukesh, D., "Succeed at Process Development", Chem. Eng.

Progress **95** (2), February 1999

[iii] Worstall, J.H., Adkison, W.L., and Durante, D.R., "Speed Your Process Development", Chem. Eng. Progress **94** (12), December 1998

[iv] Lenz, DH, Guiliano, BA, Terzian HD, Juelke, CV, and Hills,WA, "CREATING THE "WORLDS'S LARGEST COATER" FOR A NEW REACTIVE PEROXYGEN PRODUCT: A PROCESS INNOVATION CASE STUDY", Spring 2000 AIChE Meeting, Atlanta, GA, March 5-9

[v] Covey, S., "The 7 Habits of Highly Effective People"

[vi] Zeldman, M., "Keeping Technical Projects on Target", 3rd Ed., © 1998, EMZEE Associates, Inc.

[vii.] Perry's, 5th Edition, "Phase Dispersion", pages 18-58 to 18-67

[viii.] Nukiyama, S., and Tanasawa,.Y., "Experiments on the Atomization of Liquids in an Air Stream", TRANS. SOC. MECH. ENGRS. (Japan), 4, 5, 6, 1938-1940

[ix.] Laskowski, J.J., and Ranz, W.E., "Spray Quenching", AICHE J, 16,5, 802 (1970)

[x.] Wigg, L.D., "Drop-Size Prediction for Two-Fluid Atomizers", J. INST. FUEL, 37, 500 (1964)

[xi.] Ibid., Glatt Process Training Seminar

(45a) Pilot and Full-Scale Application of Technologies for Mercury Control

Noah Meeks, Research & Environmental Affairs, Southern Company Services, Inc., Birmingham, AL

With the final MATS rule going into effect in 2015, electric utilities must now consider mercury control as a needed technology in their emissions train. This talk wiill cover recent full-scale injections of activated carbon and alkali injection, as well as pilot-scale studies of non-halogenated carbons or non-carbon sorbents. We will also discuss recent mixed matrix membrane pilot studies. **(Most recent data will be included in the abstract closer to the meeting)

(45b) Acid Tunneling Vs Acidizing : A Case Study of Rudatain Oilfield-Kuwait

Adel Elsharkawy, Petroleum Engineering, Kuwait University, Kuwait, Kuwait and Mohamed Al-Hadher, Pet Eng, Kuwait Oil Company, Safat, Kuwait

Acid Tunneling Vs Acidizing : A Case Study of Rudatain Oilfield-Kuwait

By Mohammed Al-Hadher and Adel Elsharkawy

Kuwait University, College of Engineering & Petroleum

Po Box 5969 Safat 13060 Kuwait

Abstract

Acidizing is one of the most important techniques used in oil industry in order to enhance the reservoirs and increase oil production. There are three types of acidizing techniques which are matrix acid stimulation, acid fracturing, and acid tunneling. In this paper a comparison these types of acidizing is discussed. The case study of RA-0007 acid tunneling completed in Kuwait Oil Company is presented. It represents the first dendritic with extensive worm holes created by acid tunneling for maximum reservoir contact. The whole effect is aptly demonstrated by the change in PI of the well before and after the acid tunneling as mentioned before. Analysis of the well designed and implemented job showed that two tunnels were created successfully in MA-C formation through a two days rig-less operation. The results indicated an improvement in the well productivity index (PI) which had been increased from 1.3 to 3.87 bpd/Psi.

(45e) Sustained Casing Pressure Modeling and Well Integrity

Tony Rocha-Valadez, Mary Kay O'Connor Process Safety Center, Artie McFerrin Department of Chemical Engineering, Texas A&M University, College Station, TX, A. Rashid Hasan, Harold Vance Department of Petroleum Engineering, Texas A&M University, College Station, TX, C. Shah Kabir, EPT, Hess Corporation, Houston, TX and Sam M. Mannan, Artie McFerrin Department of Chemical Engineering, Mary Kay O'Connor Process Safety Center, Texas A&M University, College Station, TX

SUSTAINED CASING PRESSURE MODELING AND WELL INTEGRITY

T. Rocha-Valadez[1*], A.R. Hasan[2], C.S. Kabir [3] and S.M. Mannan[1]

[1]Mary Kay O'Connor Process Safety Center, Artie McFerrin Department of Chemical Engineering, Texas A&M University, College Station, Texas 77843-3122, US., email: *tonyrochav@tamu.edu, mannan@tamu.edu.

[2] Harold Vance Department of Petroleum Engineering, Texas A&M University, College Station, Texas 77843-3122, US, email: rashid.hasan@pe.tamu.edu.

[3]Hess Corporation, 1501 McKinney Street, Houston, TX 77010, US, email: skabir@hess.com.

With an increasing demand for hydrocarbons and other sources of energy, the methods for delivering such energy needs have become complex. In the case of the oil and gas industry, exploration and production have gone to deeper waters and deeper horizons than ever before, resulting in drilling several miles into the earth's core. Deeper well penetration means encountering increasingly high formation pressures and temperatures. For example, in deepwater drilling, well construction and production management deal with significant heat transfer between the produced fluids and its surroundings. The high-rate production and the attendant heat transfer demand close attention in terms of operational safety and potential environmental impact. One phenomenon that occurs in producing wells is sustained casing pressure (SCP), which presents a risk to both process safety and the environment.

In oilfield terminology, the casing is a steel pipe cemented to the surrounding formation. Multiple concentric casings are involved in a typical well construction. The casing immediately surrounding the production tubing is known as the production casing, which is most vulnerable to thermal and mechanical stresses and fluid leakage. The cement sheaths provide zonal isolation, support the casing, and protect it from corrosive fluids. Ideally, pressures monitored at the wellhead for multiple casings should change minimally after being bled off through a needle valve at the surface. When the pressure rebuilds after being bled off, that casing is said to exhibit SCP [1].

SCP occurs because of stresses that the system experiences owing to fluids production up the tubing. Besides the normal source of SCP, a poor cement job may allow hydrocarbon migration from the producing or other horizons and accumulation into one or more of the casings. The risk of not attaining well integrity can range from the activation of rupture discs to the blowout of hydrocarbons. Because of the safety and environmental dangers of SCP, regulations require remedial treatments for a casing string that exhibits significant SCP problems. Unfortunately, available methodologies to test for SCP are often qualitative and empirical in nature [2, 3], rendering them inapplicable in many situations. No practical systematic testing procedure has emerged to quantify SCP.

In this paper, we present a model for SCP that is rooted in the transport processes of the system. The model assumes gas leaking through the cement barrier, and then migrating up the annular-liquid column. It accounts for the fluid-temperature profile along the wellbore and liquid compressibility. Some robust assumptions about the transport processes allow us to write the governing equation as a first-order, linear-differential equation that is solved with appropriate boundary conditions. The resulting algebraic expression for the casing pressure rise as a function of time paves the way for all applications. A comparative study of the model's performance with field data suggests excellent agreement for wells where the casing pressure increase is owes gas influx into the annulus. The proposed model could form the basis for standardization of SCP testing. This study also offers some test procedures to assess SCP-related safety concerns.

References

1. Bourgoyne Jr., A.T.J., S.L. Scott, and W. Manowski, ***A Review of Sustained Casing Pressure(SCP) Ocurring on the OCS***, 2000, Final Report Submiteted to MMS.

2. Xu, R. and A.K. Wojtanowicz, ***Diagnosis of Sustained Casing Pressure from Bleed-off/Buildup Testing Patterns.*** SPE, 2001: p. Productions and Operations Symposium.

3. Wojtanowicz, A.K., S. Nishikawa, and R. Xu, ***Diagnosis and Remediation of Sustained Casing Pressure in Wells***, July 2001, MMS.

(45f) Evaluation of Particle Growth in an Electrochemical Fluidised Bed Reactor

Emmanuel Ehirim, Rivers State University of Science and Technology, Port Harcourt, Nigeria

The process of electrodeposition within electrochemical reactor results in the increase of particle sizes, which increases the volume of the bed. This work entails a mathematical model to describe the effect of this phenomenon on the behaviour of a three-dimensional fluidised bed electrode operating under the condition of electron transfer. The model incorporates the pseudo-stationary behaviour in the reactor since the bed increases in volume due to the particle size growth. The Orthogonal Collocation method was used to solve the developed differential equations. The overpotential distribution and the reaction rate within the reactor have been calculated. The principal results are the effects of the electron transfer operation on the superficial displacement with time. The effects of total applied current, electrode geometry and the bed porosity on the behaviour of the system have been analysed. The result shows a very gradual increase in the bed height.

(45g) Debottlenecking Gas Dryers and Desiccant Adsorbers

Benjamin A. Schmitt, Hydrocarbon Processing Industry (HPI) - Business Unit, Johnson Screens, Inc., New Brighton, MN and Michael Ekholm, Engineering, Johnson Screens, Inc., New Brighton, MN

Debottlenecking Gas Dryers and Desiccant Adsorbers

Though Maximization of the Useable Bed Volume with Johnson Screens' SSG^{TM}

Benjamin Schmitt, PE

Senior Mechanical Engineer, Technology

Michael Ekholm, PE

Applications and Product Design Manager

Johnson Screens Inc. St. Paul, MN

AIChE Spring Meeting, April 2013

13th Topical Conference

San Antonio, TX

Introduction:

The following presentation of the Shaped Support Grid (SSG^{TM}) releases additional supporting data and updates to the initial presentation given by Robert G. Norell at the 2010 AIChE Spring Conference.

Limitations on ethylene dryers and desiccant adsorbers can create bottlenecks in the production process which can lead to a need for increased dryer and adsorber capacity. Capital intensive investments may be necessary in order to eliminate the bottleneck which may include the addition of an additional vessel. With limits on the plot plan, this may not be a workable solution. Adding to the problem is the amount of time consumed in engineering, planning and construction, which can have negative effects on the operation of the ethylene unit. Increased desiccant volume and operational efficiency can increase the unit's production with minimal capital expenditures. Johnson Screens', patent pending, SSG^{TM} can add the increased process efficiency and cycle life necessary to extend the operating capability of the unit.

Extending the operation, though debottlenecking, can be achieved by maximizing the time before breakthrough of the media bed. There are three primary functions that maximize the breakthrough time:

Proper gas distribution in both normal operation and regeneration conditions which maximizes the bed utilization.

Increased molecular sieve or desiccant bed volume are achieved through the elimination of traditional bed supports comprised of support grids and beams. The support interface for the screen configuration also allows for thermal expansion typically present in vessels with a thermal cycle as part of the process. The design allows for thermal cycling without the risks of seal or structural failure inherent in other solutions.

Minimizing the bed volume consumed by the distribution and support system increases the net desiccant volume lost when utilizing alternative bed supports.

Johnson Screens ***SSG*** is designed to provide the industry with an alternative, cost effective, approach to increasing capacity. Supporting data and preliminary case studies will be depicted to quantify the benefits that are achievable through the implementation of the ***SSG***.

Conclusions:

The use of a Johnson Screens ***SSG*** provides a reliable immediate solution to reduce capacity bottlenecks. The ***SSG*** increases the bed volume, when compared to traditional bed support arrangements comprised of grids and beams. The increase in desiccant volume will allow for increases in feed flow rate and/or adsorbtion cycle time. The ***SSG*** maximizes the flow distribution during desiccant regeneration, maximizing the utilization of the complete cross section of the vessel. This solution can be implemented in both new and existing vessels, providing consistency in design between planned increases in performance of current vessels and future expansions.

The modular design of the ***SSG*** allows for easy installation and maintenance. Connection of the ***SSG*** to the vessel is performed within the outlet nozzle, eliminating all welding on the pressure vessel and eliminating any need for recertification of the pressure vessel after installation. The design allows for ease of dismantling for inspection of the vessel head during routine maintenance and inspections. The low profile will also reduce interference with any required media exchanges. The design package of the ***SSG*** provides the end user with a holistic solution that fulfills debottlenecking needs with limited capital expenditures.

Biography of Speaker:

Benjamin Schmitt is a Senior Mechanical Engineer in Johnson Screens' Hydrocarbon Processing Industry Technology Group. Ben graduated from North Dakota State University in 2004 with a Bachelor of Science degree in Mechanical Engineering. He currently working toward his Master's in Business Administration from Hamline University. Ben is member of AIChE and is a registered professional engineer in the state of Minnesota.

AMACS

PROCESS TOWER INTERNALS

AIChE 2013 Spring Meeting Poster board

Design Considerations for effective liquid removal In Suction Drums

Presented By:

Kanti Patel, PE

Compressor down again?

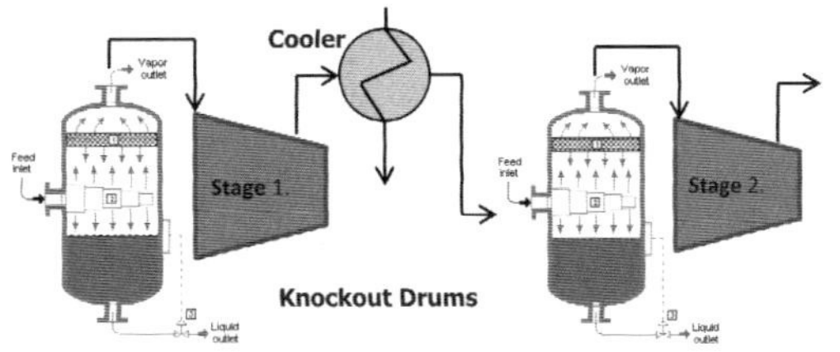

Is liquid carryover the culprit?

<u>We have heard it all before</u>:

-Premature compressor shutdown.

-Production time lost.

-High yearly compressor maintenance costs.

-Impeller damage due to pitting corrosion.

-Corrosion due to salt deposits.

AMACS

PROCESS TOWER INTERNALS

Common Design Problems with Suction Drums which promotes liquid carryover

1. Choosing the correct mist eliminator (M.E. equipment)
2. Knowing where to install the M.E. in a suction drum.
3. Consider a high mist loading in the gas when selecting your M.E.
4. Fouling considerations.
5. M.E. disengaging space, bad gas distribution & liquid slug in the feed.
6. One micron and great droplets, versus submicron mist removal.
7. High viscosity liquids.
8. Dislocated M.E. due to pressure surges.
9. Through put exceeds original design capacity.
10. Capacity range too large for suction drum.

AMACS
PROCESS TOWER INTERNALS

Choosing the correct Mist eliminator equipment

- Proper sizing

+ Use Souders-Brown equation to find optimum velocity.

$$\mathbf{Vd = K} \cdot \sqrt{\frac{Pl}{Pg} - 1}$$

+ Select the correct vapor velocity factor K. based on the operating pressure, and type of equipment.
  - Lower K-factor 0.01 for every 100 psi
  - K factor to select when using:

Mesh Pad	K= 0.1 - 0.55
Chevron Vane	K= 0.45 - 0.65
MultiPocket® Vane	K= 0.75 – 1.0

AMACS
PROCESS TOWER INTERNALS

Choosing the correct Mist eliminator equipment cont.

- Selection based on droplet size.

Table 2. Droplet sizes (water in air) typically captured with 99.9% efficiency by mist eliminator elements of various types

Element type	Size range microns
Fiber candles or panels	0.1 and larger
Mesh with co-knit yarn	2.0 μm and larger
0.006-inch knitted mesh	5.0 μm and larger
0.011-inch knitted mesh	10 μm and larger
AMACS MultiPocket ® Vane	10 μm and larger
Conventional vane arrays	15 μm and larger

AMACS
PROCESS TOWER INTERNALS

Where to install the M.E. in the suction drum

- Improperly installed M.E. can cause severe liquid carryover problems.
- General accepted spacing guidelines.

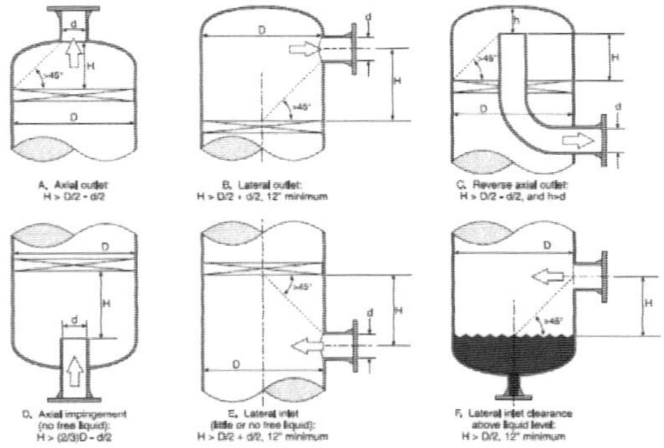

AMACS
PROCESS TOWER INTERNALS

Consider high mist loading when selecting your M.E. equipment

+ High mist loading in the gas stream can flood your M.E. equipment and cause liquid carryover.

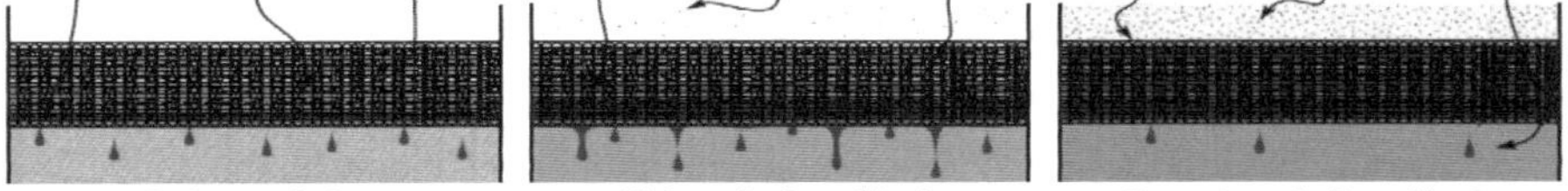

Figure 4. Envisioning mesh pad performance degradation as liquid and gas loads increase (vertical cross-sections)

AMACS
PROCESS TOWER INTERNALS

Consider high mist loading when selecting your M.E. equipment cont.

+ Limited drainage capabilities
  - mesh M.E. <0.5 gal/FT2/min
  - Vane type M.E. <2 gal/FT2/min

+ Select a combination of vane & mesh mist eliminators

AMACS
PROCESS TOWER INTERNALS

Fouling Applications

+ Fouling of mesh mist eliminator will cause liquid carryover.

+ For mild fouling install a water spray system below the mist eliminator for occasional inline cleaning (based on Delta-P)

+ For consistent fouling problems, install a Vane mist eliminator upstream of the mesh mist eliminator.
    - Vanes remove solids and bulk liquids therefore, protecting the downstream mesh mist eliminator

AMACS
PROCESS TOWER INTERNALS

Fouling Applications cont.

+ For consistent fouling problems where the solids are "sticky" install a water spray system below the vane mist eliminator.

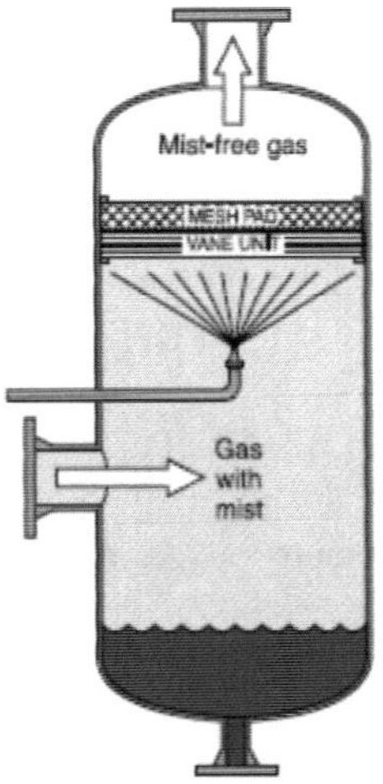

AMACS
PROCESS TOWER INTERNALS

Bad Gas distribution, limited disengaging space and liquid slugs.

+ Bad gas distribution, limited disengaging space in a suction drum can cause severe liquid carryover.

+ CFD analysis can assist in locating high velocity areas.

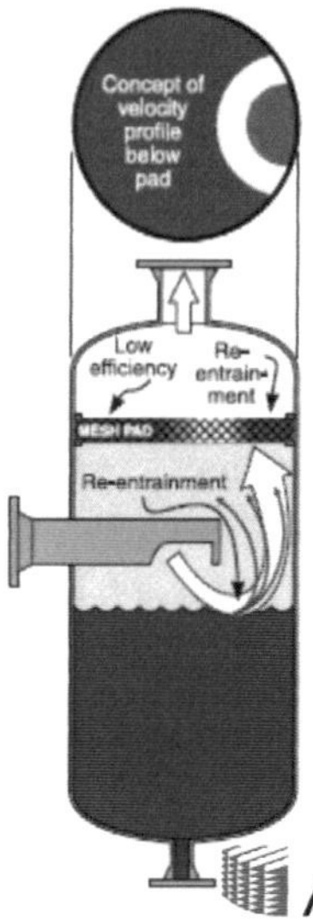

Bad Gas distribution, limited disengaging space and liquid slugs cont.

+ Use inlet diffuser to force uniform gas distribution i suction drum.

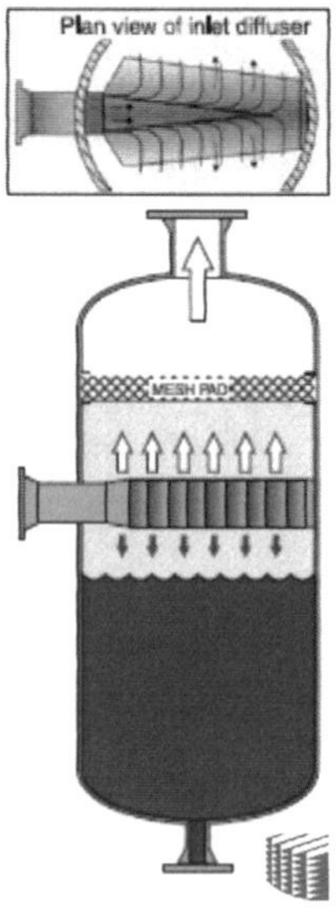

Inlet Feed Device

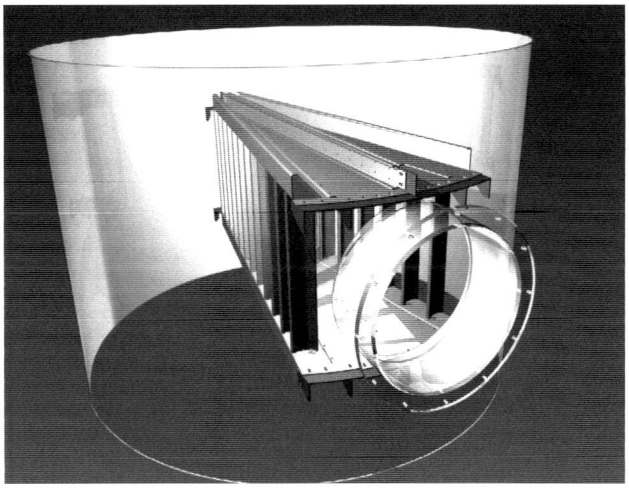

AMACS PROCESS TOWER INTERNALS

Inlet Feed Device

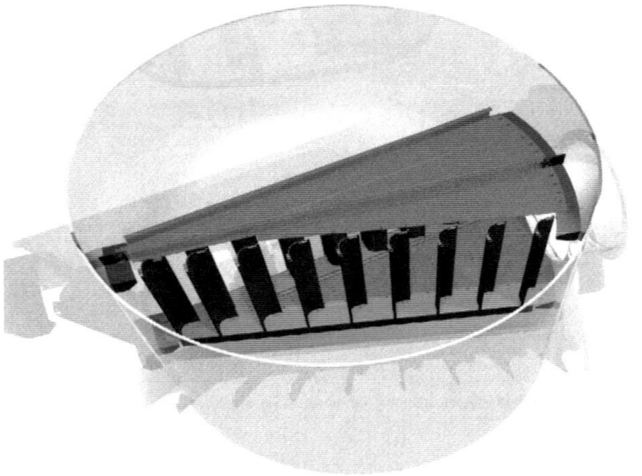

AMACS PROCESS TOWER INTERNALS

Bad Gas distribution, limited disengaging space and liquid slugs cont.

+ Use a perforated distribution plate downstream of the M.E. for additional distribution assistance.

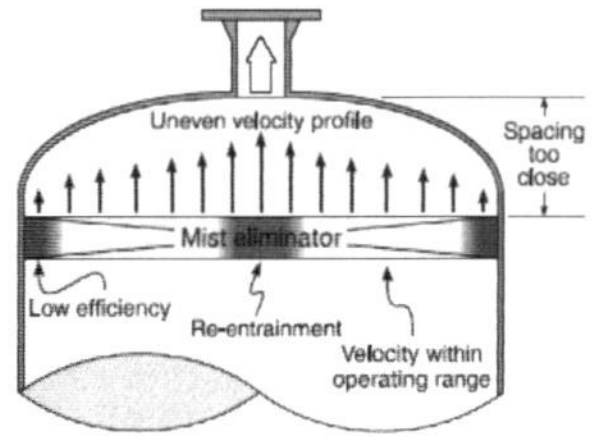

Figure 7. Example of mist eliminator performance degradation due to uneven velocity profile

+ Inlet diffusers also handle incoming liquid slug which otherwise can cause flooding of the Mist eliminator.

AMACS
PROCESS TOWER INTERNALS

Know what mist size droplets You are dealing with

+ Know if your mist is > 1 micron or submicron.
+ When gas travels from A to B, condensation in the pipeline will cause a fine mist (fog or submicron droplets)

  Examples:

  - Incoming gas from an offsite location
  - Fuel gas to furnaces/heaters with low NOx burners

AMACS
PROCESS TOWER INTERNALS

Know what mist size droplets You are dealing with cont.

+ Submicron mist is not captured by conventional M.E.

+ Submicron mist requires a coalescing filter, typically
installed in a two stage filter separator.

AMACS
PROCESS TOWER INTERNALS

Know what mist size droplets You are dealing with cont.

1c stage; submicron droplets are coalesced to > 10 micron droplets.

2c stage; greater than > 10 micron droplets are ejected from the coalescing filter and captured by a vane, which captures droplets > 8 micron.

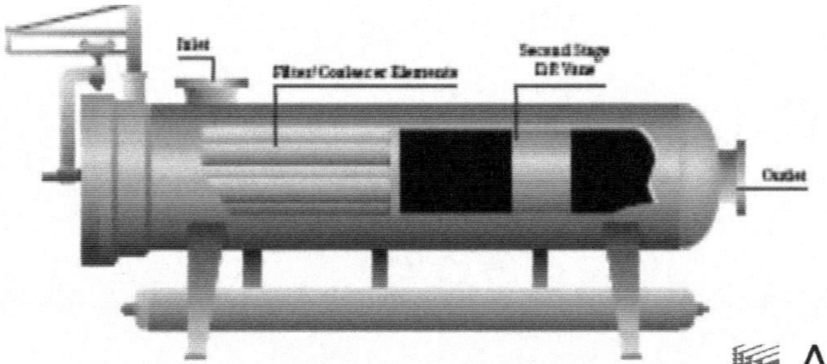

AMACS
PROCESS TOWER INTERNALS

High viscosity liquid as mist

+ Viscous liquids drain slower out of the M.E. and could cause localized flooding.
+ Use a mesh mist eliminator with an open drainage layer on the bottom to improve liquid drainage.
+ Use a lower K- value (lower face viscosity) for the M.E. To allow additional drainage time.
+ Use a vane type M.E. upstream of the mesh mist eliminator, to remove the bulk of the viscous liquid, so the M.E. can effectively handle the residual liquids.

Dislocated M.E. due to System pressure surges

+ M.E. sections can be dislodged into the downstream equipment, which then creates massive carryover.
+ Standard M.E. equipment often gets installed and secured to the support ring with tie wire.

Dislocated M.E. due to System pressure surges cont.

+ Where system pressure surges are common, there are more secure ways to install the M.E.
  - dual support rings
  - through bolts
  - install a pressure relief door

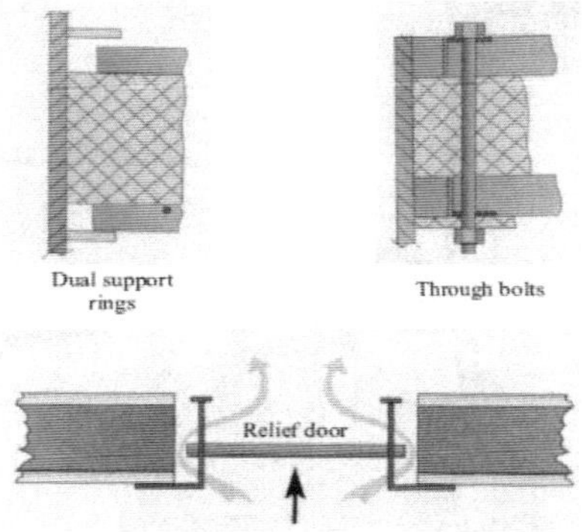

AMACS
PROCESS TOWER INTERNALS

When throughput exceeds original capacity

+ There are various solutions to increase capacity in an existing vessel.
  - Patented High capacit
    mist eliminator 0% - 1(

AMACS
PROCESS TOWER INTERNALS

When throughput exceeds original capacity cont.

-Patented MISTFIX™
cylindrical mist eliminator 0% - 25%

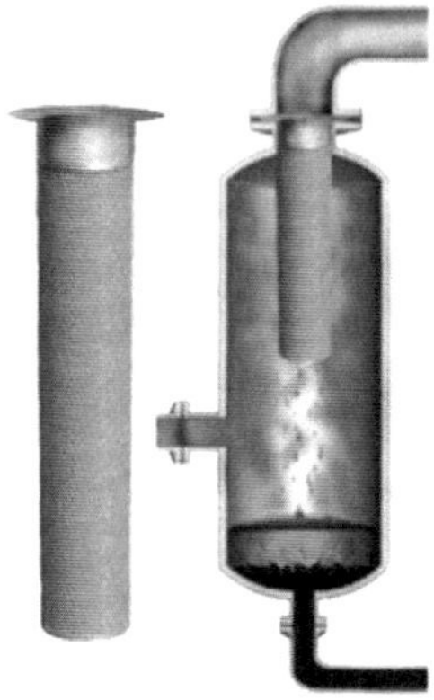

When throughput exceeds original capacity cont.

- High capacity Patented MultiPocket® vane. 0%-50%

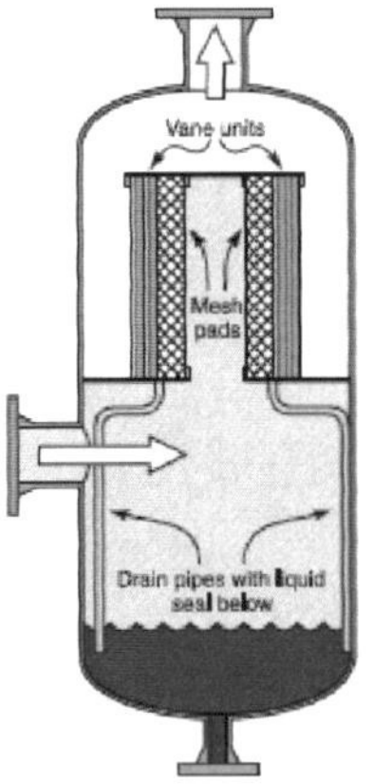

AMACS
PROCESS TOWER INTERNALS

When throughput exceeds original capacity cont.

+ MistFix™ is installed through the gas outlet, all other M.E. equipment comes in sections, installed through the manway.
+ With increased capacity comes increased bad gas distribution.
  - Use an inlet diffuser.
+ To avoid welding new support rings and housing for the high capacity new equipment, use double expansion rings, a compression fit.

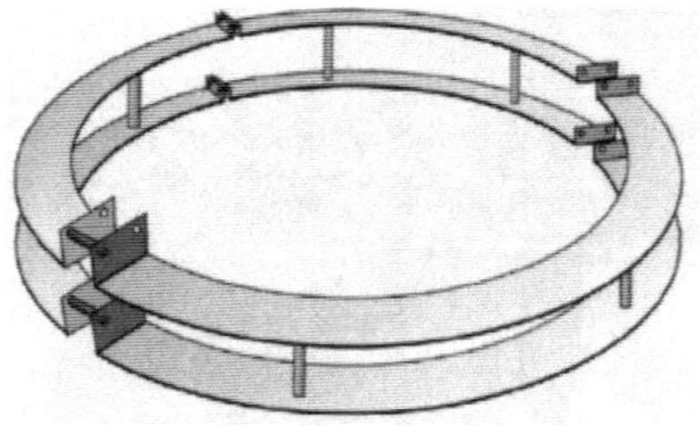

AMACS
PROCESS TOWER INTERNALS

Support Rings

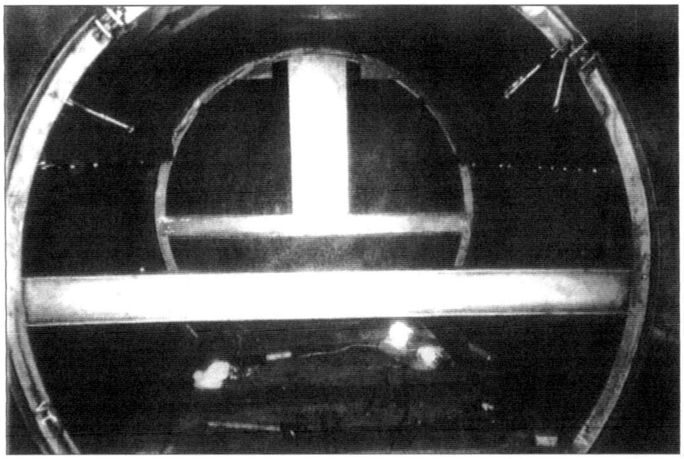

AMACS
PROCESS TOWER INTERNALS

Support Rings

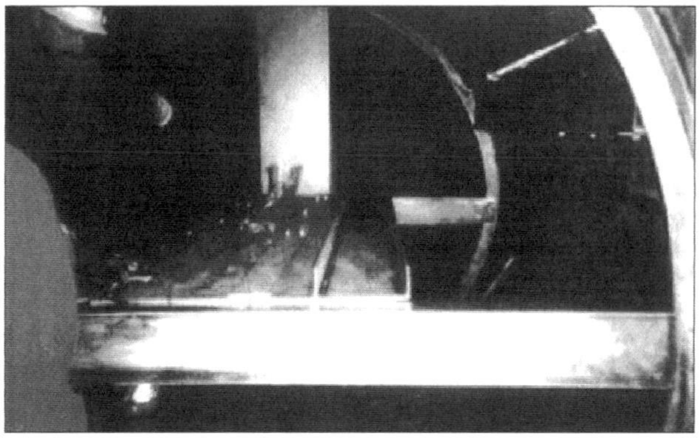

When the capacity range is too large for a suction drum

+ Typical M.E. equipment will operate between

  30% – 110% capacity.

+ When a suction drum often operates below 30% liquid carryover may occur.

When the capacity range is too large for a suction drum Cont.

To facilitate a capacity range of 10% - 110% consider a 2 phase M.E. in the same drum.

- one that operates 30% - 1
- second that operates between 10% - 30%

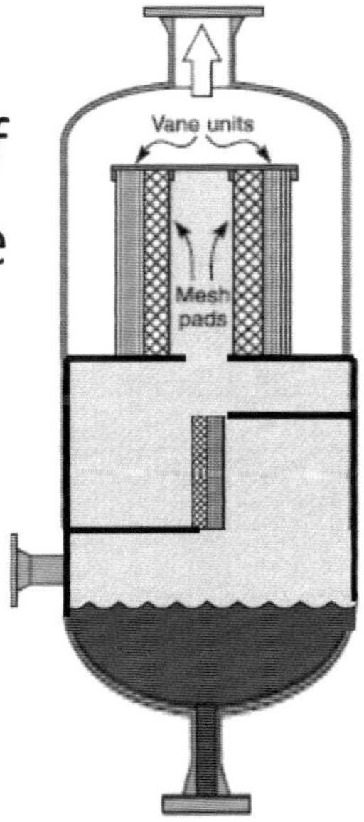

(45i) Renewable Feedstock for Steam Crackers: Catalytic Upgrading of Crude Tall Oil (CTO) Into Bio-Naphtha

Jinto Anthonykutty, Juha Linnekoski, Antero Laitinen and Ali Harlin, VTT Technical Research Centre of Finland, Espoo, Finland

Renewable feedstocks from bio resources are better alternatives to petroleum derived naphtha for the production of olefins and aromatics by steam cracking or reforming.[1, 2] Mostly, these renewable feedstocks (bio-naphtha) rich in n-alkanes are produced by catalytic upgrading methods via hydrotreating/hydroprocessing; in which oxygenates are removed from bio-oil by various deoxygenation processes. At present, hydrotreated vegetable (HVO) oil is produced in many countries on industrial scale using existing petroleum refinery methods[3]. However, employing higher priced vegetable oils could be a major concern in these processes and the main challenge is therefore finding an inexpensive and non-food chain affecting feedstock. Tall oil, obtained as a by-product from Kraft pulping process, is abundant and cheap;and hence, meets the criteria of a low cost starting material for upgrading[4]. Moreover, the utilization of tall oil in conventional oil refineries could serve as an eco-friendly method to convert bio-wastes into bio-naphtha[1].

In the present research approach, catalytic upgrading (hydrotreating) of crude tall oil (CTO) into bio-naphtha suitable for a steam cracker is discussed. CTO, which contains mainly fatty acids (42.9 wt. %) and resin acids (25.7 wt. %) was obtained from Stora-Enso Kaukopää (Finland) mill and used as received. A commercial NiMo catalyst was employed and the experiments were conducted in a continuous, down flow, fixed bed reactor at different process conditions (WHSV 1-3h^{-1}, Pressure 5 MPa and Temperature 350 - 450 ^{o}C). After the stabilization period (6 h.), the liquid sample was collected and separated into organic and aqueous phases.The product distribution in organic phase was analyzed with a Shimadzu GC/MS equipped with a Zebron MS 30m × 0.25 mm × 0.25μm column.

This current approach reports that CTO produces mainly paraffins, cycloalkanes and aromatics during upgrading process. Paraffinic hydrocarbons mainly comprise n-octadecane (20-25 wt. %) and n-heptadecane (15-20 wt. %), are obtained in maximum yield (55- 60 wt. %) in low temperature reactions. The yield of paraffinic hydrocarbons increased as a function of increasing space time in low temperature reactions. 18-Norabietane and other mono-, di- and tri- cycloalkanes are obtained from resin acids in CTO, and these cyclic structures are converted to aromatics such as naphthalene and phenanthrene in high temperature reactions (>400 ^{o}C). Stigmastane, apparently produced from sterols in CTO are also present as a major three ring component in the product stream. Additionally, gaseous products obtained and the yield increased as a function of temperature. The trend observed with the distribution of gaseous products reveal that decarboxylation and cracking reactions are favored at higher temperatures, while hydrogenation and hydrodeoxygenation (HDO) are suppressed.

The elemental composition of CTO shows that it contains relatively high sulfur content (1800 mg/kg), could be useful to keep the catalyst's activity with time on stream (TOS). However, the considerable amount (87 mg/kg) of alkali metals (Na, K, Ca and P) in CTO might be critical for the catalyst activity with TOS, will be discussed in this contribution.

The product composition obtained with CTO is assessed in comparison with compositions and characteristics of refinery naphtha feeds, and will be discussed in detail. Furthermore, this contribution will discuss the achieved deoxygenation rate and success criteria for the product composition of upgraded CTO.

References

1. Gornay, J.; Coniglio, L.; Billaud, F.; Wild, G. Steam Cracking and Steam Reforming of Waste Cooking Oil in a Tubular Stainless Steel Reactor with Wall Effects. ***energy&fuels***, **2009**, 23(11), 5663-5676.
2. Pyl, S.P.; Schietekat, C.M.; Reyniers, M.F.; Abhari, R.; Marin, G.B.; Van Geem, K.M. Biomass to olefins: Cracking of renewable naphtha. ***Chemical Engineering Journal***, **2011**, 176-177, 178-187.
3. Sunde, K.; Brekke, A.; Solberg, B. Environmental Impacts and Costs of Hydrotreated Vegetable Oils, Transesterified Lipids and Woody BTL- A Review. ***Energies***, **2011**, 4(6), 845-877.
4. Coll, R.; Udas, S.; Jacoby, W.A. Conversion of the Rosin Acid Fraction of Crude Tall Oil into Fuels and Chemicals. ***energy&fuels***, **2001**, 15, 1166-1172.

AIChE Paper Number 45j
High Efficiency Ethylene Cracking Furnaces

Rajaram Ramesh

Product Development Manager - Ethylene
Technip
Netherlands

Marco W. M. van Goethem, PhD

Product Development Manager – Modeling and Testing
Technip
Netherlands

25th Ethylene Producers Conference
AIChE 2013 Spring National Meeting
San Antonio, Texas, April 29 – May 3, 2013

Abstract

Environmental and economic pressures generate a driving force to produce products more efficiently. A direct effect of more efficient processes is that the emission of green house gases, particularly carbon dioxide decreases. In households nowadays "high efficiency furnaces" are available with efficiencies in the range of 94-97%. This paper describes the outcome of a feasibility study on applying the high efficiency furnace concept to steam cracking furnaces. The high efficiency is achieved by enhanced heat recovery from the flue gases before they are released to the atmosphere at ca. 65°C.

Technip studied the application of this concept for cracking furnaces based on gas as well as liquid feedstocks. Natural gas was used as basis for the energy price. Since energy prices have a strong impact on project economics, the evaluations were done for an EU average natural gas price and a US natural gas price.

Two alternative heat sinks have been considered: preheating of combustion air and preheating of cold hydrocarbon feed. Both are viable for liquid cracking furnaces. In the case of gas cracking furnaces feed preheating results in a pinch and this makes it not viable.

Capital costs are based on Technip's cost database and quotations from air preheater vendors. Special attention has been paid to the material selection of the affected convection section banks.

This study shows that high efficiency ethylene cracking furnaces are economically viable. For Europe the study indicates an attractive payback period. In the US the payback period is longer due to the lower energy prices. If energy prices increase in the future, the concept's economic viability will become better. Inclusion of CO_2 credits will make the payback period more attractive.

Furnace Efficiency

In this paper furnace efficiency has been calculated as net thermal efficiency as interpreted by API560[1)]. Furnace efficiency is determined by drawing an envelope around the furnace and including in it the radiant section, convection section and any heat exchanger equipment that recover heat from the flue gas (air preheater, economizer, steam superheater, etc.). The efficiency is the ratio of the useful heat absorbed to the fired heat of the fuel and any other external heat inputs. Ambient temperature is a reference temperature for efficiency calculations and any heat inflow or outflow above this reference is taken into consideration as a heat input or heat loss.
Furnace efficiency can be determined as follows:

$$\frac{heat\ absorption\ in\ radiant\ section\ +\ convection\ section}{(firing\ rate\ \times\ fuel\ heating\ value) + sensible\ heat\ of\ fuel\ and\ air\ above\ ambient\ temp. + other\ external\ heat\ inputs}$$

Alternately, $\text{Furnace efficiency} = 100\% - \text{Total heat loss, where:}$

$$\text{Total heat loss} = \frac{stack\ loss\ +\ radiation\ heat\ loss}{(firing\ rate\ \times\ fuel\ heating\ value) + sensible\ heat\ of\ fuel\ and\ air\ above\ ambient\ temp. + any\ external\ heat\ inputs}$$

In a furnace where fuel and combustion air enter the furnace at ambient temperature and there are no other external heat inputs, the first expression can be simplified to the following:

$$Furnace\ efficiency = \frac{heat\ absorption\ in\ radiant\ section\ +\ convection\ section}{fuel\ firing\ rate\ \times\ fuel\ heating\ value}$$

Please note that net thermal efficiency, as defined by API 560, is based on the lower heating value of the fuel.

Study objectives

Technip undertook this study with the goal to reduce fuel firing in cracking furnaces by at least two percent. Although this figure does not look very ambitious, it must be remembered that since efficiencies in current designs are already in the mid nineties (upper part of the technology S-curve), further improvements will not be easy and can only be incremental.

Since stack loss is the main source of inefficiency (radiation heat loss is comparatively smaller and more difficult / expensive to reduce) it was decided to make stack loss reduction as the main focus of the study.

By a relatively straight forward calculation it can be shown that the two percent efficiency rise target would require the flue gas to be cooled to approximately sixty or seventy degrees Celsius.

The main challenge in reducing stack temperature to this level is finding an appropriate heat sink for the recovered heat. In a cracking furnace there are two suitable heat sinks, combustion air and hydrocarbon feed, and both were studied. The following sections discuss the results for liquid furnaces and gas furnaces.

Basis of Study

We have done the study on a naphtha furnace and an ethane furnace. We have chosen these two feedstocks since a large number of furnaces operate on these two feeds and we can draw conclusions that can be applied approximately to other liquid and gas feeds. The study has been performed on the following bases:

- The reference liquid and gas furnace designs used in this study are based on actual furnaces from Technip's past projects. The furnaces were simulated using Technip's proprietary tool SPYRO which is used in the design and simulation of cracking furnaces.
- The budgetary capital cost need to achieve the target furnace efficiency was calculated as an incremental cost. It was estimated from vendor quotes and Technip's internal cost database.
- The cost estimates are for a new project, not a revamp, and are for a West European location.
- Operating costs are based on cost of base utilities and derived utilities. Base utilities are natural gas and demineralized water. These are taken from market data.
- Since natural gas prices are very different in different parts of the world, the operating costs and economic analysis have been done for European and United States gas prices.
- EU natural gas price was fixed at €25.1 /MWh of higher heating value (HHV) which is equivalent to €28.0 /MWh of lower heating value (LHV). The prices are based on EUROSTAT data for 1st half of 2010 and were obtained from the publications of the Department of Energy & Climate Change in the UK. They correspond to the EU-15 average for Large Industrial Consumers (i.e. 2.13 pence/kWh HHV).
- US Natural gas prices were obtained from the US Department of Energy. The average value was 6 \$/MM BTU (HHV) or 17 €/MWh (LHV) for 1st half of 2010. The average HHV used is 1029 BTU per cubic foot or 1.029 MM BTU per thousand cubic feet.
- Cost of demineralized water was assumed at € 2.00 /ton.
- The prices of other utilities (boiler feed water, steam and electricity) have been calculated from those of the two base utilities. Electricity was priced assuming 35% overall efficiency of power generation.
- The economic calculations do not include any CO2 credit for the fuel saved.
- Assumed exchange rate: 1 US dollar = 0.75 euros
- The plant is assumed to operate 8400 hours per year

The resultant utility prices used in the study are summarized below.

UTILITY COSTS

		EU basis	US basis	
FUEL as NATURAL GAS	Base	**28.0**	**17.0**	€ / MWh
DEMINERALISED WATER	Base	2.0	1.5	€ / ton
VHP STEAM (net value)	Derived	23.7	14.4	€ / ton
MP STEAM (net value)	Derived	13.9	8.5	€ / ton
LP STEAM	Derived	9.7	5.9	€ / ton
ELECTRICAL POWER	Derived	80.1	48.6	€ / MWh

Study Results

Naphtha Furnace using Air Preheat

Combustion air can be preheated in a plate type air preheater (APH). These are cross flow type exchangers and have low pressure drops. Combustion air is at a lower temperature than any of the process or utilities streams entering the convection section; therefore, from the standpoint of efficiency, APHs are ideally suited to be the last heat recovery section before the stack.

Fig 1 shows the Base Case Naphtha furnace while Fig 3 shows the same furnace with an APH. Please also refer to Table 1 for a performance comparison of the APH design with the Base Case naphtha furnace. The following points are of interest:

1) The overall efficiency of the furnace will improve from 94.4 to 97.0% for the APH option. This is mainly due to the lower stack temperature but also partly due to the smaller excess air, which is explained below.

2) This option will require a forced draft combustion air fan to overcome the pressure drop in the APH. As a result the furnace operation will change from induced draft mode to balanced draft. The availability of air in forced draft will enable more accurate control of the excess air and it will be possible to operate the furnace with lower excess air (5% instead of 8%). This leads to a lower flue gas flow, a lower stack loss and therefore better efficiency.

3) There will be a drop of 4.8% in fuel firing which is greater than the efficiency increase. This is because of the lower absorbed heat. The furnace is designed and controlled in such a way as to supply all the required process heat and to absorb the surplus heat into utilities to the extent possible. When the furnace operates more efficiently and the heat demanded by the process media remains unchanged, as in the APH case, there will be less surplus heat available for VHP steam. As a result the VHP steam generation will go down from 79.7 to 76.0 t/h.

4) The heat transfer surface in some parts of the APH will be below the water dew point of the flue gas (around 60°C). This can lead to water condensation and create two possible issues: 1) accumulation of water in the APH and 2) corrosion due to the presence of liquid water, oxygen and carbon dioxide.

   Two measures have been included in the design to counter these problems:
   - Placement of unit such that flue gas flows downward to aid easy draining
   - Use of corten steel to resist corrosion

5) We have evaluated the impact of the air preheat option on NOx emissions using API 535[2)]. The rise in air temperature and the drop in excess air will have opposing effects on the NOx production. The net effect is that the NOx concentration will be close to the Base Case.

   If a DeNOx unit is present, as is often the case in the US, the APH option may help to slightly extend the life of the DeNOx catalyst. This is because the total NOx load will be smaller than the Base Case due to the smaller flue gas flow.

Table 2 summarizes the results of the economic analysis of this option. We have used payback period to compare different options. Payback period is the ratio of the incremental capital cost above the Base Case CAPEX divided by the savings in yearly operating cost.

1) The capital cost (CAPEX) of the APH option includes the cost of the APH, the combustion air fan, extra ducting and structural steel. The operating cost (OPEX) savings is the savings in furnace fuel but this must be offset against the value of the lost VHP steam production (which notionally must be made up in a utility boiler elsewhere) and the additional electricity for the two fans.

2) It can be seen that the payback period is three years based on European gas prices and five years based on American gas prices.

Naphtha Furnace using Feed preheat

Naphtha is normally stored at ambient temperature and is preheated to 60°C using quench water before it is routed to the furnaces. Most cracking furnaces are equipped with a Feed Preheater (FPH) bank in the convection section. If flue gas is utilized to preheat naphtha from ambient conditions the stack temperature can be lowered to the desired range of sixty to seventy degrees Celsius. Till now designers and operators have not taken full advantage of this mainly because of the higher CAPEX and fear of dew point corrosion.

One of the consequences of this feature will be that the quench water duty previously absorbed by the naphtha will now shift to the air coolers in the quench water circuit. The additional electrical power that the air cooler fans will require will be about 10 KW per furnace.

Please refer to Table 3 for an overview of the performance of a naphtha furnace with and without an extended feed preheater. Please note that the Base Case naphtha furnace is the same as the one considered in the APH option.

1) The overall efficiency of the furnace can be improved from 94.4 to 96.0% for this option.

2) Unlike the APH option a combustion air fan will **not** be required. The furnace will continue to operate in induced draft mode.

3) The fuel firing and the VHP steam generation will drop marginally.

4) The upper tube rows of the FPH will be below the water dew point of the flue gas and some condensation will occur. As a precaution we have considered that the top four rows will be of stainless steel.

5) The flue gas NOx concentration will be unaffected by this option. However the total hourly NOx emissions and load on the DeNOx unit, if any, will decrease to the same extent as the fuel firing, which is around 0.6%.

Table 4 summarizes the results of the economic analysis of this option.

1) The capital cost (CAPEX) of this option covers the incremental cost of the extended FPH, extra ducting and structural steel. The OPEX savings is the reduction in furnace fuel after adjusting for the reduction in VHP steam production and the additional electricity for the ID fan and the Quench Water Air Cooler fans.

2) In the Base Case the naphtha feed piping between the external feed preheater and the convection section is insulated for heat conservation. In the FPH option naphtha will be fed at ambient temperature to the furnace and so these lines do not have to be insulated. This will yield some capital cost saving but this has not been taken into account in our analysis.

3) The payback period excluding CO2 credit will be three and a half years with European natural gas and five and a half years with US natural gas.

Ethane Furnace using Air Preheat

Fig 2 shows the Base Case Ethane furnace that was used in the study. The APH arrangement on an ethane furnace will be quite similar to the one on the naphtha furnace. Table 5 shows the impact that the APH would have on the furnace performance. The main points are as follows:

1) The overall efficiency of the furnace will improve from 91.7 to 96.5% for the APH option. This is mainly due to the lower stack temperature but also partly due to the smaller excess air.

2) In this furnace also the APH will require a forced draft combustion air fan. This will allow the excess air to be reduced from 8% to 5% which will result in lower stack losses and a better efficiency.

3) The fuel firing will go down by 11.0%, from 8.2 to 7.3 t/h. The absorbed heat will drop from 116.8 to 110.6 MW. As a result of these the VHP steam generation will go down from 97.0 to 88.8 t/h.

4) A portion of the APH will operate below the flue gas water dew point. Consequently the unit must be designed to prevent water accumulation and corrosion.

5) The impact on NOx was evaluated here also. The rise in air temperature will have a greater impact on NOx than the fall in excess air. We estimate that as a result the flue gas NOx concentration will increase slightly (around two percent). This can be a cause for concern if the furnace does not have a DeNOx unit and is operating close to the limits imposed by authorities. However since the total firing will go down, the hourly NOx production will drop. Plants can consider discussing with authorities to obtain a higher allowable NOx limit,

   If the furnace has a DeNOx unit, the chemical load on the unit will reduce, which should help to obtain a longer catalyst life.

The operating cost savings and payback for this option have been summarized in Table 6.

1) The capital cost includes the cost of the APH, the combustion air fan, extra ducting and structural steel. The operating cost savings is the savings in fuel firing with an adjustment for the value of the reduced VHP steam generation and the extra electricity for the ID fan and FD fan.

2) The payback period with natural gas priced at European rates will be one and a half years. The corresponding figure for US gas prices will be three years.

Ethane Furnace using Feed Preheat

Using feed at ambient temperature does not significantly reduce the stack temperature in a gas furnace. This is because the feed preheat curve has a steeper slope than the flue gas cooling curve, causing a pinch on the hot side of the Feed Preheater. Fig 4 shows the heating and cooling curves of ethane, boiler feed water and flue gas in the convection section. The hatched shape on the plot shows the maximum additional heat duty that can be recovered into feed, over and above the Base Case. The plot clearly shows that when the ethane feed temperature is lowered from the base value of 55°C to 30°C, the stack temperature will drop from 170°C to about 150°C. The flue gas cannot be cooled any more than this modest amount.

For this reason the feed preheat option was not investigated any further.

Conclusions and Summary

The key results from the different study options are listed together below.

Cracking furnace feed	Naphtha		Naphtha		Ethane	
Cold sink	Combustion air		Hydrocarbon feed		Combustion air	
	EU nat. gas	US nat. gas	EU nat. gas	US nat. gas	EU nat. gas	US nat. gas
OPEX savings	€ 0.60 mill	€ 0.36 mill	€ 0.17 mill	€ 0.10 mill	€ 1.27 mill	€ 0.77 mill
Payback time	3 years	5 years	3.5 years	5.5 years	1.5 years	3 years
Salient features of each option	Small number of additional equipment		No additional equipment		Small number of additional equipment	
	Higher CAPEX		Lower CAPEX		Higher CAPEX	
	More OPEX savings		Less OPEX savings		Best OPEX savings	
	Slight effect on NOx concentration		No effect on NOx		Around 2% higher NOx concentration	

All three scenarios have attractive payback periods and will be worth considering in high energy cost countries such as Europe and Asia. In US and other energy rich nations, the economic incentive will be much lower. In the US, where ethane cracking is common and increasing even more, improvement of furnace efficiency with air preheat can be worth considering as the payback time is less than 3 years.

If the options are compared against each other, in our opinion the Air preheating option has more potential, irrespective of feedstock.

The Feed Preheating option in liquid furnaces is cheaper but it also gives proportionally modest savings. Moreover, as earlier mentioned, the feed preheat duty is being shifted from one waste heat source (quench water) to another (flue gas). This option will not lead to a net improvement or cash savings in the plant as a whole unless there is a shortage of quench water and there are other, better uses for the saved quench water duty

Acknowledgements

The authors wish to express their thanks to their colleagues Mr Ashok Laghate, Ms Susan Cooper, Mr Sachin Singare and others in Technip Benelux who contributed to this study.

References

1) API 560: Fired Heaters for General Refinery Service, 4th edition

2) API 535: Burners for Fired Heaters in General Refinery Service, 2nd edition

Table 1
Naphtha Furnace with Combustion Air Preheating – Performance Summary

		Base Case	Air Preheat Case
Naphtha feed	t/h	60.8	60.8
Ambient conditions	-	15°C, 85% R.H.	15°C, 85% R.H.
Air temp. to burners	°C	15	77
Stack temperature	°C	108	61
Excess air	%	8	5
Absorbed heat	MW	113.2	110.4
Overall efficiency	%	94.4	97.0
Fuel flow	t/h	8.4	8.0
Fired heat	MW	119.9	113.9
VHP steam production	t/h	79.66	76.01
NOx concentration		Base	Close to Base

Table 2
Economics of Naphtha Furnace with Combustion Air Preheating

		Air Preheat Case (EU location)	Air Preheat Case (US location)
Savings in fuel firing	million €/yr	1.42	0.86
Cost of lost VHP steam capacity	million €/yr	0.73	0.44
Cost of additional electricity	million €/yr	0.10	0.06
Net OPEX saving	million €/yr	0.60	0.36
Payback	Years	3	5

Table 3

Naphtha Furnace with Feed Preheating – Performance Summary

		Base Case	Feed Preheat Case
Naphtha feed	t/h	60.8	60.8
Feed temperature	-	60°C	30°C
Combustion air temp.	°C	15	15
Stack temperature	°C	108	71
Excess air	%	8	8
Absorbed heat	MW	113.2	114.3
Overall efficiency	%	94.4	96.0
Fuel flow	t/h	8.4	8.35
Fired heat	MW	119.9	119.1
VHP steam production	t/h	79.66	79.63
NOx concentration		Base	Base

Table 4

Economics of Naphtha Furnace with Feed Preheating

		Feed Preheat Case (EU location)	Feed Preheat Case (US location)
Savings in fuel firing	million €/yr	0.19	0.12
Cost of lost VHP steam capacity	million €/yr	< 0.01	< 0.01
Cost of additional electricity	million €/yr	0.01	< 0.01
Net OPEX saving	million €/yr	0.17	0.10
Payback	Years	3.5	5.5

Table 5
Ethane Furnace with Air Preheating – Performance Summary

		Base Case	Air Preheat Case
Ethane feed	t/h	51.9	51.9
Ambient conditions	-	15°C, 85% R.H.	15°C, 85% R.H.
Air temp. to burners	°C	15	129
Stack temperature	°C	169	64
Excess air	%	10	5
Absorbed heat	MW	116.8	110.6
Overall efficiency	%	91.7	96.5
Fuel flow	t/h	8.2	7.3
Fired heat	MW	127.3	114.5
VHP steam production	t/h	97.0	88.75
NOx concentration		Base	Base + 2%

Table 6
Economics of Ethane Furnace with Air Preheating

		Air Preheat Case (EU location)	Air Preheat Case (US location)
Savings in fuel firing	million €/yr	3.01	1.83
Cost of lost VHP steam capacity	million €/yr	1,64	1.00
Cost of additional electricity	million €/yr	0.10	0.06
Net OPEX saving	million €/yr	1.27	0.77
Payback	Years	1.5	3

Fig 1
Naphtha cracking furnace

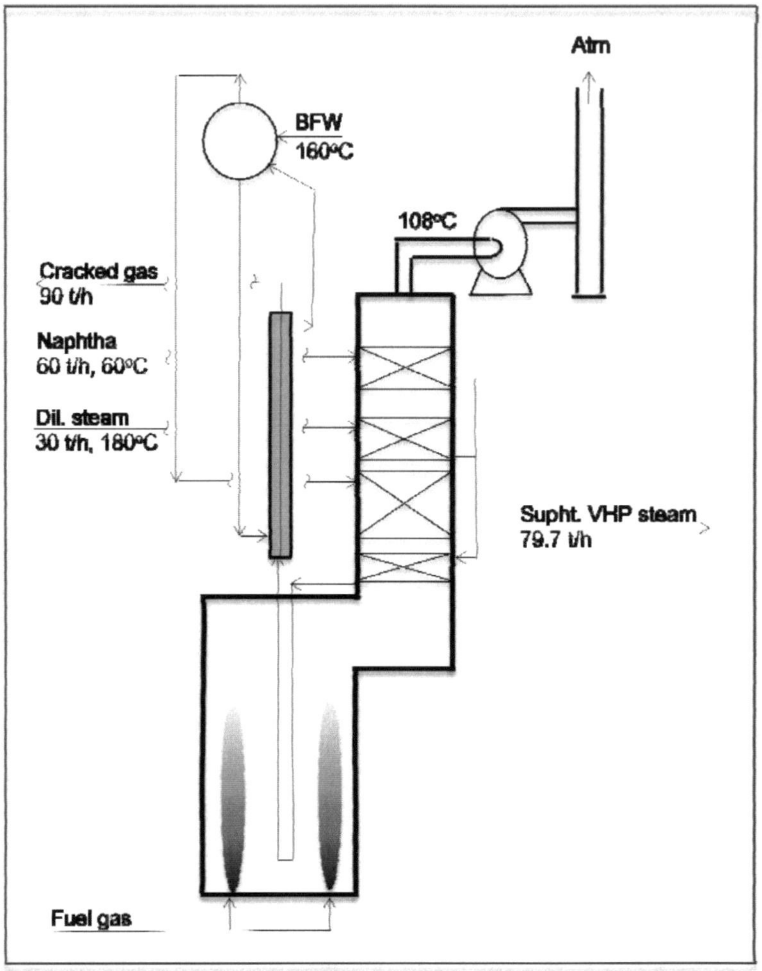

Fig 2
Ethane cracking furnace

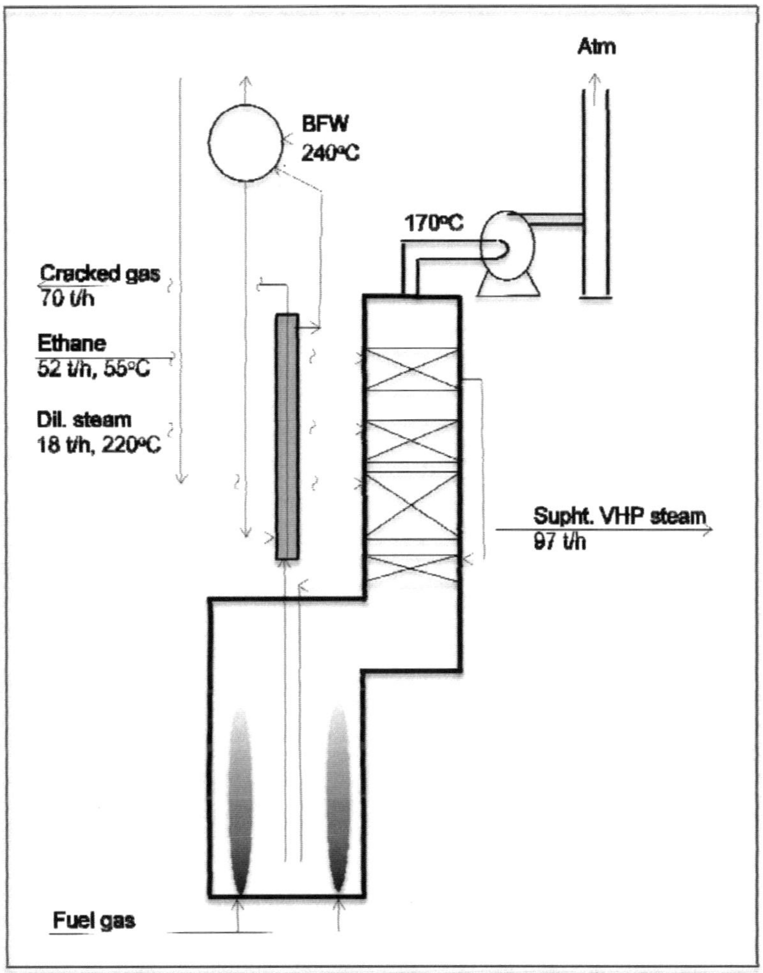

Fig 3

Naphtha cracking furnace – Air Preheater option

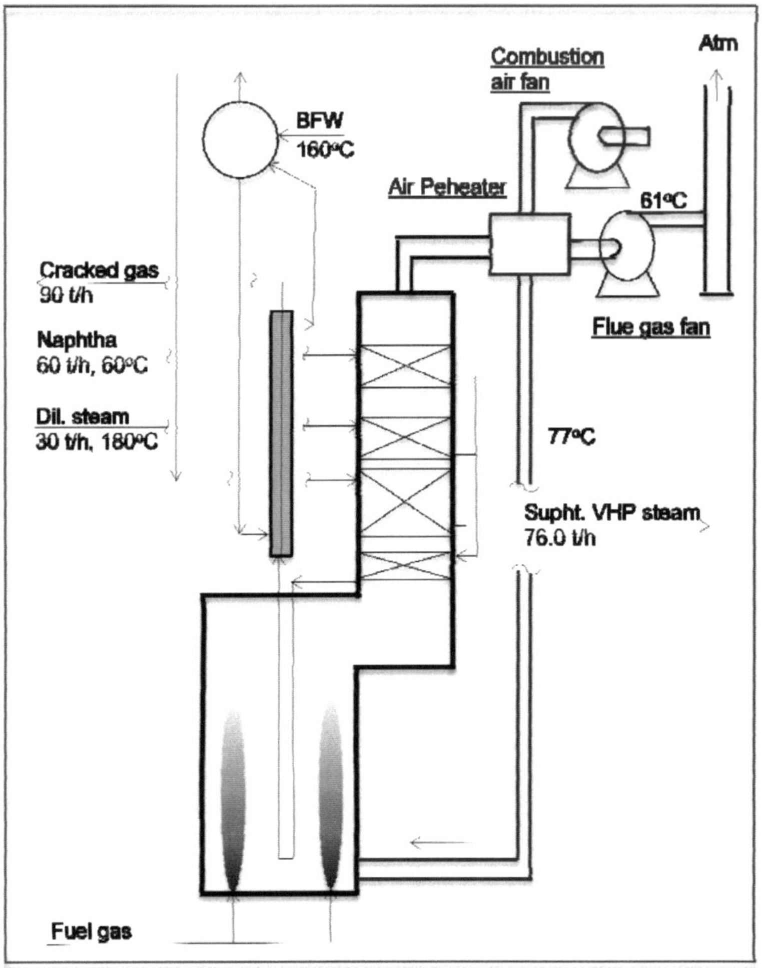

Fig 4

<u>Ethane cracking furnace – Feed Preheater option</u>

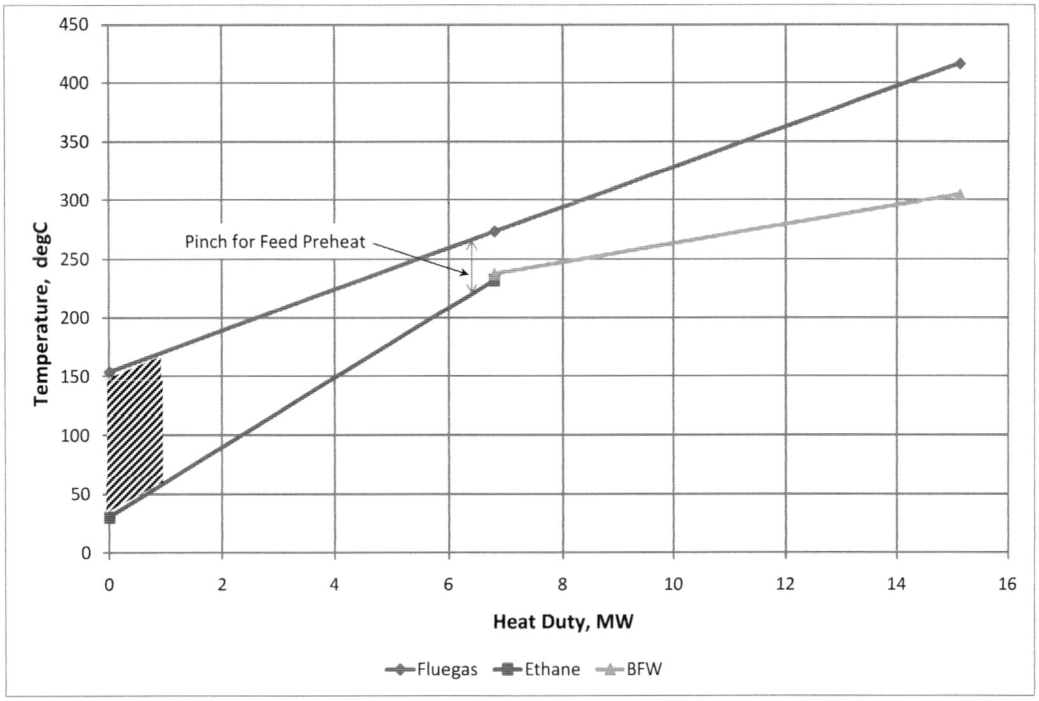

Modernize Your Control Functionality
Easy Steps to Start Process Optimization

13 AIChE
Spring Meeting, San Antonio, TX

Presenters

- Laurie Ben
- James Beall

EMERSON
Process Management

Agenda

- Introduction
- Examples – Best Practices & Lessons Learned
  - Split Range Duty
  - Smith Predictor
  - Gain Scheduling/Adaptive Control
  - Signal Characterization
- Summary

Introduction

- Modernizing Control System – Decision Done Check!
- 'Do I' or 'Don't I' take advantage of new technology
  - Implemented older cumbersome techniques?
  - Complex custom strategies that handle automation challenges?
  - Hard to implement & maintain?
- Standard features that are field proven, practical, and easy to implement
  - Easier to implement & maintain
  - More efficient and effective control strategies
  - Achieve improvements in process performance
  - Lays the foundation for optimization

Agenda

- Introduction
- Examples – Best Practices & Lessons Learned
  - Split Range Duty
    - Control Challenge
    - Legacy Solution
    - Newer Technology Solution
    - Summary
  - Smith Predictor
  - Gain Scheduling/Adaptive Control
  - Signal Characterization
- Summary

Split Range Duty – Control Challenge

- Use steam and hot oil as energy source for distillation column for temperature controller.
- Hot oil is “free energy” and should be maximized
- Control Scenarios
  - Maximize hot oil and supplement with steam
  - Base load steam (if meter bad) and supplement with hot oil
  - Base load hot oil flow (if bad meter) and supplement with steam
- Constraints
  - Minimum and maximum hot oil valve position and flow rate
  - Minimum steam flow rate

Split Range Duty – Legacy Solution

– Old legacy system implemented this feature using a velocity PID with incremental output and a "micro expert rules system"

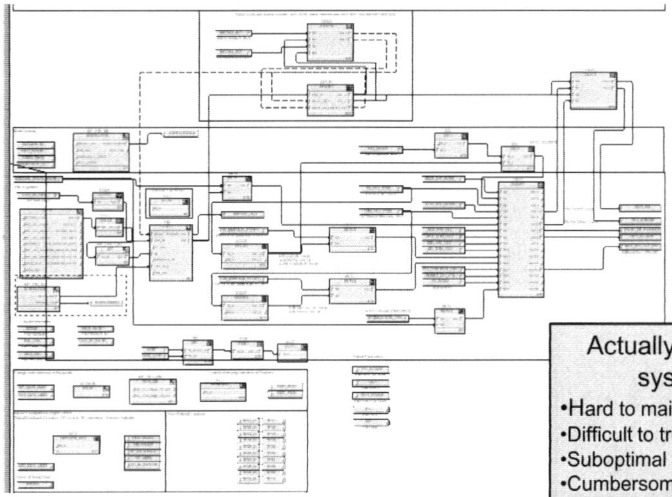

Actually Duplicated in New system, but still...

- Hard to maintain and modify
- Difficult to troubleshoot
- Suboptimal performance
- Cumbersome/Kludge/Ugly

Split Range Duty – New Technology Solution

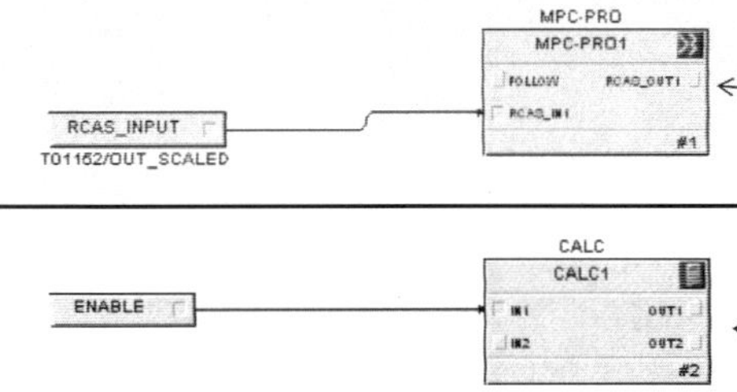

Embedded MPC Function Block provides all control functionality! EASY!!!

Simple calculations to calculate limits

Split Range Duty – New Technology Solution

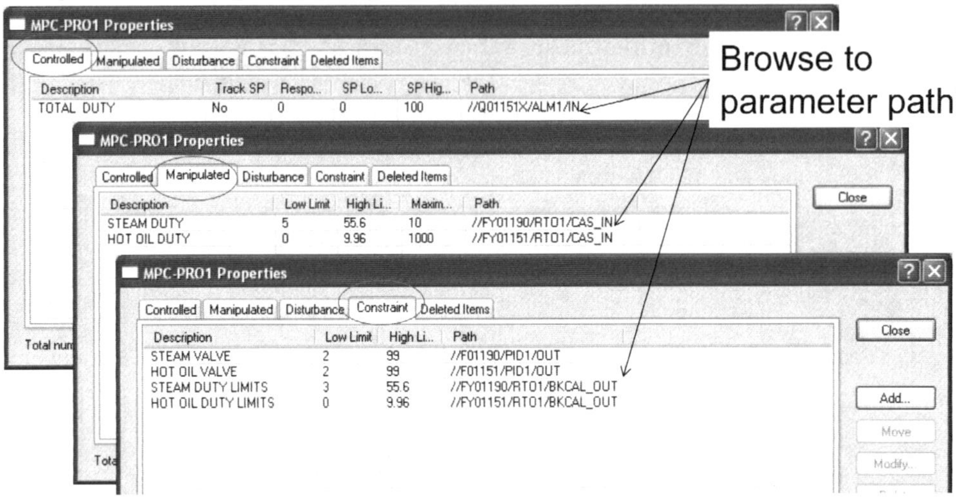

Split Range Duty – New Technology Solution

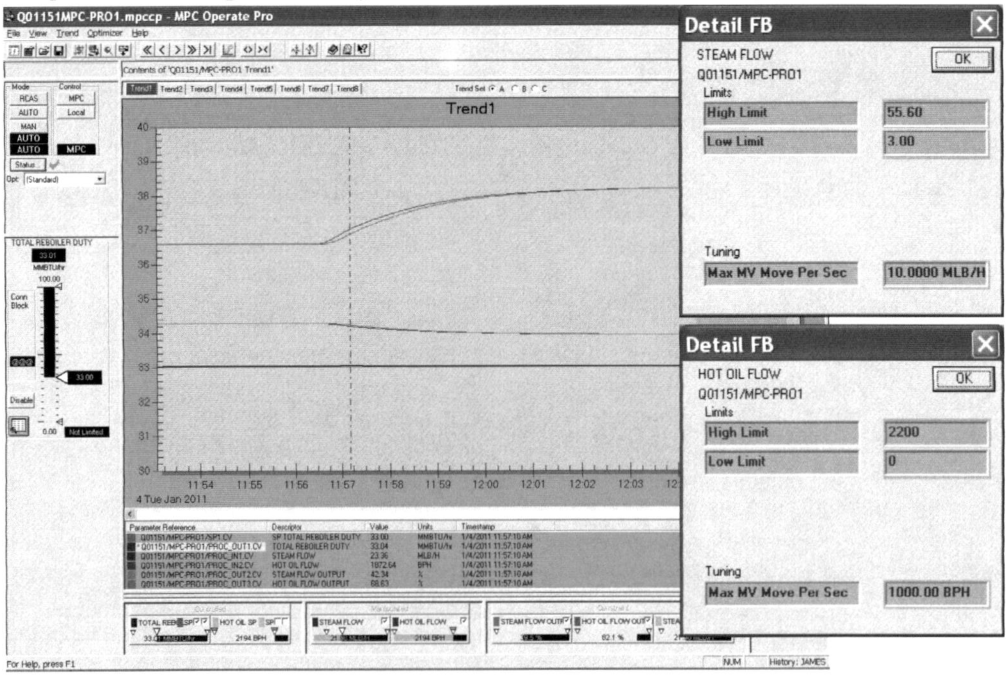

Summary Split Range Duty Control

Embedded MPC (New Technology) vs Legacy Solution

- Better performing solution
  - Bumpless transfers and constraint control
- Easier to Maintain
  - Change limits and settings
  - Simpler implementation
  - Easy to understand, train others, document
- Easily expands to handle future constraints
- Reduce energy costs!

Agenda

- Introduction
- Examples – Best Practices & Lessons Learned
  - Split Range Duty
  - Smith Predictor
    - Control Challenge
    - Legacy Solution
    - Newer Technology Solution
    - Summary
  - Gain Scheduling/Adaptive Control
  - Signal Characterization
- Summary

Smith-Predictor PID – Control Challenge

- Control of processes that are "dead time dominant" (dead time > time constant)
- Typically custom configured
  - some legacy control systems have preconfigured options, e.g. PROVOX
- Algorithm uses a process model
  - (gain, dead time and time constant)
- Challenge: as process dynamics change from model used, performance degrades rapidly

Smith-Predictor PID – Legacy Solution

- Traditional Smith-Predictor

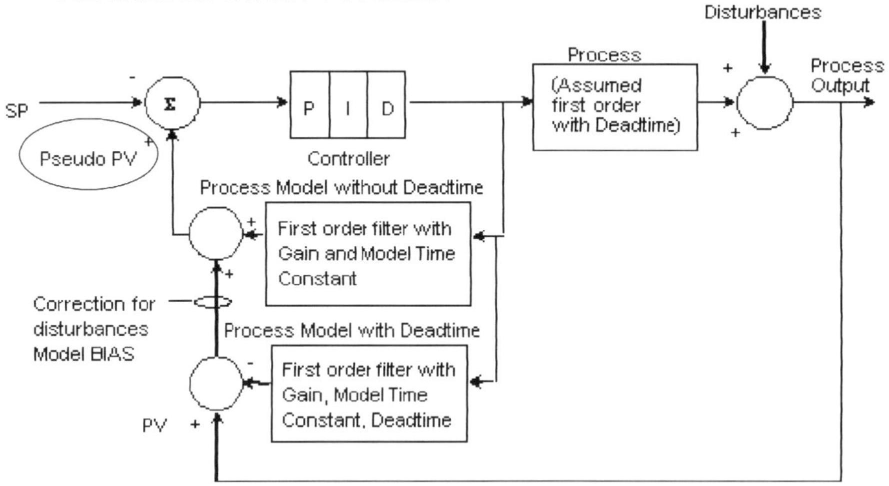

•High maintenance when process dynamics change
•Performance degrades rapidly as process dynamics change from configured model

Smith-Predictor PID – New Technology Solution #1

Easy to implement the Smith-Predictor but…

- High Maintenance when process dynamics change
- Performance degrades rapidly as process dynamics change from configured model

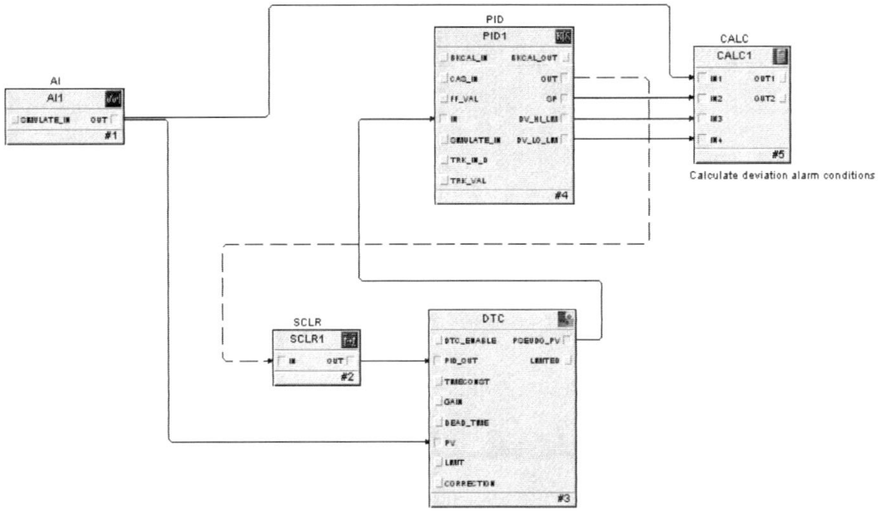

Smith-Predictor PID – New Technology Solution #2

- Embedded MPC Solution

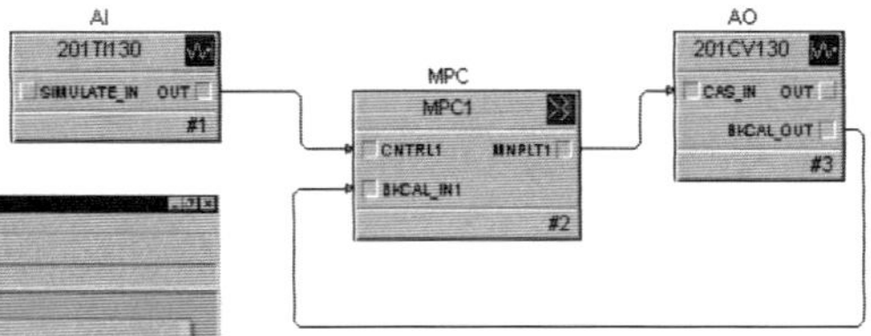

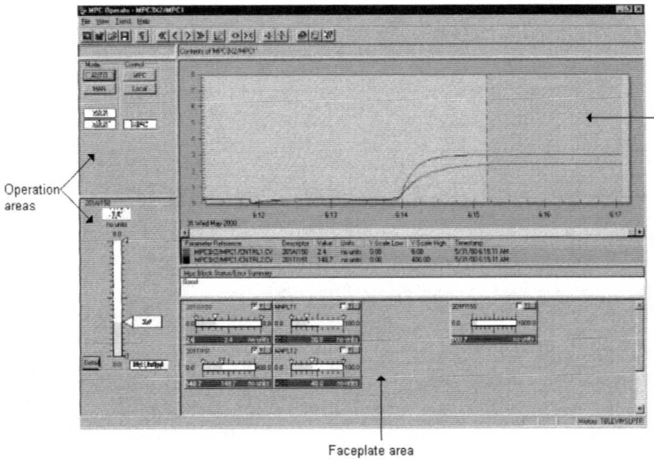

- Provides Future Prediction
- Provides MPC Faceplate – no configuration needed
- More robust under changing process dynamics

Smith-Predictor PID – New Technology Solution #2

- Embedded MPC Solution

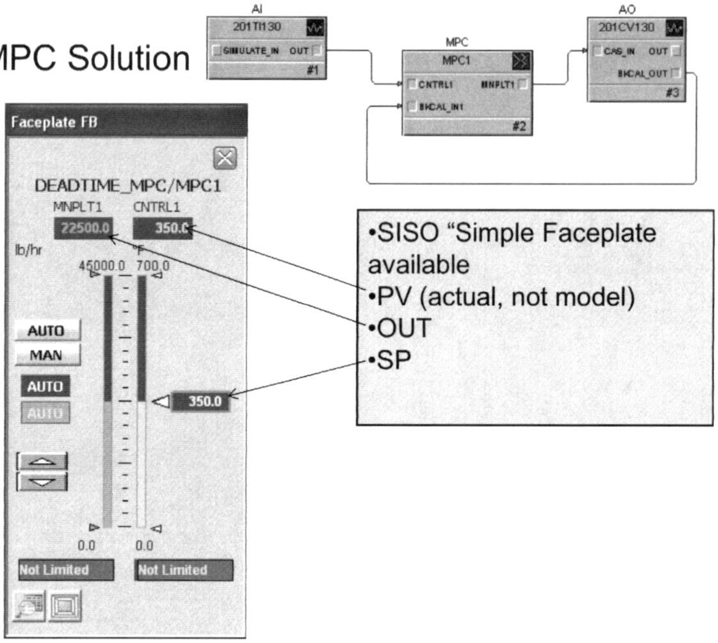

Summary Smith-Predictor PID

New Technology Solutions

- Standard Smith Predictor (#1) and Embedded MPC (#2)
- Both easy to implement and understand!

New Technology Solution #2 (Embedded MPC)

- More robust solution under a changing process model
- Provides visual display of the "future" PV
- Save money through better control performance

Agenda

- Introduction
- Examples – Best Practices & Lessons Learned
  - Split Range Duty
  - Smith Predictor
  - Gain Scheduling/Adaptive Control
    - Control Challenge
    - Legacy Solution
    - New Technology Solution
    - Summary
  - Signal Characterization
- Summary

Gain Scheduling – Control Challenge

- Legacy systems have limited capability to modify tuning based on process conditions
- Others had add-on application software, e.g. PROVOX "Gain Scheduler" system provided 3 sets of Gain, Reset, and Rate applied in 3 separate regions of a "state variable"
  - State variable = PV, OUT or "other"
  - Transition between tuning sets interpolated over "dead band"
- Beware: some suspend loop execution when tuning parameters change, so don't change the tuning every scan

Gain Scheduling – Legacy Solution

- Custom code
- Multiple points, registers, displays
- Restricted (slower) update rate

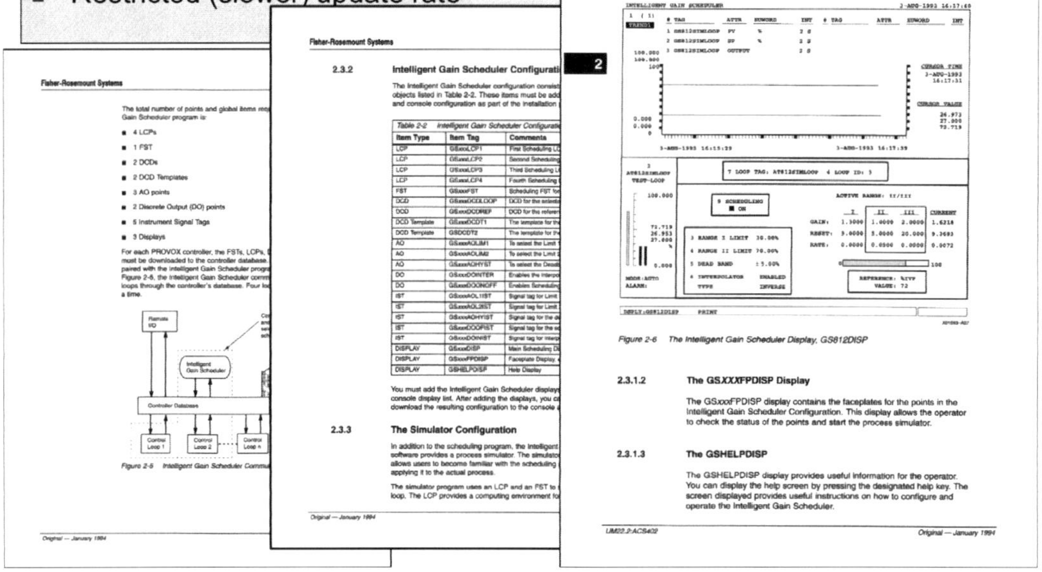

Gain Scheduling – New Technology Solution #1

Standard Modules

- Standard configuration
- Preconfigured faceplate & detailed display
- Easy to implement
- Tuning updates at loop execution

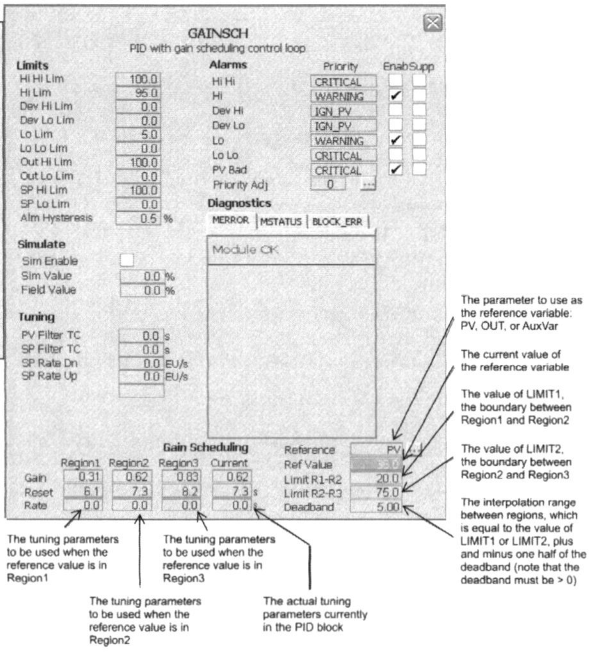

Gain Scheduling – New Technology Solution #2

New Technology "Adaptive Control"

- "Learns" process models from normal operating moves
- Allows up to 5 tuning regions of the "state variable"
- Partial Adaptive Control
  - Use "approved model" only & selected tuning rule in each region
  - Lambda PI or IMC PID; "Closed loop time constant" or "Lambda Factor".
- Optional Full Adaptive Control
  - continuous "Learning" of process model in each region
  - updates tuning while in each region per selected tuning rule
  - set allowable deviation (%) from approved model in each region

New Technology – Adaptive Control

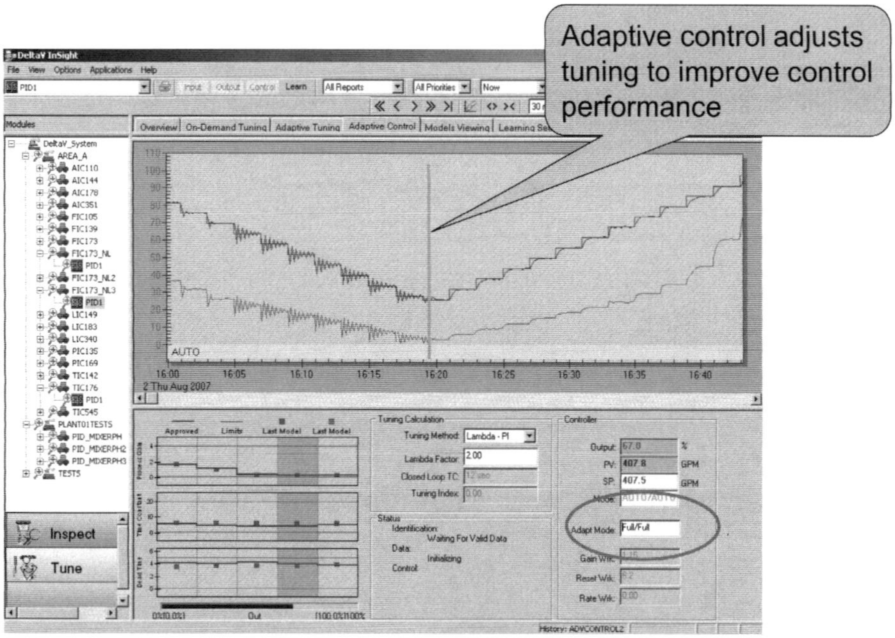

Summary Gain Scheduling

New Technology Solution(s) vs. Legacy Solution

- Gain Scheduling easy to implement and understand!
  - Standard configuration
  - Preconfigured faceplate & detailed display
  - Tuning updates at loop execution
- Adaptive Control provides additional benefit and ease of implementation
  - Learns models during normal operation – NO STEP TESTs!
  - More Regions – Up to 5
  - Automatic tuning updates when process model changes
- Improved control performance saves money!

Summary Gain Scheduling

New Technology Solution(s) vs. Legacy Solution

- Gain Scheduling easy to implement and understand!
  - Standard configuration
  - Preconfigured faceplate & detailed display
  - Tuning updates at loop execution
- Adaptive Control provides additional benefit and ease of implementation
  - Learns models during normal operation – NO STEP TESTs!
  - More Regions – Up to 5
  - Automatic tuning updates when process model changes
- Improved control performance saves money!

Agenda

- Introduction
- Examples – Best Practices & Lessons Learned
  - Split Range Duty
  - Smith Predictor
  - Gain Scheduling/Adaptive Control
  - Signal Characterization
    - Control Challenge
    - Legacy Solution
    - New Technology Solution
    - Summary
- Summary

Signal Characterization - Control Challenge

- Input Characterization (AI)
  - Combustion control – characterize fuel and air
  - Level – characterize pv (e.g. level to volume)
- Output characterization (AO)
  - modify installed flow characteristics of valves, dampers or other final control elements
  - Use to modify loop gain, e.g. output path to slave loop
- Legacy System Challenges
  - Lack of resolution (few points)
  - External black box characterizers may shift, lose documentation
  - Potential operator/maintenance confusion on signal going to final control element

Signal Characterization – Legacy Solution

- Legacy system using external analog characterizer

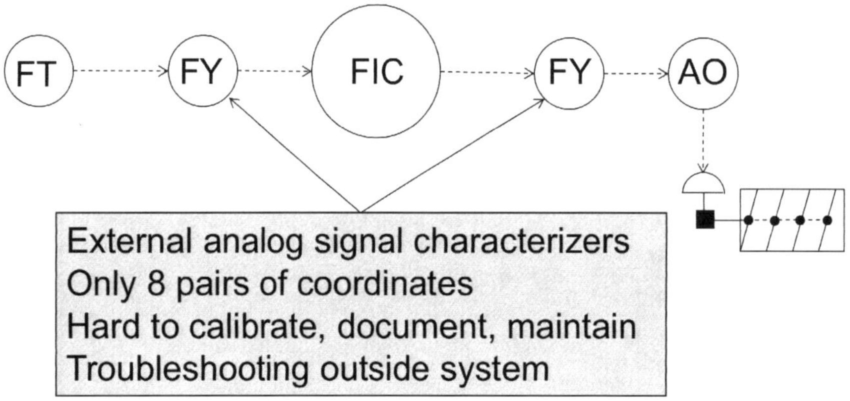

Signal Characterization – New Technology Solution #1

- Standard Module implementation provides 21 point signal characterizer function block (SGCR)

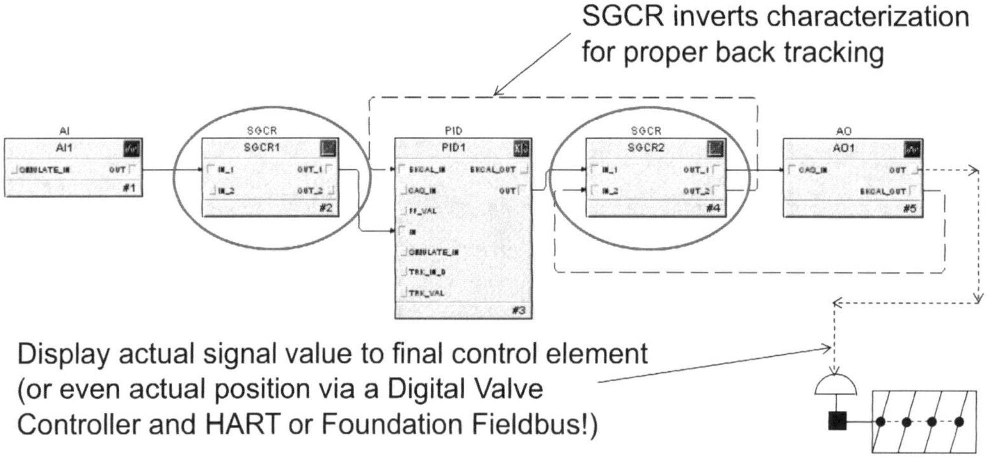

Signal Characterization – New Technology Solution #2

- Adaptive Control
  - Adjust tuning to compensate for final control element characteristics

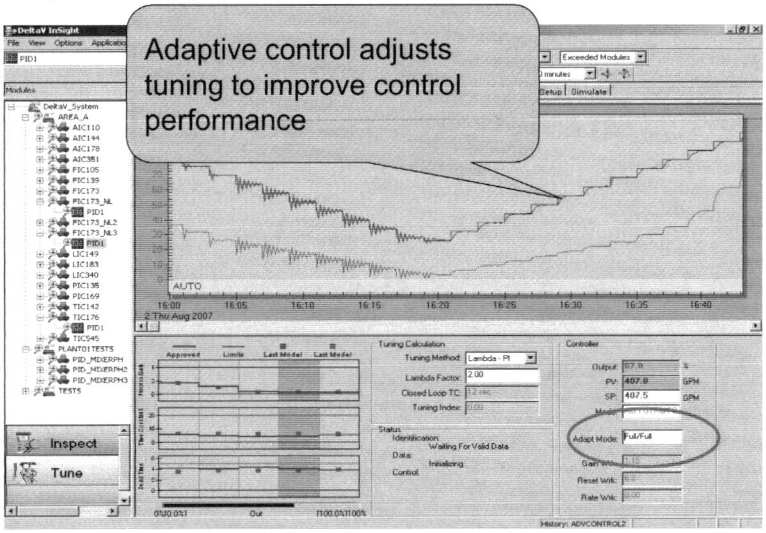

Summary Signal Characterization

- Newer Technology Signal Characterization function block
  - Better resolution (21 pairs of coordinates)
  - Self documenting (exists in control system)
  - No calibration required vs. 'external black box'
  - Provides "inverse" characterization for analog output characterizations
  - Reduce confusion: display signal being sent to final control element
- Applying Adaptive Tuning may provide a superior solution for AO characterization of final control element

Agenda

- Introduction
- Examples – Best Practices & Lessons Learned
  - Split Range Duty
  - Smith Predictor
  - Gain Scheduling/Adaptive Control
  - Controller Characterization
- Summary

Summary

- Practical application of control features that deal with common challenges
- Highlighted best practices & lessons learned
  - Split Range Duty
  - Smith Predictor
  - Gain Scheduling/Adaptive Control
  - Signal Characterization
- Do look at new functionality available when modernizing
  - Allows jump start to process optimization
  - Leads to improved process performance
  - Easier to implement and maintain over the life investment

Contacts & Questions?

James.Beall@Emerson.com
903-235-7935

Laurie.Ben@Emerson.com
512-834-7237

Paralleled Double-Effect Distillation: Simulative Case Studies

*Pei Wang, Leilei Dai, Ming Shi , Fan Wang, Hong Gao, Jinsheng Sun**
School of Chemical Engineering and Technology, Tianjin University, No. 92 Wei Jin Road, Tianjin 300072, P. R. China

Abstract

The increasing scarcity and growing demand for energy have led to an increasingly attention to the energy-efficient methods. Double-effect distillation is a common method to achieve remarkable energy conservation, which has three kinds of typical configurations including the feed splitting (FS), the light split/forward (LSF) and the light split/reverse (LSR). In spite of a good pathway for energy saving, better options are expected to cut down energy expenditure by a bigger margin than double-effect distillation does.

In this paper, case studies of paralleled or combined double-effect distillation schemes are proposed on the basis of current double-effect distillation. In order to verify the feasibility of this method, the improved methanol distillation scheme and organosilicon distillation scheme are optimized under this principle as examples. Exergy losses and the energy consumptions are measured as a tool to compare the two prevalent schemes with their corresponding optimized counterparts, respectively. The results of the total exergy loss and the energy consumption of these schemes show that the optimized schemes can cut the corner of the exergy loss by 22.8% and 40.41% and the energy consumption by 15.61% and 44.28%, respectively, in comparison to their popular cousins.

Keywords: double-effect distillation, paralleled double-effect distillation, methanol distillation, organosilicon monomer distillation, distillation process simulation, energy saving, exergy analysis

1. Introduction

Distillation, in chemical process and related industries, is by far the most widely applied for the separation of liquid mixture. In spite of its well-known benefit and widespread application, the major drawback is its high energy consumption. In classical distillation process, heat is transferred to the reboiler of distillation column. Along with the distillation column to separate the mixture, a large number of heat are directly discharge into the surrounding environment and cooling water. Araujo AB [1] has done some research to achieve that distillation has a relatively low thermodynamic efficiency and it requires the input of high quality energy in the reboiler to push the separation of liquid mixture. Therefore, the focus of energy conservation is to consider how to recover or reuse the discharged heat, or how to reduce the quantity of transferred heat to the column. In the early 20th century, many scholars have devoted to the study of saving energy in distillation, and now most of correlative techniques are widely used in distillation process [2,3]. These energy saving methods involve multi-effect distillation, heat pump distillation, heat integrated distillation and etc [3-8]. But without extra rotating machines by electricity or other power producers, multi-effect distillation has been more favored in research and application [9].

Many researches confirmed multi-effect distillation to be one of the effective methods to reduce the energy consumption [10-12]. The energy saving effect of those schemes used multi-effect distillation is associated with the number of the effects and the characteristics of separation system. Generally speaking, those schemes can offer energy savings around 30% with respect to the original or classical multicomponent distillation scheme [23, 27, 28]. Considering the interaction between investment of plants and energy saving effect, the double-effect distillation has been the most commonly used in not only research but also field application.

The use of the double-effect distillation technology for optimizing the conventional distillation schemes can cut a large amount of the energy requirement in comparison with the correspondingly prevalent schemes [13]. Although the system is very difficult to control in the sense that it is nonlinear, multivariable and interacting [14]. More and more published studies [15-19] have been focused on further questing the way of energy saving on the basis of the double-effect distillation schemes. However, the overhead vapor of one high pressured column only can integrate with the reboiler of one low pressured column in those published studies. In view of this, in this paper, the paralleled or combined double-effect distillation is proposed to broaden the thought for the application of the double-effect distillation, in which the overhead vapor from one of the high pressured column is divided into two or more streams to successively integrate with the reboiler of the other columns [9].

2. *Theory*

2.1 Double-effect or heat integrated distillation system

A double-effect distillation system is made up of two adjacent distillation columns running in different pressure, which can lead to a temperature difference between the top of the high pressured column and the bottom of the low pressured column. Through the thermodynamic connection, the reboiler of the low pressure column plays a role of the condenser for the high pressure column. According to the direction of the mass flow and the heat integration, the double-effect distillation can fall into three kinds of configuration: the feed splitting (FS), the light split/forward (LSF) and the light split/reverse (LSR), as showed in Figure. 1.

One of the configurations with a high thermodynamic efficiency is the feed splitting (FS), in which the feed is divided roughly equal between the two columns running in different pressure, and the products can be obtained from both of the top and bottom of the two columns, as illustrated in figure. 1 (a). The overhead vapor form the high pressured column is used to heat the reboiler from the low pressured column. There is only energy exchange between the two columns, so the heat duties of the feed splitting (FS) are approximately equal to the heat duty for single column with the same conditions divided by the number of effect.

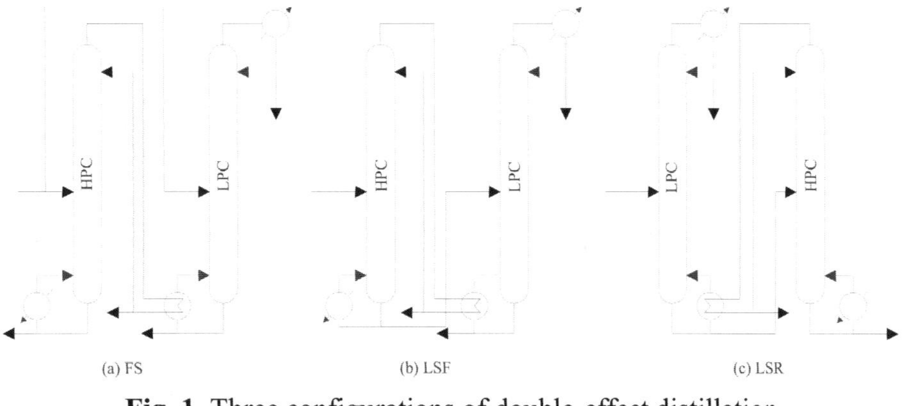

Fig. 1. Three configurations of double-effect distillation

Figure. 1 (b) and (c) respectively shows the configuration of the light split/forward heat integration (LSF) and the light split/reverse heat integration (LSR). Different from the feed splitting (FS), the total feed stream was sent to one column in the other two configurations. In the light split/forward (LSF), the total feed is pumped to the high-pressure column (HPC) and the bottom stream of HPC flows into the low-pressure column (LPC). In this way, all of the heavy components are from LPC and the light components from the top both of the two columns. Figure. 1 (b) indicates the direction of mass flow and heat integration is the same. The last configuration is the light split/reverse (LSR), as shown in figure. 1 (c), all of the feed is sent to the low-pressure column (LPC). It can remove the two kinds of light components from the top both of the two columns and one kind of heavy components from the bottom of HPC. Moreover, the direction of heat integration is in the reverse with the mass flow. These two kinds of configurations have not only energy exchange but also substantial exchange, therefore, when the operating parameters of one column are changed, the other column will also change.

Theoretically, the three kinds of configurations are empirically 30% less energy consumption than the single column distillation. As for choosing what kind of the configurations of the double-effect distillation, it depends on the properties of the multi-component mixture, the pressure of saturated steam, the temperature of cooling water, structure parameters of existing distillation column and so on.

2.2 Exergy analysis of the distillation process

According to the First Law of Thermodynamics, also called the Law of Energetic Balances, no matter what happens, the total amount of energy has not yet changed. However, when the various types of energy generate the mutual conversion, the quality of the energy is always decreased. The separation process in distillation is irreversible. In order to separate the mixture in distillation column, it needs the driving force which must consume a certain amount of external work, by this means, the decline of the quality of energy is inevitable. Therefore, we should choose a reasonable driving force in distillation process to avoid unnecessary loss of the quality of energy. It defined the concept of exergy (E_x) to compare the quality of the different types of energy. Exergy, which is a measure for evaluating the extent of available energy or the quality of energy, is based on the Second Law of Thermodynamics. Exergy analysis reveals the essence of the energy, and can be used as an important tool to evaluate the effect of energy saving. And it usually combines with process optimization studies, and as its guidance, to identify cost saving opportunities and amend the existing scheme.

Szargut [29] has summarized a comprehensive definition of the exergy, that is: 'Exergy is the amount of work obtainable when some matter is brought to a state thermodynamic equilibrium with the common components of the natural surroundings by means of reversible processes, involving interaction only with the above mentioned components of nature'. The state of one stream recorded as 1, according to the definition of exergy, the exergy of the stream can be expressed as:

$$E_x = (H_1 - H_0) - T_0(S_1 - S_0) \quad (2)$$

where, H_1 and S_1 is the enthalpy and the entropy of the stream 1, and T_0 is the temperature of the surrounding and H_0 and S_0 represent the enthalpy and the entropy of the surrounding.

The state of another stream recorded as 2, when the system stream change from state 1 (T_1, P_1) to state 2 (T_2, P_2), the variation of exergy (ΔE_x) is given by

$$\Delta E_x = E_{x,1} - E_{x,2} = \Delta H - T_0 \Delta S \quad (3)$$

In the conventional distillation process, it must input the heat or energy to drive the occurrence of separation process. The bottom of the column input the heat (Q_R) at this temperature (T_R), and the top of the column was sent to the heat (Q_C) at this temperature (T_C). The net work consumption (W) in distillation is

$$W = Q_R(1 - \frac{T_0}{T_R}) - Q_C(1 - \frac{T_0}{T_C}) \quad (4)$$

When the enthalpy difference between the product and the feed is very small, $Q_R=Q_C=Q$ can be approximately gotten, then Equation (4) is rewritten to be:

$$W = QT_0(\frac{1}{T_C} - \frac{1}{T_R}) \quad (5)$$

Then, the total exergy loss (Ex_{loss}) of the distillation column can be calculated by

$$Ex_{Loss} = W - \Delta E_x \quad (6)$$

3. *Prototype case: the prevalent distillation schemes*

3.1 The improved four-column methanol distillation scheme

In the methanol industry, it has gradually appeared two-column, three-column, four-column and five-column distillation scheme [20-23]. With the increase of the number of distillation column, the corresponding distillation scheme can not only increase the efficiency of the process, but also cut the energy consumption. At present, the four-column scheme has a dominant status to produce the purified methanol. However, the improved four-column distillation scheme can cut some more energy requirement with respect to the traditional four-column distillation scheme. The improved four-column distillation scheme is seen as the comparative reference in the paper.

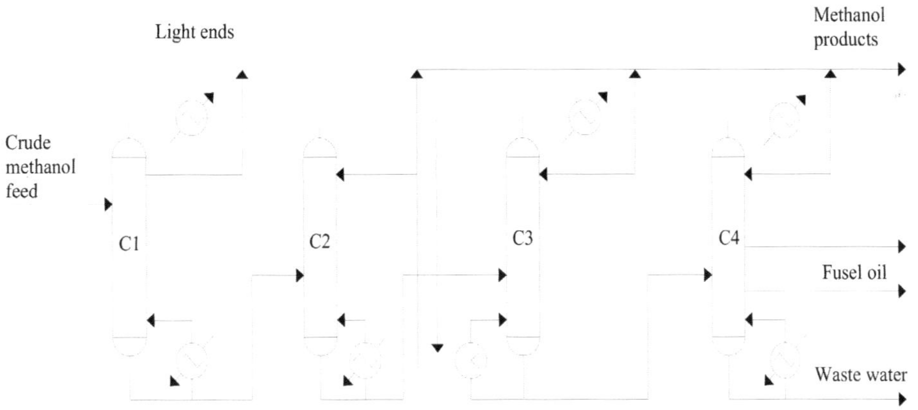

Fig. 2. The improved four-column double-effect methanol distillation scheme (C1, light ends column; C2, pressured column; C3, atmospheric column; C4, water column)

In the improved four-column distillation scheme, as shown in figure. 2 [24], the crude methanol flows into the light ends column (C1) in 87.5 ℃ for removing the light ends and the remnants of dissolved gas from the top of C1. Besides, in order to raise the methanol recovery rates, the light ends column (C1) designs a two-stage condenser system. The crude methanol was distillated successively from the pressured column (C2) and the atmospheric column (C3), and most of the methanol products can be received at the top both of C2 and C3. The bottom stream of C1, C2 and C3 is pumped respectively as the feed into C2, C3 and the water column (C4). Residual methanol, fusel oil and waste water are extracted from the corresponding position of the top, side stream and the bottom of the water column (C4).

In the improved scheme, the operation pressure of C2 is 0.9MPa, and the other three columns run under the atmospheric pressure. Due to the different of operating pressure, the overhead vapor of C2 can undertake thermal coupling connection with the reboiler of C3.

3.2 The current organosilicon monomer distillation scheme

According to the survey, there are several kinds of organosilicon distillation schemes. However, the 9-column organosilicon distillation scheme [9], in China, has an important role in the organosilicon industrial, as illustrated in Figure. 3, it is used as the base case.

Figure. 3 shows that the crude monomer is delivered to the heavies removal column (C1') and the bottom stream of C1' has a lot of the unsteady substances of high boiling points. These unsteady substances of high boiling points are pumped to the high boiling point column (C9') as the feed, and the products from the top of C9' continue to serve as the crude monomer by circumfluence. By this way, it can accomplish the separation between the unsteady substances of high boiling points and the useful substances at a relatively low temperature in the high boiling point column (C9') and avoid the unsteady substances of high boiling points from thermal cracking to generate many unknown substances which would greatly affect the subsequent separation of product. The low removal column (C2') provides trimethylchlorosilane (Me3), methyldichlorosilane (MeH), and other light components for lights removal column (C3'), MeH column (C4'), azeotrope removal column (C5') and Me3 column (C6'), which can finally achieve two products (MeH and Me3) and three wastes. It also provides methyltrichlorosilane (Me1) and dimethylchlorosilane (Me2) and other heavy components for Me1 column (C7') and Me2 column (C8'), which can finally get two products (Me1 and Me2).

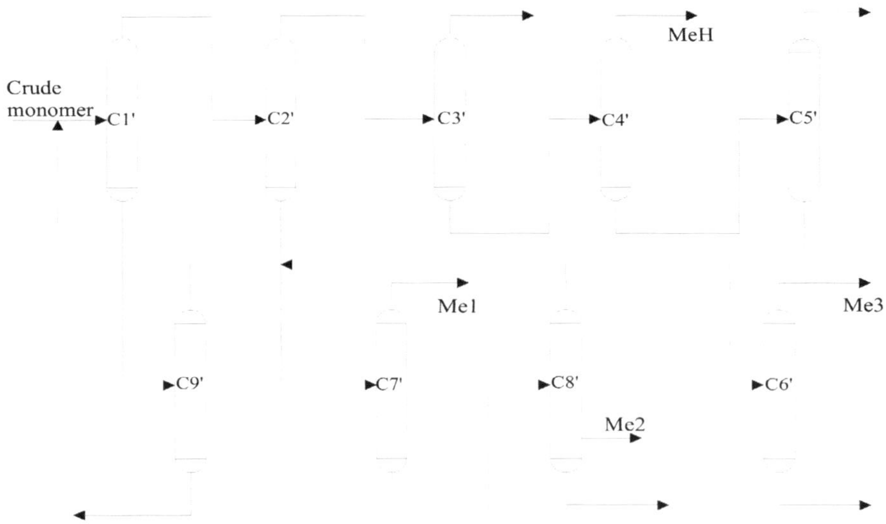

Fig. 3. The prevailing 9-column distillation scheme (C1', heavies removal column; C2', low removal column; C3', lights removal column; C4', MeH column; C5', azeotrope removal column; C6', Me3 column; C7', Me1 column; C8', Me2 column; C9', high boiling point column)

In this scheme, the operating pressure of all nine columns is relatively low, and the highest operating pressure is 260 kPa in the lights removal column (C3'). Although it has a different operating pressure between C3' and the other columns, but the temperature difference is not enough to make them use double-effect or heat integrated distillation.

4. Simulation and analysis both of the prevalent schemes and the optimized schemes

4.1 The novel five-column heat integrated methanol distillation scheme

Although the energy saving of the improved four-column methanol distillation scheme can reach 10% compared with the traditional four-column scheme, it still exist several shortages, which will be illustrated by the results of computer simulation to the improved four-column methanol distillation scheme. The flow rate of the crude methanol is 239230 kg/hr, and the temperature is 87.5 ℃, and the composition of the crude methanol is exhibited in table 1. The inlet and outlet temperature of the cooling water are 30 ℃ and 40 ℃.The methanol content is specified as 99.99 wt% in the methanol products and 10ppm in the wastewater of the water column. Table 2 shows the operating parameters of each column in the improved four-column scheme.

Table 1. Crude methanol feed data

Component	Mass fraction%	Component	Mass fraction%
H_2	0.0100	Pentanol	0.0338
N_2	0.0710	Formic acid	0.0030
Methane	0.0100	Acetic acid	0.0030
CO	0.0360	Methyl formate	0.0092
CO_2	0.6640	Acetone	0.0193
Methanol	95.0000	Methyl ethyl ketone	0.0030
Dimethyl ether	0.0309	Diisopropylether	0.0056
Trimethylamine	0.0010	Water	3.7100
Ethanol	0.0900	i-Pentane	0.0562
i-Propanol	0.0024	n-Pentane	0.0169
Propanol	0.0264	n-Hexane	0.0487
i-Butanol	0.0338	n-Heptane	0.0009
Butanol	0.0185	n-Octane	0.0050
3-Pentanol	0.1407	n-Nonane	0.0050

Table 2. The operating parameters of improved four-column scheme

The operating parameters	C1	C2	C3	C4
Top pressure (kPa)	150	900	110	110
Top temperature (℃)	81.0	132.6	67.7	67.7
Bottom temperature (℃)	83.8	139.1	96.5	104.5
Reflux ratio	0.58	2.44	1.68	11.57
Number of theoretical stage	33	52	58	61

The results of simulation indicate that the hot utility is still in a high level because the reboilers of C1, C2 and (C4) still need the saturated steam to transmit the heat. Table 2 shows the reflux ratio of C4 is 11.57 and this large reflux ratio can cause the increasing of the energy requirement of the water column (C4). Besides, according to the flowsheet of this scheme, the diameter of column is so large that the emergence of many urgent cases.

In view of above these defects in the improved four-column scheme, the novel five-column heat integrated methanol distillation scheme (shown as Fig. 4) is designed in this paper to reduce the impact of these factors as far as possible by using the paralleled or combined double-effect distillation. Juntao Zhang [23] and Jinsheng Sun [25] are also proposed a five-column methanol distillation scheme by using triple-effect distillation and double-effect distillation respectively, and the effect of energy saving of the two schemes is outstanding.

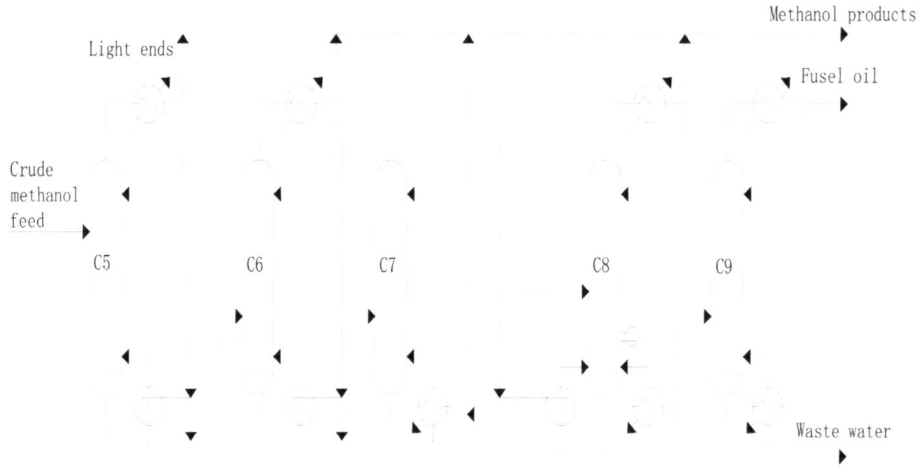

Fig. 4. The five-column heat integrated methanol distillation scheme (C5, light ends column; C6, prepositive atmospheric column; C7, pressured column; C8, postpositive atmospheric column; C9, water column)

According to the principle of the paralleled double-effect distillation, the four columns are unable to constitute the heat integration network in the improved four-column scheme. Through repeated verification, the prepositive atmospheric column (C6) is added after the light ends column (C5), meanwhile, the operating pressure of the pressured column (C7) is reduced to increase its throughput. In this way, the overhead vapor of C7 is split into three streams to respectively integrate with the light ends column (C5), the prepositive atmospheric column (C6) and the postpositive atmospheric column (C8). Due to use the paralleled double-effect distillation, the duties of the hot utility have decreased in a great margin. However, there are still part of heat from the background process is absorbed by the cooling water. So an inter-reboiler with a low temperature level is designed in the postpositive atmospheric column (C8) and it can share the heat with the background process streams, besides, we have proved the feasibility by using the column grand composite curve analysis of columns. Figure. 4 shows that the product of the top of the water column (C9) is fusel oil instead of purified methanol, which can not only reduce the reflux ratio of C9, but also make the diameter of column smaller to reduce the difficulty of operation.

Due to produce the fusel oil instead of purified methanol in the water column (C9) in the novel five-column distillation scheme, the methanol content of the feed for the water column have to limit at 0.016 wt%. In the five-column scheme, the operating pressure and the number of theoretical stage for each column are determined through the existing process, the onsite data and the requirement of the heat coupling. The proportion of the methanol products from the C6, C7 and C8 is determined by their own double-effect or heat integrated distillation system. The simulation results of the novel five-column scheme are listed in table 3.

Table 3. The operating parameters of five-column scheme

Column	C5	C6	C7	C8	C9
Top pressure (kPa)	160	110	400	110	110
Top temperature (℃)	83	66.6	104	66.6	86.5
Bottom temperature (℃)	85.6	84	118.8	104.5	103.2
Reflux ratio	1.46	1.49	2.23	1.87	4
Number of theoretical stage	33	54	54	65	29

4.2 Energy consumption and exergy loss analyses for the two methanol schemes

In the improved four-column scheme, the pressured column and the atmospheric column constitute the distillation system of heat coupling, there are three columns (C1, C2, C4) need the hot utility, and the entire duty of reboiler is 145.4×10^3 kw. As described above, the light ends column (C5) and the prepositive atmospheric column (C6) don't need the saturated steam in the five-column scheme and the total duty of reboiler is reduced to 122.7×10^3 kw. The detailed data of the reboiler duty as shown in table 4.

Table 4. The duty of reboiler in the two schemes

Column	C1	C2	C4	C7	C8	C9
Duty of reboiler (10^4 kw)	3.30	10.04	1.20	8.97	2.86	0.44

Obviously, the novel five-column scheme is much better than the improved four-column scheme in energy saving and the energy consumption of the novel scheme can be reduced by 15.61%. However, the recovery rate of methanol of the novel scheme is approximately equal with the improved four-column scheme.

Only the energy consumption analysis is not revealed the energy saving essence of the novel five-column scheme. Exergy analyses are used to further compare the two schemes, and the exergy flow rates are calculated through using the computer simulation data. The activity coefficient of the liquid is calculated using the Alcohol Package thermodynamic model. The calculation of the net work consumption of distillation column should pay more attention to the temperature of the cooling water. When the cooling water exchanges heat with other streams, the temperature of the cooling water is momentarily changed, so it is replaced the temperature of cooling water with the arithmetic average of inlet and outlet temperature of cooling water. Figure. 5 show the overall exergy loss of each column in the two methanol distillation schemes, in addition to the overall exergy loss of C6. Due to optimize the four-column scheme, the prepositive atmospheric column (C6) is added in the transformed scheme, and the overall exergy loss of C6 is 9157.6 kw. The exergy loss of each column, besides the pressured column (C7), in the novel five-column scheme is obviously less than their counterparts in the improved four-column scheme.

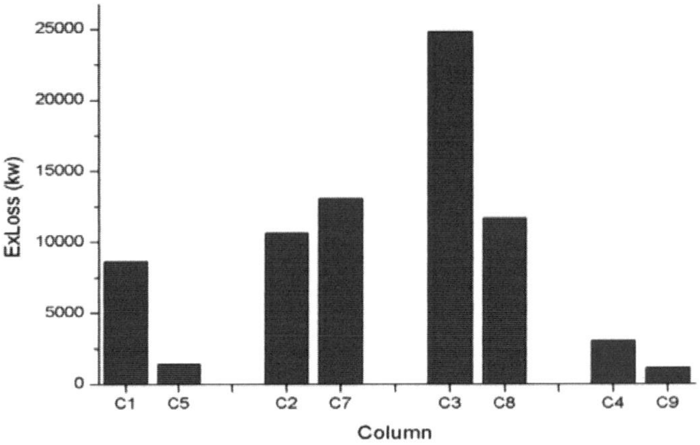

Fig. 5. The overall exergy loss analysis of each column in the two schemes

The total exergy loss of the improved four-column scheme and the transformed scheme is, respectively, 47100.1 kw and 36355.1 kw. The total exergy loss of the optimized scheme is dropped by 22.8%. Through the comparison between the exergy loss and energy consumption of the two schemes, it is indicated that the novel five-column scheme should be applied to produce the purified methanol in industry.

4.3 The novel heat coupling organosilicon distillation scheme

The primary source of the crude organosilicon monomers is the catalytic reaction products between chloromethane and silicon at high temperature and the table 5 displays the normal boiling points and mass percentage of the main components of the crude organosilicon monomers. Methyltrichlorosilane (Me1), dimethylchlorosilane (Me2), trimethylchlorosilane (Me3) and methyldichlorosilane (MeH) are the most valuable components in the crude organosilicon monomers. The purification requirements on organosilicon of these materials are shown in table 6.

Table 5. The crude organosilicon monomer feed data

Component	Normal boiling point at 1 atm/℃	Mass percentage %
LB		1
Methyldichlorosilane	41.8	4
Silicon tetrachloride	57.6	0.5
Trimethylchlorosilane	57.9	2.5
Methyltrichlorosilane	66.4	11
Dimethyldichlorosilane	70.2	80
HB		1
Temperature (℃)	40	
Pressure（kPa）	100	
Flow rate (kg/h)	26200	

Table 6. Purification demand on organosilicon

Monomer	Purification demand
MeH	99.5%
Me3	99.5%
Me1	99.5%
Me2	99.95%

The conventional nine-column scheme is simulated with the aid of existing process and onsite data to achieve the operating parameters of each column, as illustrated in table 7. The bottom temperature of the heavies removal column (C1') and the high boiling point column (C9') are obviously much bigger than that of the other seven columns in table 7. According to the principle of the double-effect distillation, the top temperature of the high pressured column is relatively high in comparison with the bottom temperature of the low pressured column, and the high pressured column can integrate with the low pressured column. Although the top temperature of C1' is less than the bottom temperature of the other seven columns, the overhead vapor of C1' still has a great opportunity to conduct heat with the other seven columns with the increasing of its top temperature.

Table 7. The simulation results of prevailing 9-column

Column	C1'	C2'	C3'	C4'	C5'	C6'	C7'	C8'	C9'
Top temperature (℃)	70.8	65.3	57.3	66.7	72.1	70.6	71.2	74.9	72.7
Bottom temperature (℃)	124.1	89.1	80.4	85.2	74.9	75.4	75.3	75.3	126.5
Reboiler duty (MW)	11.67	5.61	1.13	0.51	1.19	0.46	12.27	16.08	0.02
Condenser duty (MW)	-12.47	-5.53	-1.08	-0.53	-1.20	-0.47	-12.69	-16.12	-0.02
Top pressure (kPa)	110	180	260	220	170	150	110	110	110

According to the way of the paralleled double-effect distillation, the increasing operating pressure in the heavies removal column (C1') can make the top temperature of C1' ascending to integrate with the other seven columns. The idea is still in study to overcome some difficulties. However, we have got another feasible method, in which the operating pressure of C1' and the Me1 column (C7') is raised to 360 kPa and 280 kPa, respectively. In this way, the top temperature of C1' and C7' is increased to ensure they can integrate with the other columns, besides the high boiling point column, and the parallel double-effect or heat integrated organosilicon monomer distillation scheme is designed as shown in figure. 6 [26]. The feasibility of this heat coupling scheme has confirmed through three x-y diagrams including the heavies component and Me1, the heavies component and Me2, Me1 and Me2 [9].

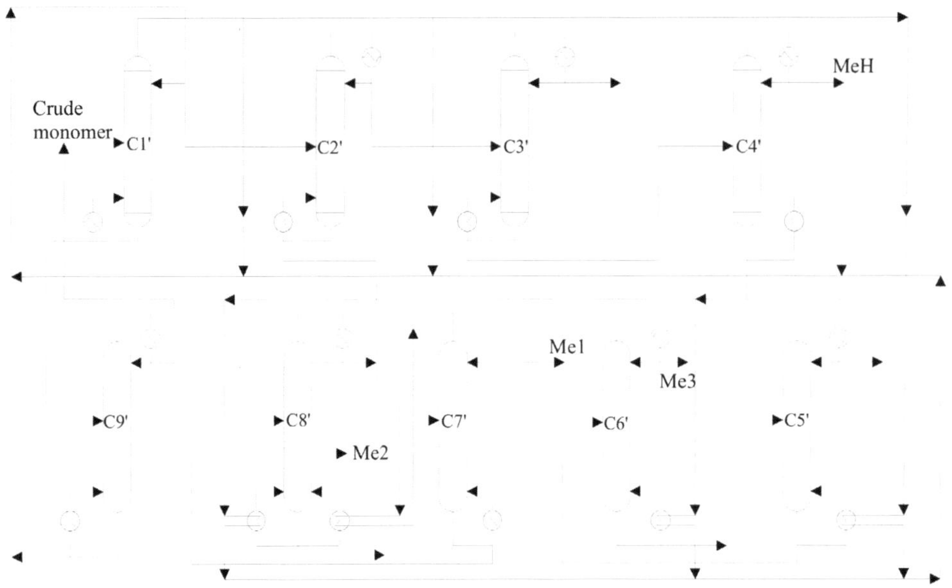

Fig. 6. The heat coupling 9-column scheme (C1', heavies removal column; C2', low removal column; C3', lights removal column; C4', MeH column; C5', azeotrope removal column; C6', Me3 column; C7', Me1 column; C8', Me2 column; C9', high boiling point column)

Figure. 6 shows the heavies removal column (C1') make the heat coupling with the low removal column (C2'), the lights removal column (C3'), the MeH column (C4'), the azeotrope removal column (C5'), the Me3 column (C6') and the Me2 column (C8'), successively. In order to gain a better energy saving effect, C8' is designed with two reboilers, the one integrated with C1' and the other integrated with the Me1 column (C7'). The operating parameters and the duty of reboiler and condenser of each column in the heat integrated 9-column are enumerated in table 8.

Table 8. The simulation results of heat integrated 9-column

Column	C1'	C2'	C3'	C4'	C5'	C6'	C7'	C8'	C9'
Top temperature (℃)	114.3	65.3	57.3	66.7	72.1	70.6	105.1	74.9	72.7
Bottom temperature (℃)	124.9	89.9	78.8	84.5	75.4	75.4	110.9	76.6	121.4
Reboiler duty (MW)	13.06	5.62	1.13	0.51	1.19	0.46	13.83	14.12	0.38
Condenser duty (MW)	-13.72	-5.54	-1.08	-0.53	-1.20	-0.47	-14.02	-14.43	-0.41
Top pressure (kPa)	360	180	260	220	170	150	280	110	110

4.4 Energy consumption and exergy loss analyses for the two organosilicon schemes

Figure. 4 indicate that each column of the prevalent nine-column scheme is not integrated with the other columns and the reboilers of all of these columns are required the saturated steam (600 kPa). According to the data from the table 7, the total duty of reboiler is 48.94 MW for the prevailing 9-column scheme. In the new nine-column scheme, C1' integrate with C2', C3', C4', C5', C6' and C8' successively and C7' integrate with C8', so C2', C3', C4', C5', C6' and C8' don't consume the saturated steam. The total duty of reboiler is 27.27 MW. Obviously, the heat integrated 9-column scheme has an absolute advantage over the prevailing 9-column scheme in energy saving, and the energy savings can reach 44.28%.

As the same with the methanol distillation scheme, the inlet and outlet temperature of cooling water are 30 ℃ and 40 ℃, and the total exergy loss is calculated through the results of computer simulation of the two schemes. The overall exergy loss comparison of each column in the two schemes is illustrated in figure. 7. In the normal scheme, the throughput of C9' is so small that the overall exergy loss of C9' is very near to zero, so it only have the exergy loss of C9' of the new scheme in figure. 7. The overall exergy loss of each column in the optimized scheme is declined in comparison with that of the corresponding column in the normal scheme except the high boiling point column (C9'). Because the crude organosilicon monomers contains some unstable components, in order to avoid these components happening thermal cracking in the heavies removal column (C1') in the novel scheme, more liquid from the bottom of C1' is pumped to C9' as the feed. That can cause the ascent of the throughput and the overall exergy loss of C9'.

The total exergy loss is 14035.5 kw and 8363.2 kw, correspondingly, in the conventional scheme and the novel scheme, and the exergy loss of the new scheme can be reduced by 40.41%. Through the above analysis and comparison, the heat integrated nine-column organosilicon distillation scheme should be a substitute for the prevalent nine-column scheme.

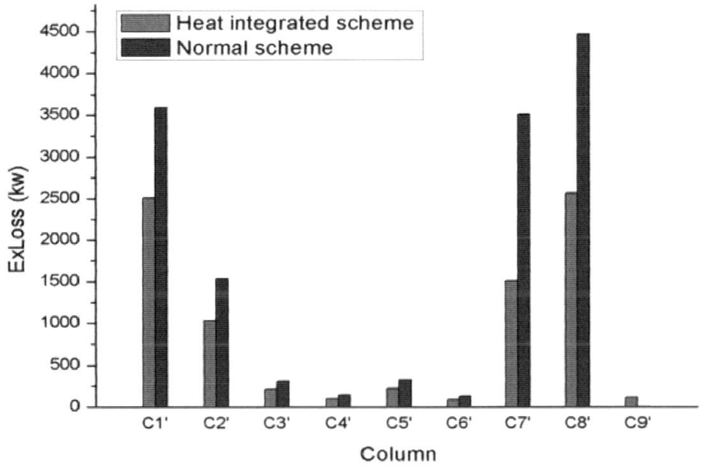

Fig. 7. The exergy loss comparison between the heat integrated scheme and the normal scheme

5. *Conclusions*

Paralleled double-effect or heat integrated distillation for energy saving is proposed, exampled by heat integrated five-column methanol distillation scheme and the heat coupling 9-column organosilicon monomer distillation scheme, with the aid of computer simulation. Besides, the optimized methanol distillation scheme and the transformed organosilicon scheme are comprehensively analyzed through the energy consumption and the exergy loss in comparison with the correspondingly prevalent scheme.

Adding a prepositive atmospheric column and decreasing the operating pressure of the pressured column in the novel five-column scheme can make the pressured column to simultaneously integrate with the light ends column, the prepositive atmospheric column and the postpositive atmospheric column. Besides, an inter-reboiler is added to integrate with the background process streams. The energy consumption and the exergy loss of the new scheme are 15.61% and 22.8% lower than the improved scheme.

Increasing the operating pressure of the heavies removal column and the Me1 column can make them to integrate with the other columns, in addition to the high boiling point column, to constitute a paralleled double-effect distillation system in the optimized nine-column scheme. The energy consumption and the exergy loss of the novel scheme are 44.28% and 40.41% less than the normal scheme.

The two cases expound that the key of paralleled double-effect distillation process lies in the reasonable adjustment of the energy distribution, the operating parameters and the configuration of some columns. It not only reduces the quantity of energy consumption, but also avoids the decrease of the quality of energy. For factories with a large numbers of distillation schemes of high energy requirement, the paralleled schemes lead to improve economic benefit.

References

[1] Araujo AB, Brito RP, Vasconcelos LS, "Exergetic analysis of distillation processes-a case study," Energy 2007;32:1185-93.

[2] Michele Mascia, Francesca Ferrara, Annalisa Vacca, Giuseppe Tola, Massimiliano Errico, "Design of heat integrated distillation systems for a light ends separation plant," Applied Thermal Engineering 2007;27:1205-1211.

[3] Amiya K. Jana, "Heat integrated distillation operation," Applied Energy 2010;87:1477-1494.

[4] Hilde K. Engelien, Sigurd Skogestad, "Multi-effect distillation applied to an industrial case study," Chemical Engineering and Processing 2005;44:819-826.

[5] James G. Gebbie, Michael K. Jensen, Piotr A. Domanski, "Experimental transient performance of a heat pump equipped with a distillation column," International Journal of Refrigeration 2007;30:499-505.

[6] Massimiliano Errico,Giuseppe Tola, Michele Mascia, "Energy saving in a crude distillation unit by a preflash implementation," Applied Thermal Engineering2009;29:1642-1647.

[7] Massimiliano Erricoa, Giuseppe Tola, Ben-Guang Rong, Daniele Demurtas, Ilkka Turunen, "Energy saving and capital cost evaluation in distillation column sequences with a divided wall column," chemical engineering research and design 2009;87:1649-1657.

[8] Ivar J. Halvorsen, Sigurd Skogestad, "Energy efficient distillation," Journal of Natural Gas Science and Engineering 2011;3:571-580.

[9] Jinsheng Sun, Fan Wang, Tingting Ma, Hong Gao, Yanzhen Liu, Fang Cai, "Exergy analysis of a parallel double-effect organosilicon monomer distillation scheme," Energy 2012;47:498-504.

[10] Hilde K. Engelien, Sigurd Skogestad, "Multi-effect distillation applied to an industrial case study," Chemical Engineering and Processing 2005;44:819-826.

[11] Kazuo Matsuda, Kenichi Kawazuishi, Yasuki Kansha, Chihiro Fushimi, Masaki Nagao, Hiroshi Kunikiyo, Fusao Masuda, Atsushi Tsutsumi, "Advanced energy saving in distillation process with self-heat recuperation technology," Energy 2011;36:4640-4645.

[12] M.A. Sharaf, A.S. Nafey, Lourdes García-Rodríguez, "Exergy and thermo-economic analyses of a combined solar organic cycle with multi-effect distillation (MED) desalination process," Desalination2011;272:135-147.

[13] Weiwen Wang, Wenjian Wang, "Energy Saving in Methylchlorosilane Distillation," International Conference on Computer Distributed Control and Intelligent Enviromental Monitoring 2012:290-293.

[14] Myungwan Han, Sunwon Park, "Multivariable control of double-effect distillation

configurations," J. Proc. Cont. 1996:247-253.
[15] Hilde K. Engelien, Sigurd Skogestad, "Multi-effect distillation applied to an industrial case study," Chemical Engineering and Processing2005;44:819-826.
[16] Debadrita Maiti, Amiya K. Jana, Amar Nath Samanta, "A novel heat integrated batch distillation scheme," Applied Energy2011;88:5221-5225.
[17] Anton A. Kiss, Servando J. Flores Landaeta, Carlos A. Infante Ferreira, "Towards energy efficient distillation technologies-Making the right choice," Energy2012;47:531-542.
[18] Diego C. Alarcón-Padilla, Lourdes García-Rodríguez, Julián Blanco-Gálvez, "Experimental assessment of connection of an absorption heat pump to a multi-effect distillation unit," Desalination2010;250:500-505.
[19] Diego C. Alarcón-Padilla, Lourdes García-Rodríguez, Julián Blanco-Gálvez, "Design recommendations for a multi-effect distillation plant connected to a double-effect absorption heat pump: A solar desalination case study," Desalination2010;262:11-14.
[20] Fiedler E, Grossman G. Methanol, "Ullmann's Encyclopedia of Industrial Chemistry," Weinheim : Wiley-VCH Verlag/Gmbh & Co; 2002.
[21] Meyers RA, "Handbook of synfuels Technology," New York : McGraw-Hill; 1984.
[22] Chu YZ, Qin LP, "Optimum design of methanol distillation process and columns," Chem Ind Eng Prog 2008;(27):1659-62.
[23] Juntao Zhang, Shengrong Liang, Xiao Feng, "A novel multi-effect methanol distillation process," Chem Eng Pro 2010;(49):1031-37.
[24] Jinsheng Sun, Fan Wang, Tingting Ma, Hong Gao, Peng Wu, Lili Liu, "Energy and exergy analysis of a five-column methanol distillation scheme," Energy2012;45:696-703.
[25] J. S. Sun, P. Wu, Y. H. Wang, H. Gao, L.L. Liu.CN Patent102039058 A. 2011.
[26] J. S. Sun, P. Wu, Y. H. Wang, Y. X. Su, H. Gao, J. Li, D. Yan. CN Patent 101798322 A. 2010.
[27] Michele Mascia, Francesca Ferrara, Annalisa Vacca, Giuseppe Tola, Massimiliano Errico, "Design of heat integrated distillation systems for a light ends separation plant," Applied Thermal Engineering 2007;27:1205-1211.
[28] Olga A. Flores, J. Carlos Ca´ rdenas, Salvador Herna´ ndez, Vicente Rico-Ramı´rez, "Thermodynamic Analysis of Thermally Coupled Distillation Sequences," Ind. Eng. Chem. Res. 2003;42:5940-5945.
[29] Szargut J, "Exergy Analysis of Thermal, Chemical and Metallurgical Processes," New York: Hemisphere Publishing Corporation;1988:55-152.

(45n) In Situ Transesterification of Jatropha Curcas Seed Particles with Alkaline Phase Transfer Catalyst

Subbarao Duvvuri[1], Shuhaimi Mahadzir[2] and Sintayehu Hailegiorgis[2], (1)Department of Chemical Engineering, Universiti Teknologi PETRONAS, Tronoh, Malaysia, (2)Chemical Engineering, Universiti Teknologi PETRONAS, Tronoh, Malaysia

Economic viability of biodiesel technology is at stake due to the high cost of edible oil feedstock and slow reaction rates due to limited solubility of oil in methanol. In situ transesterification of non edible oil seed particles can minimize the cost of feedstock. Microwave pretreatment of seeds can make oil molecules more reactive. Use of phase transfer catalysts can enhance the reaction rates. In the present work, these three concepts are utilized together to investigate alkaline in situ transesterification of microwave pretreated jatropha curcas seed particles in presence of benzyltrimethylammonium hydroxide (BTMAOH) as a phase transfer catalyst (PTC). It was observed that addition of PTC to alkaline catalyst has improved conversion and reduced reaction time. Microwave pretreatment of jatropha seed particles reduced reaction time drastically. The quality of biodiesel produced was within the limits of the international standard requirements of ATME D6751 and EN-14214. Thus, combined use of microwave pretreatment of seed particles, phase transfer catalysis in the presence of alkaline catalyst can increase the rate of in situ transesterification and product yield.

(45o) Catalyzed-Assisted Manufacture of Olefins (CAMOL): Updated for Use in Naphtha Service

Maximilian Walter, Petrochemicals, BASF SE, Ludwigshafen, Germany and **Suzanne Rech**, BASF Qtech, BASF Corporation, Iselin, NJ

Since the commercialization of the olefins plant, operators have searched for ways to be more competitive. One area always targeted includes keeping their furnaces in operation longer between decoking steps. Over the years, many anti-coking methods/technologies have become commercially available using different approaches to prevent coils from coking. One of the latest available technologies is the novel CAMOL™ (Catalyzed-assisted Manufacture of Olefins) Technology from BASF Qtech. As reported in the introduction of the technology to the EPC in 2008 & 2010, the CAMOL™technology differs from other available anti-coking technologies in that the technology not only eliminates the catalytic (or filamentous) coke make through a physical barrier, but also gasifies the thermal (or amorphous) coke for an even greater coking resistance. The coating process has generated a coating with almost identical thermo-mechanical properties as the base metal substrate (alloy). This allows bonding to the base metal similar to being welded with a minimum risk of spallation when going through thermal cycling. Commercially, the CAMOL Technology has been available since 2006 and is now in operation in furnaces of various designs.

With CAMOL™, operators have experienced extended run lengths, reduced energy requirement and GHG emissions, lower average TMT operating temperatures, reduced DMDS consumption, and less carburization. In addition, the technology has exhibited high operating temperature stability (>1130°C/ 2065°F).

At BASF SE in Ludwigshafen, Germany, the technology has been in operation since early 2010 in mostly naphtha service. After overcoming some initial challenges associated with a catalytic surface technology, such as reactivating the catalytic species after decoking, the coil is now successfully operating with run lengths significantly higher than uncoated coils. This paper will report on the results and learnings from the experiences gathered over the past three years along with the advancements made since the last report in 2010.

(45p) Extending Furnace Run Length and Coil Life by Optimizing Decoking

Vinod Mishra, Ingenero, Mumbai, India and **James Brigman**, Ingenero, Houston, TX

Abstract:

In the Ethylene manufacturing process, coke deposited in the radiant coils and Transfer Line Exchangers (TLEs) during normal operation is removed periodically by decoking the furnaces using a mixture of steam and air. There is a significant variation in operating parameters adopted by different plants for decoking. These are often either too conservative or aggressive, leading to production loss from increased off-stream time, TLE limitation and deterioration in coil life. This paper describes how the decoking procedure was modified based on furnace and TLE operating conditions and the decoking steps changed to obtain the best possible Ethylene yield, without adversely impacting coil life and TLE performance. This optimization has now been done in several major ethylene facilities and furnace run length has more than doubled and production has significantly increased.

SOLVING OPERATIONAL PROBLEMS IN A NAPHTHA SPLITTER

Flávio Weissheimer, GE Water & Process Technologies do Brasil Ltda, São Paulo/Brasil
Giorgio Franceschetti, Braskem, Santo André/Brasil
Rafael Henrique Camparin, GE Water & Process Technologies do Brasil Ltda, São Paulo/Brasil
Patrick Lucas, GE Water & Process Technologies, The Woodlands/US

Abstract

This paper sets out how the polymerization process of Reboilers of the (A-100) Crude Naphtha Fractionation Unit of Braskem UNIB 3 ABC can be controlled by applying polymerization inhibitors developed by GE Water & Process Technologies Ltda and Braskem´s engineering solutions.

These heat exchangers have a high degree of polymerization because of the content of olefinic compounds present in the feed composition which, when in contact with high temperatures in the heat exchange areas of reboilers, tend to polymerize. The joint solution of GE technology and Braskem engineering solutions applied to this system ensures regular operation of such equipment for periods of time greater than those hitherto recorded without manifesting loss of heat transfer efficiency.

In the 2012 shutdown while the opening of these reboilers was being inspected, it was observed that pipes had not been clogged nor was surface fouling present, which is a very positive state when compared to the historical records of previous campaigns.

The strategy adopted by GE/Braskem to ensure a longer campaign for these pieces of equipment was based on the quality control of feeding by adjusting dosages and monitoring performance using statistical analysis tools.

The use of statistical models based on Multivariable Regression Analysis (MRA) to evaluate performance, which is validated by inspecting reboilers internally, has been an efficient way to determine the degree to which the system has been fouled and also to schedule shut-downs for cleaning the mechanical parts.

Description of the Process

The (A-100) Naphta Fractionation Unit includes two distillation columns, the objective of which is to separate the light (C_6), medium (C_6 - C_8) and heavy (C_{9+}) cuts, included in the feed supply of crude naphtha.

The light cut and a portion of the heavy cut are sent to pyrolysis furnaces so as to produce ethylene. The medium cut is sent to the Hydrotreating and Catalytic Reforming Unit, with the objective of increasing the aromaticity of the system so as to produce BTX.

The A-100 processes 100 tons/h of crude naphtha, and approximately 35% of this current represents the average cut. Figure 1 below shows the schematic flow diagram of this fractionation process.

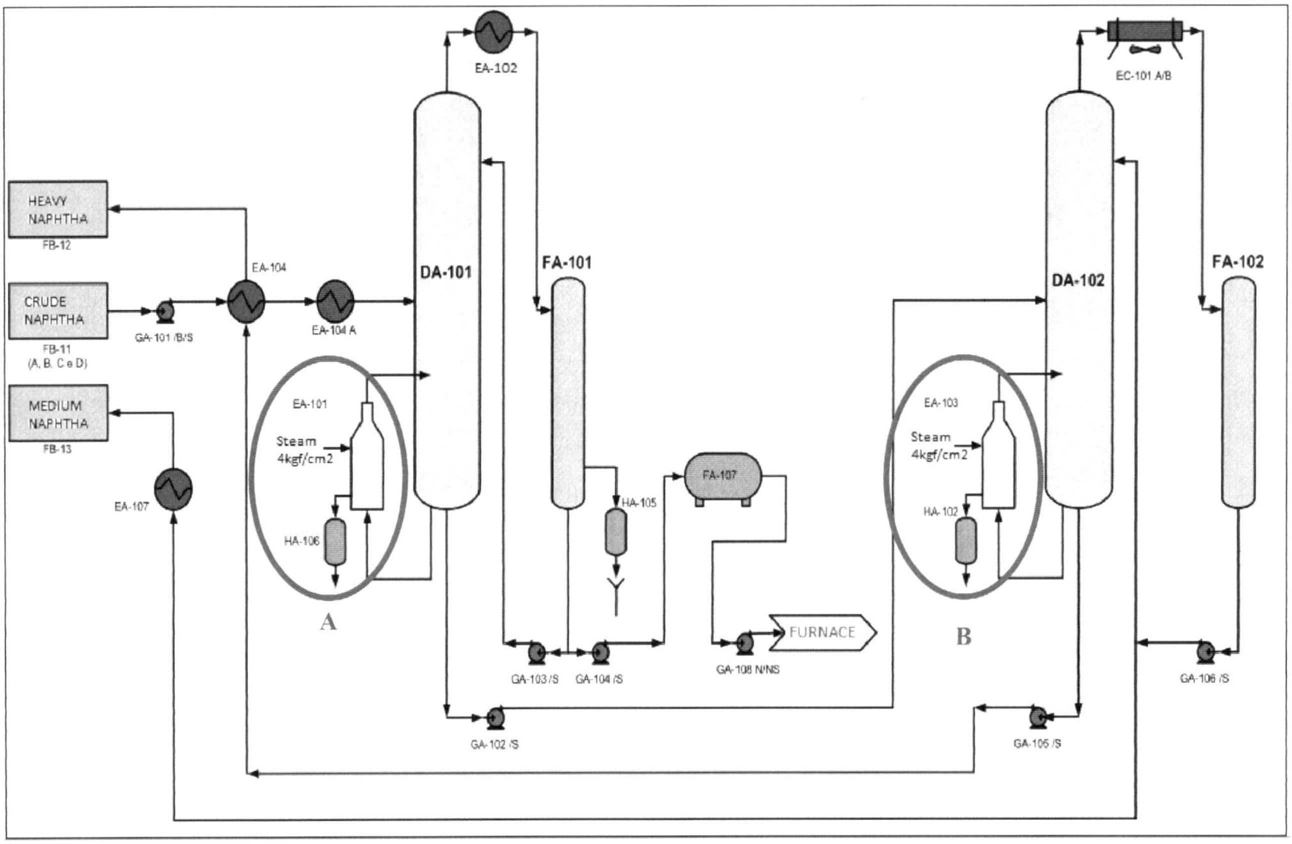

Figure 1. Process Flow Diagram (A-100)

The Reboilers, indicated by the letters A and B in Figure 1 above, operate with an injection of low steam (4kgf/cm2) to promote vaporization of the bottom stream of the A-100 distillation columns. These devices operate over a temperature range of about 150°C, which is sufficient to promote the polymerization of certain olefinic compounds present in the feed of crude naphtha.

Poor performance of these Reboilers can lead to production stoppages for maintenance and cleaning, which very often affects up to five days of production. The possibility of raising the times of campaigns and predicting the degree of fouling in downstream units helps to minimize production losses due to unscheduled stoppages.

This study will discuss the results of Reboiler A, which is the most critical equipment, with a view to synthesizing data, but gains in equipment B were also significant.

Evaluation System

Historical Record of the Campaign

Reboiler A was installed in 1972 but examination of the historical records did not reveal that the sheaf of tubes had ever been replaced. This exchanger has currently 11 tubes plugged into it.

The Campaign times of this equipment are shown in Table 1, with the last campaign being highlighted in blue so as to set it apart from the other ones. Table 1 gives a summary in the order of the timings of the campaigns, the average load processed in the period, the total load processed, the reason for operational stoppages etc.

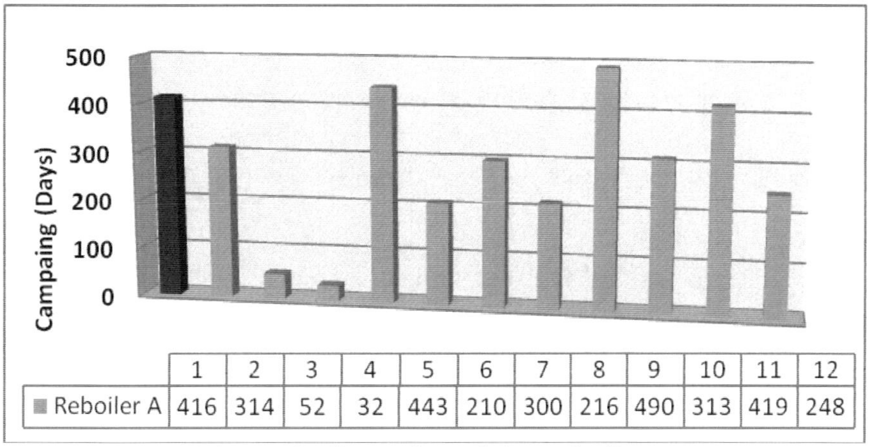

Figure 2. Campaign of the Reboiler A

Table 1. Historical Record of Reboiler A Campaigns

Date	Campaign *(days)*	Capacity *(ton/h)*	Capacity *(ton/campaign)*	Occurrence	Tubes Clogged *(%)*
13/06/12	416.0			Programmed Shutdown	0
17/03/11	314.0	93.1	701601.6	Heat exchange performance	30
03/05/10	52.0	90.8	113318.4	Stuck tube	-
11/03/10	32.0	71.1	54604.8	Heat exchange performance	50
09/02/10	443.0	72.0	765716.6	Heat exchange performance	20
16/11/08	210.0	98.1	494575.2	Heat exchange performance	-
16/04/08	300.0	104.0	748800.0	Programmed Shutdown	-
16/06/07	216.0	92.8	481075.2	Programmed Shutdown	-
10/11/06	490.0	-	-	Programmed Shutdown	-
30/06/05	313.0	-	-	Stuck tube	-
17/08/04	419.0	-	-	-	-
18/06/03	248.0	-	-	-	-
10/10/02	-	-	-	-	-

Conditions of Opening

In this review, a comparison is presented between two operational shut-downs of A-100, which occurred in March 2011 and May 2012, in order to observe the improvement in performance from the last campaign associated with the change in the GE Treatment strategy.

Campaign I: Production shutdown on 03/2011

Table 2 below summarizes the features observed during the inspection of the state of the inlet of this piece of this equipment during the shutdown of March 2011:

Table 2. Internal inspection of Reboiler A in 03/2011

Bottom Cover	Surface residue
Bottom Mirror	No deposits or internal blockage of the tubes were observed.
Top Mirror	Approximately 30% of the tubes were clogged, especially in the region of the vapor inlet, which has higher temperatures. Magnetic and non-magnetic deposits.

May/2011	**May/2011**

Figure 3. The mirror of Reboiler A before it was cleaned.	**Figure 4.** The mirror of Reboiler A at the vapor inlet before it was cleaned.

Note: It is difficult to carry out internal cleansing by Hydro jetting (pump pressure during cleaning is 20,000 psi). This reboiler has 11 tubes plugged into it, but none of them are plugged in during this shutdown.

Campaign II: Production shutdown on 03/2011

Table 3 below summarizes the features observed during the inspection of the state of the inlet of this piece of equipment during the shutdown in May 2012:

Table 3. Internal inspection of Reboiler A in 05/2012

Bottom Cover	Surface residue
Bottom Mirror	No deposits or internal blockage of the tubes were observed.
Top Mirror	Clogged tubes were not identified, not even in the region of the inlet vapor. There was only a thin layer of deposit that was easy to remove and which was evenly distributed over the upper surface of the mirror.
Spool	A thin deposit on the inner surface of the caisson.

May/2012	**May/2012**

Figure 5. The mirror of Reboiler A before it was cleaned.	**Figure 6.** The mirror of Reboiler A before it was cleaned.

Note: Hydro jetting was done quickly, without any obstructions due to deposits having been identified. This equipment has 11 tubes plugged into it, but none of them were during this shutdown.

Analysis of Deposits

The characterization of the deposits found in Reboiler A showed an 86.0% loss on ignition, thus indicating the strong presence of organic compounds in their composition, with the functional groups of carboxylic acid and ester standing out. The inorganic analysis of the sample indicated a high concentration of iron oxides and also sulfur oxides in the form of sulfides on the cover and spool of this reboiler. This is summarized in Table 4 below:

Table 4. Analysis of the Deposits of Reboiler A

Organic Compounds - LOI (%)	86.0
Water of hydration	Smaller
Sulfur compounds	Smaller
Carboxylic Acid and Ester	Greater
Dichloromethane soluble (%)	2.0
Elemental Sulfur	Greater
Dichloromethane insoluble (%)	98.0
Water of hydration	Smaller
Inorganics	Smaller
Carboxylic Acid and Ester	Greater
Composition of Dichloromethane insolubles (%)	
Carbon	56.0
Hydrogen	5.0
Nitrogen	3.0
Sulfur	8.0
Inorganic Compounds (%)	
Iron, Fe2O3 + Fe3O4	82.0
Sulfide, S2-	16.0
Phosphate, P2O5	1.0
Sodium, Na2O	1.0

Evaluation of Performance

The evaluation of performance of Reboiler A was prepared using the MRA tool, which is a GE good practice to predict fouling in critical systems. This statistical tool enables a dependent variable to be determined, in this case the percentage of opening the steam valve, by means of the numerical regression of the other operational variables of the column. This method generates a mathematical equation to predict the percentage of inlet steam when the equipment is clean, given that the equation obtained by regression includes data of this condition.

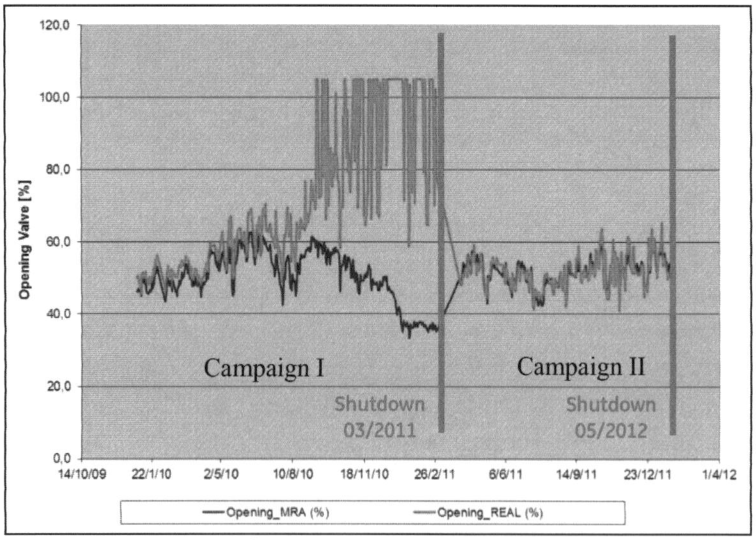

Figure 7. Performance Evaluation of Reboiler A

In Figure 7 the dark blue line represents the percentage of inlet steam when the equipment is clean, as per the equation defined by the MRA. The pink line represents the actual percentage of inlet steam. The sharp upward shift evidenced in the period between November 2010 and March 2011 shows the fouling of the system during Campaign I (prior to the 2011 operational shutdown).

After the 2011 shutdown, it can be seen that the two curves follow the same trend, without a shift, indicating that the unit has the equivalent performance of a clean piece of equipment. The change in the GE treatment strategy, taking the control of the dosages of the product Petroflo 20Y25 into account, was seen to have been effective in increasing the campaign time of this equipment.

Conclusion

The change of strategy in applying dispersant and controlling the dosages after the March 2011 shutdown, presented positive results, given the states of the inlet seen during the inspection of reboilers A and B in May 2012..

The low degree of fouling of these reboilers evidenced in the 2012 shutdown led to the production process being considered reliable, as the A-100 reboilers are no longer the pieces of equipment responsible for the loading limit of the unit. The downtime during the cleaning of the mechanical parts of the reboilers was about 10 times less than that observed in the previous shutdown, thus reducing Braskem´s maintenance costs.

In addition, the MRA statistical tool was able to produce a reliable representation of the degree of fouling of the system. This tool can be effective for scheduling shutdowns more than 03 months in advance.

References

1. Setal Lummus, "Manual de Operação da Unidade de Processamento de Nafta Bruta (Área 100)", Santo André, Brazil.
2. Ponce de Leon, Carolina; Weissheimer, Flavio and Camparin, Rafael H. (2011), "Proposta Técnica de Fornecimento de Especialidades Químicas, Equipamentos e Serviços para Tratamento Químico para a Unidade de Processamento de Nafta Bruta (Área 100)", Santo André, Brazil.
3. Kleinbaum, David G., Kupper, Lawrence L., Muller, Keith E. (2008) "Applied Regression Analysis and Other Multivariable Methods", Belmont, USA.

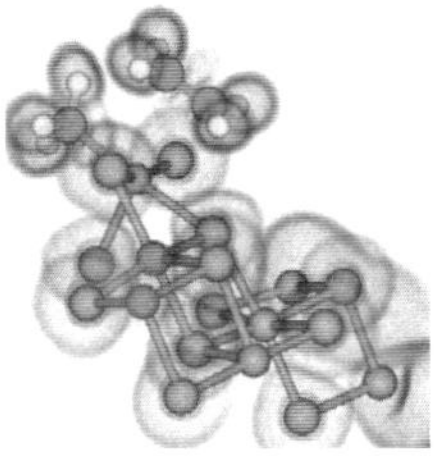

SmartProcess® Application Strategy Achieves Faster and More Efficient Control of an Ethylene Complex

Patrick Truesdale
Principal Consultant
Industry Solution Group
Emerson Process Management

April 29, 2013
San Antonio, TX

Speaker Introduction

- Principal Consultant
  Emerson Process Systems & Solutions
- BS Electrical Eng, PE
- 40+ Years experience with Refinery & Petrochem Process Control
- Joined Emerson in 2007
- Previous employers: duPont, Bonner & Moore, SETPOINT, AspenTech, Comcept Technology

Patrick Truesdale

E-mail: patrick.truesdale@emerson.com

Agenda

- Performance Package Review
- SmartProcess Ethylene Overview
  - Furnaces
  - Distillation
- Benefits

13 AIChE®
Spring Meeting, San Antonio, TX

EMERSON. Process Management

SmartProcess Ethylene Packages

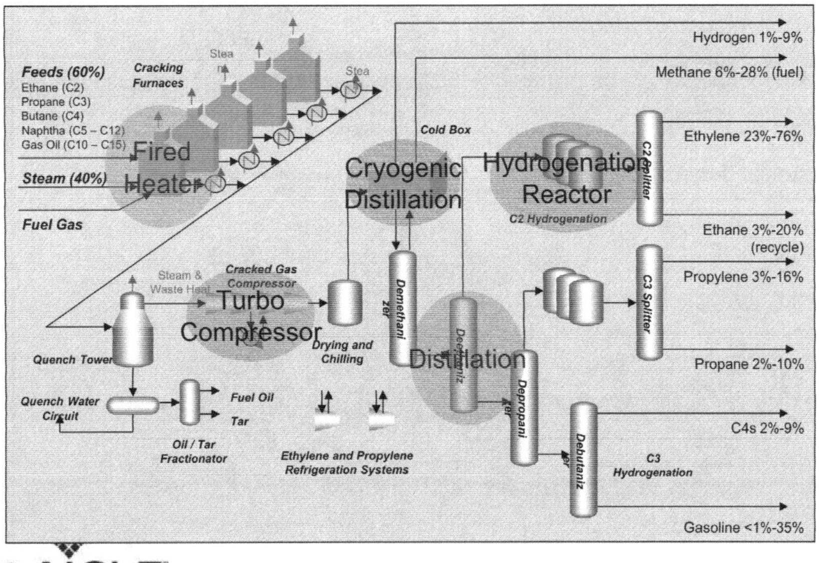

13 AIChE®
Spring Meeting, San Antonio, TX

EMERSON. Process Management

SmartProcess Scope

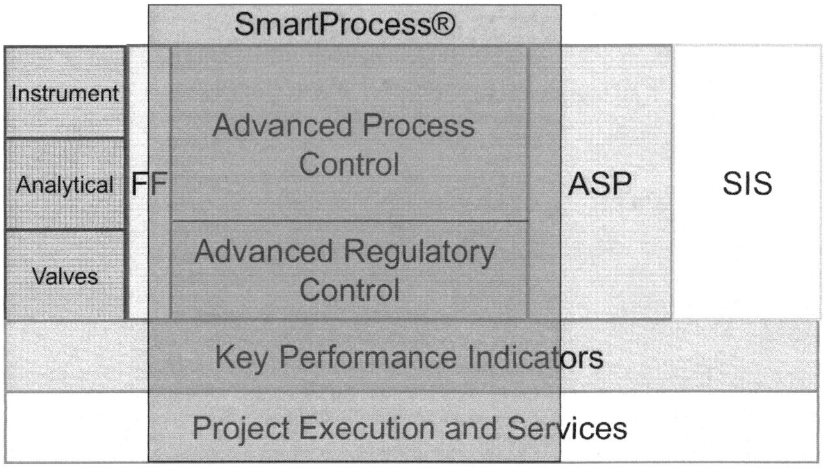

Basic Ethylene Furnace Variables

Key Operating Objectives:

- **Constant severity**
- **Maintain HC/Stm ratios**
- **Min excess air**
- **Maintain within constraints**
- **Max charge (if needed)**

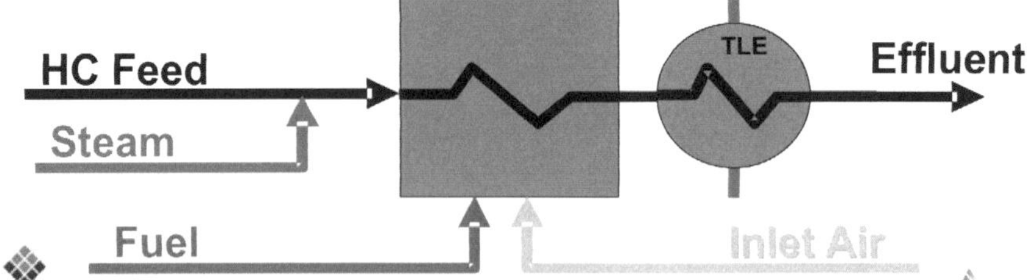

SmartProcess Furnace Module

- Design allows for different furnace configurations
  - Process Zones
  - Firing Zones
  - Multiple TLE analyzers
- Advanced Regulatory
  - Charge Control
  - Steam Control
  - Severity Calculation
  - Combustion Control
  - KPI's
- Advanced
  - Pass Balancing
  - Heat Balancing
  - Severity Control
  - O2 Control

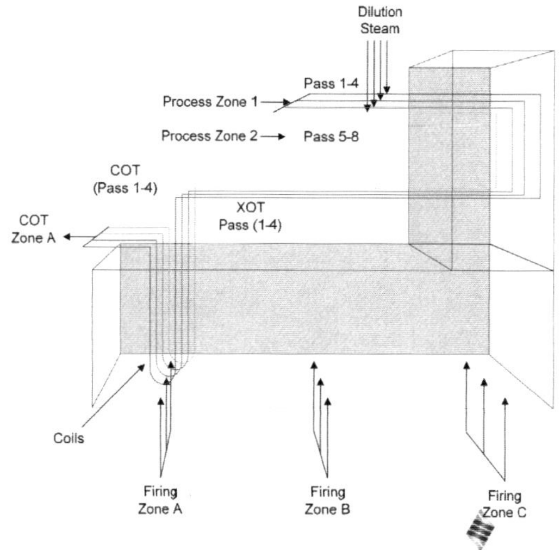

13 AIChE
Spring Meeting, San Antonio, TX

EMERSON. Process Management

Standard Furnace Module Design

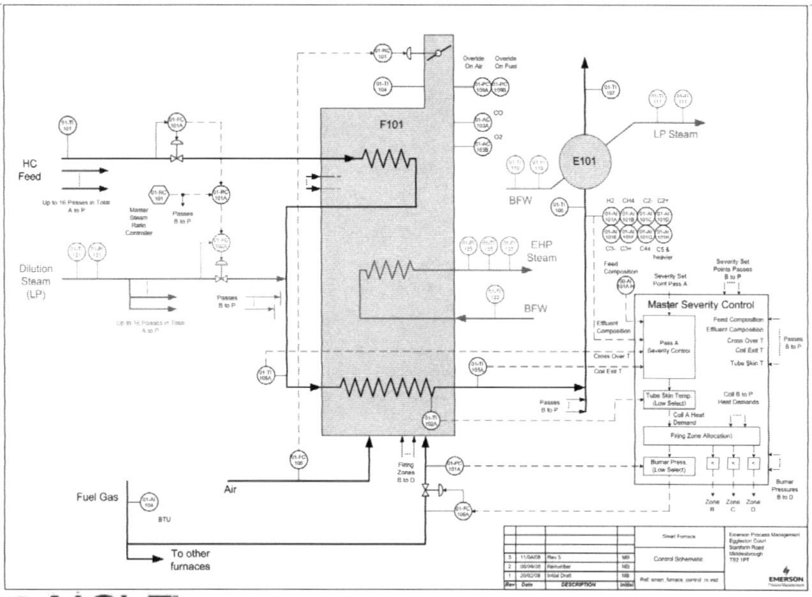

13 AIChE
Spring Meeting, San Antonio, TX

EMERSON. Process Management

Charge Rate Control

Functions:
- Pass Balancing
- Constraints
- Feed Maximization

Functions:
- Ramping
- Limits & Alarms
- Multiple Feed Types
- Furnace Status
- Pass Offsets

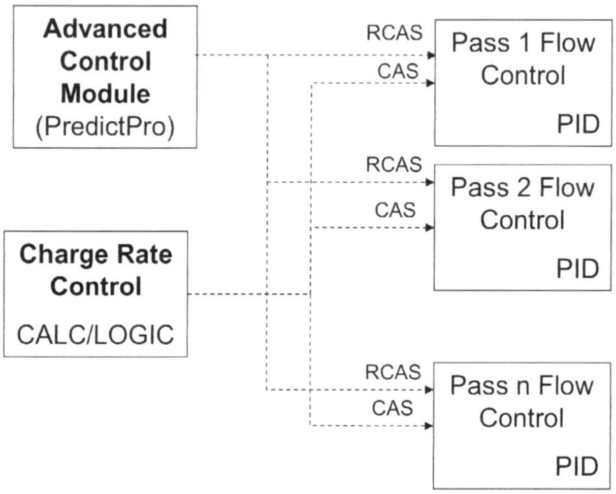

Steam Ratio Control

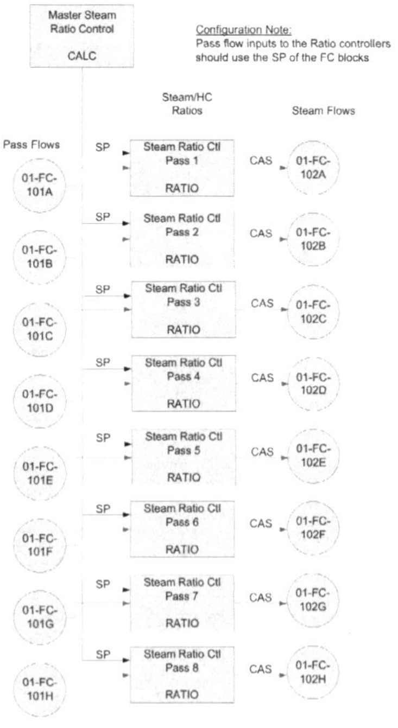

13 AIChE Spring Meeting, San Antonio, TX
© 2013 Emerson Process Management

Severity Control

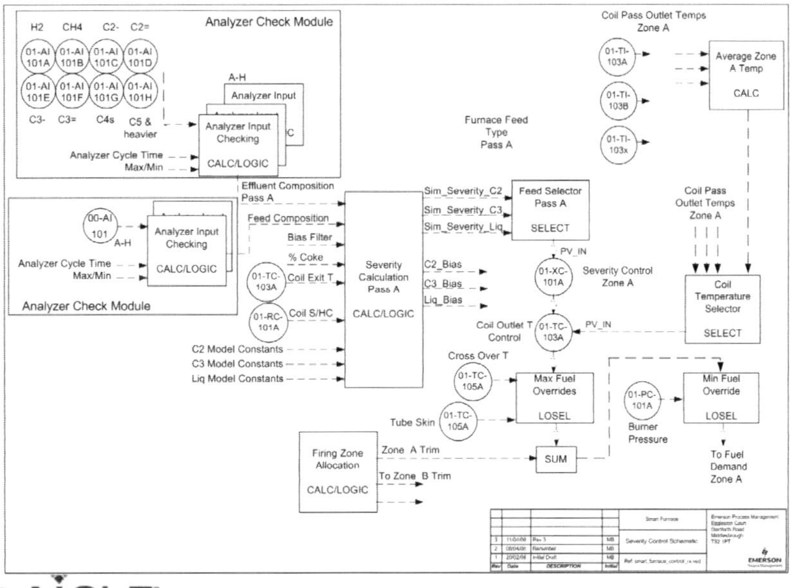

Automated Furnace Sequences

- Standard module templates to be used for Furnace Startup, Shutdown and Decoking sequences.
- Utilizes preconfigured Sequence Function Charts (SFC) running in DeltaV controllers or App Station
- Standard sequence functions can be customized to match site requirements
- Manual operator steps with confirmation can be mixed with automated steps
- Sequence execution control graphics and faceplate templates

Key Performance Calculations

- Furnace efficiency
  - Heat loss method
- Furnace Duty
- TLE Duty, Fouling factor
- Stack gas estimated composition
- Stack gas daily total tons exhaust for CO, CO2, S, N, etc.
- Stack gas NOx emissions (optional – based on CEMS)

13AIChE
Spring Meeting, San Antonio, TX

EMERSON. Process Management

Distillation - Classic Multivariable Problem

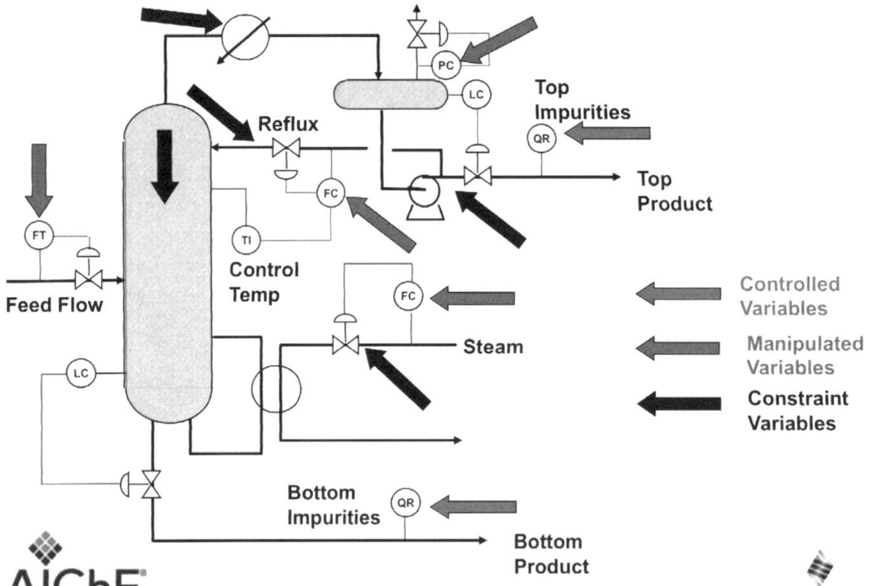

13AIChE
Spring Meeting, San Antonio, TX

EMERSON. Process Management

Distillation Control Module

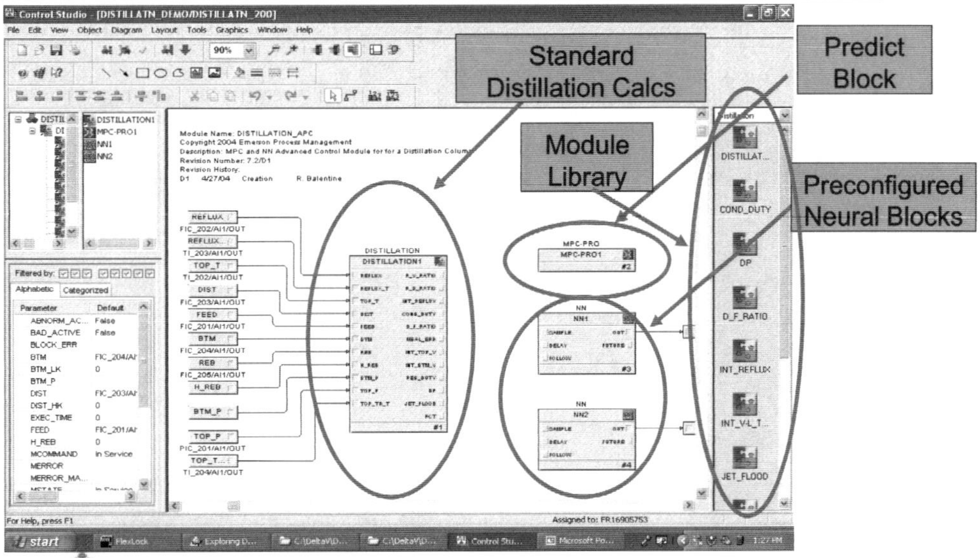

SmartProcess Distillation

- DeltaV PredictPro with up to 4 MV's
- DeltaV Neural for up to 2 product qualities
- Standard set of calculations & constraints
  - Internal reflux
  - Pressure compensated temperatures
  - Jet flooding
  - Reboiler & condenser duty
- Engineering Services to design, implement and commission

SmartProcess Applications

- Licensed industry applications for common processes that are:
  - Pre-engineered, tested, documented
  - Designed by APC Consultants with extensive practical experience
  - Supported & enhanced
- Bundled combination of:
  - DeltaV libraries: composite blocks, modules & templates
  - DeltaV simulation and demonstration example
  - APC tools (e.g. Predict, Neural)
  - Application code (e.g. VB, .Net)
  - Other databases and apps (e.g. SQL*Server, Excel, TankMaster)
  - Sample DeltaV Graphics
  - Documentation
- Templates to be used to configure for specific site requirements
- May be implemented by Emerson or Customer

13 AIChE
Spring Meeting, San Antonio, TX

EMERSON. Process Management
Page 17

Benefits

13 AIChE
Spring Meeting, San Antonio, TX
© 2013 Emerson Process Management

EMERSON. Process Management
Page 18

Ethylene Furnace High Temperature Imaging Cameras to Monitor Burner Flames ,Tube Temperature and Tube Failure Detection

Thomas M. Canty, P.E.
President
J.M. Canty Inc
Buffalo, NY USA

Abstract

The development of explosion proof, high temperature imaging cameras which operate in ethylene furnaces without cooling air now allows monitoring of ethylene furnaces to improve energy/operating efficiency along with providing important safety benefits. Imaging cameras work off the visual wavelength which provides 2 dimensional view and temperature measurement of the tube temperature which is more accurate than thermocouples and IR probes. In addition visual images of the furnace burners lets you detect flame detachment, smoking burners and flame shape in order to increase burner efficiency and save energy. The safety aspects of the Camera imaging systems is significant as it allows visual detection of pin hole leaks and tube failure in the furnaces.

Introduction

Ethylene furnaces, or any type of cracking unit, rely on thermal management in order to make product efficiently and to operate in a safe manner. Having 'eyes' in the furnace can help to identify problems in their early stages of development. High temperature camera technology is able to survive the environment and give a wide angle view of the tubes from which temperature and tube blockage may be determined. Achievable benefits from a camera system include:

1. Temperature analysis capacity for all areas of the image, not just single point. Multiple zone capability for user customization of analysis.
2. Flame detection and size measurement
3. Detection of tube movement and ability for software to automatically move measurement zones as tubes distorts.

Imaging Fundamentals

Basics of an imaging system include a high resolution sensor, proper illumination, appropriate hardware for the process and a robust software analysis package. For this particular application illumination is created by the process and need only be managed optimally for best results. A high density pixel array yields resolution of the image and measurement as well. The hardware and software are the more challenging aspects of the application and will be discussed in some detail.

Hardware

The design of the camera system is unique in that it's internally exposed lens assemblies are **designed to survive without cooling air**. When installed properly, the combination of system materials and heat transfer coefficients at the camera tip limit heat transfer to the system and allow it to run hot without chance of damage. This feature, when system is installed in a clean, gas burning environment, does not disrupt the temperature or air balances of the furnace because no cooling air is being introduced to the furnace. In addition, the design includes a disposable, quarts shield which protects the process exposed lenses from direct contact with the internal atmosphere. Should any particulate contact the system, the quartz shield is there to protect the optical lenses. Replacement of the protective quartz shield is inexpensive. Replacement of damaged lenses is very expensive and so the Canty design guards against that possibility.

Internally, the system is a two piece assembly consisting of the internal insertion lens and the external camera package. As mentioned earlier in this document, Canty employs a fused glass window as part of the electronic housing package. This has a dual purpose; Firstly, the camera housing assembly is a separate entity which carries WP and EXP ratings. Secondly, the camera housing bolts to the internal lens system and transmits the optical rays through the fused glass window to the CCD. This mounting also creates a hermetic barrier between the process and the electronic housing thereby preventing migration of hot, process gasses outside the pressure boundary of the system.

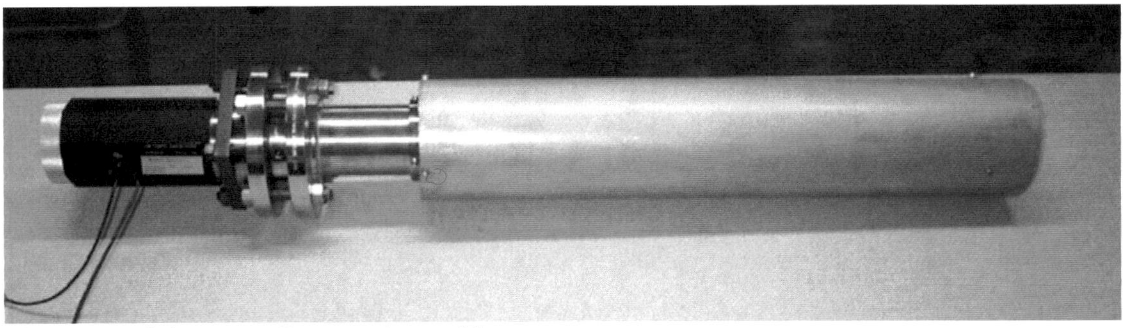

Figure 1. High Temp Camera Assembly

The camera electronics come with a full array of controls that allow it to be optimized to the software function to achieve accurate, consistent data output.

As previously mentioned, the installation is an important feature of the design. When installed correctly, the camera sits inside the refractory (see Figure 2) where its exposure to the process environment is minimized thus minimizing heat transfer into the internal lens.

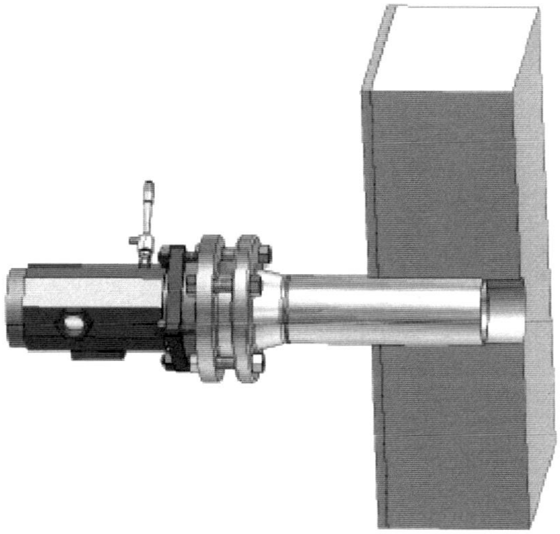

Figure 2. Camera Installation Schematic

Software
With respect to software design, the Canty software is fully flexible in analyzing any portion of the captured image. Multiple analysis zones can be drawn to monitor discreet areas of the view and each zone can run its own particular algorithm depending on need. The benefit here is that multiple tubes can be continually surveyed for changes and trends. The analysis data is also not single point, but is comprised of all the data points in the analysis zone. By choosing proper software filters various information can be obtained from the full array of data such as hot or cold spots. This data can also be archived in total or as snap shots at determined increments of time. The on screen data output can also be updated in the short or near term and can thus be used to judge what is happening in the furnace over time with just a glance at the screen. For instance, using the histogram feature with a 1x daily output a long term trend can be presented on screen in real time.

The basic theory of temperature measurement is based on the radiative emissions of a body at temperature – as the body temperature increases, the radiation curve shifts towards the shorter wavelengths and that is why as a body heats up the eye cannot detect it until it reaches 1000F which is the approximate visible point where emitted radiation is dark red in color. As the temperature increases, the color graduates to bright red, then orange then toward white. A pyrometer functions by measuring a narrow band of wavelength intensity and comparing it to the intensity of a black body to determine the temperature. From the graph below it can be seen that if a single wavelength is monitored the intensity of that measurement changes as the body temperature changes (curve shifts) which then corresponds to a different temperature on the calibration scale.

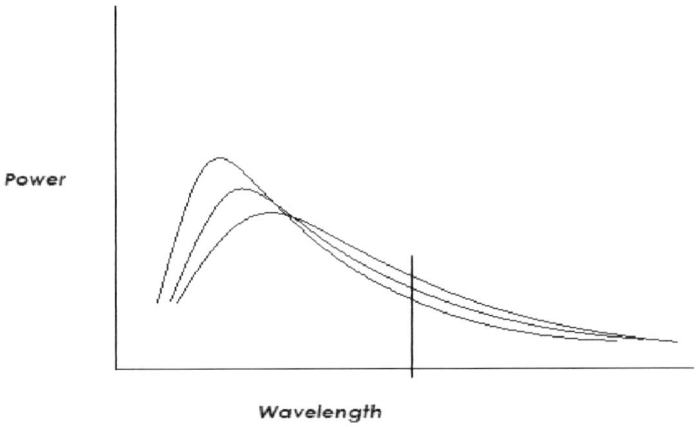

Figure 3. Power Curve at Varying Temperatures

The same is true of a camera system measuring the increasing intensity of an object as it radiates from approx 1000 F (dim, dark red color) on up the temperature scale to 3000+ F for instance which would be a bright white. From the graph above one can see that as color goes from red toward white the peak intensity is always increasing and it is the measure of this intensity which indicates the object temperature. Using an RGB camera the measurement can also be made by monitoring the individual color components of the signal impinging on the chip. For industrial uses, the measurement of intensity usually suffices.

Standard instrumentation used in industry include thermocouple type devices (contact) and thermal imaging and pyrometer systems (non contact). In severe temperatures and atmospheres, the contact devices deteriorate quickly and provide gradually less reliable information leaving the operator running blind. They also create a discontinuity on the surface they are welded to which, in the case of a furnace tube, can lead to premature tube failure.

Non contact systems have difficulties with calibration drift and emissivity adjustments to account for the fact that many items being measured are not black body emitters. In addition, these technologies are single point measurement systems and are best used as hand held devices which can be difficult to use and provide subjective measurement results based on operator preferences for targeting.

Vision technology offers several benefits these other technologies do not. The visual aspect of the system allows the operator to view the process and confirm the readings received. It also allows flexibility to set up measurement zones on screen at any time to assess any or all areas of the process in view should a change be required. Since vision systems are generally fixed installations, an added benefit is that the information is taken from the same process points and features within a given measurement zone at all times which means there is no error in sighting of the instrument.

The vision system is also unique in that it does not suffer calibration drift as almost all instruments do. Once the calibration is set the readings repeat themselves within a couple degrees F. The calibration is based on thermocouple or pyrometer readings initially, however after that calibration is finished it does not have to be redone in the future as long as the system remains in its location.

Process history and performance can also be best documented by a vision system. First off, the camera can analyze any area of process that is within the systems view. This affords the operator the capability of viewing many points of interest instead of just single point temperature readings. As an example, if the system is monitoring a tube bank it can assess the temperature of the tubes for the full length of view. If there is a local flow problem in a tube it will become overheated in one section. The camera can locate that section and indicate its temperature rise over the rest of the tube or adjacent tubes. This information allows the tube location to be identified as well as the position in the tube that has encountered the upset condition. Over time this historical data, recorded in video or still image format by the control software, may be significant in improving the process. In addition to the ability to monitor the full area of the view, the full area of view can also be segmented into multiple zones of measurement, i.e. each tube having its own measurement zone for automated process control capabilities. This capability allows the tube to be operated at optimum efficiency since the addition of excess heat can be avoided thereby reducing energy draw to operate the furnace.

Thermal Analysis

Tube Temperature

The following figures illustrate the temperature analysis capability for tubes in a live test situation. Measurement zones are placed over the tubes (figure 4) as the furnace is heated and cooled. The temperature plots are shown in figures 5 and 6. The results are as expected and uniform from tube to tube, although a small number are included in this demonstration.

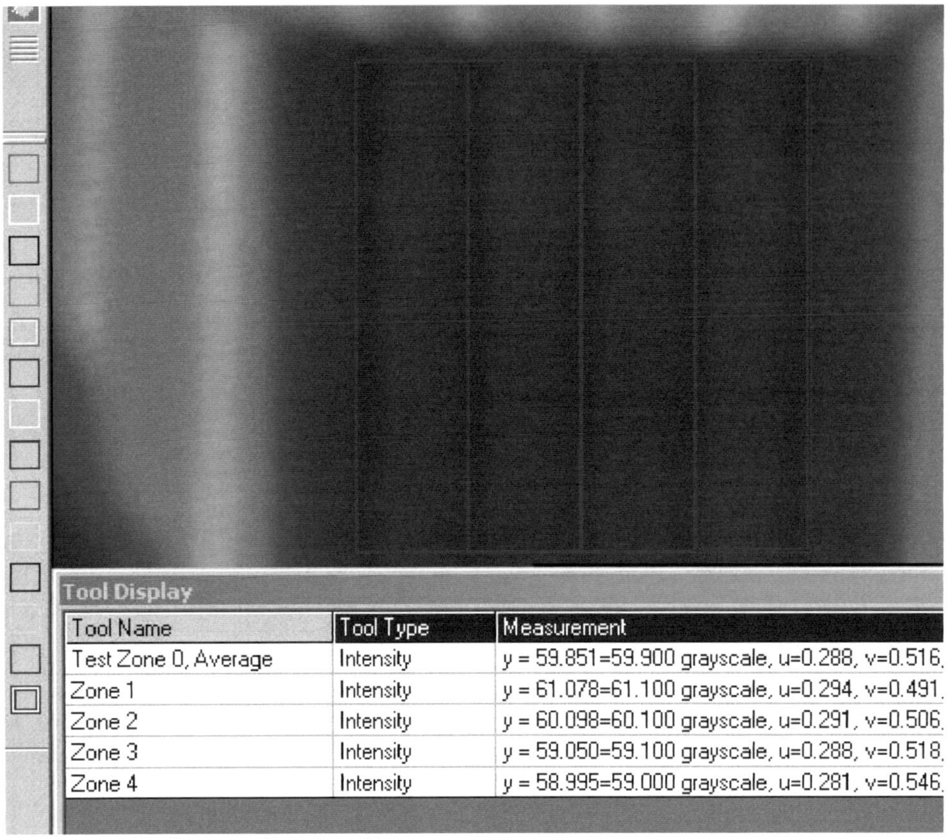

Figure 4. Software Screen Shot

As thermal loads are fluctuated up and down the following graphs track those changes. The cooling cycles are left to natural convection and thus are not linear in nature.

For the following curves, the vertical scale is gray scale intensity which is measured 0 – 255 (pitch black to bright white). The hotter the tube, the higher, or more intense, the gray scale reading. This value in process would be calibrated to the temperature. The horizontal scale is the image number, or time.

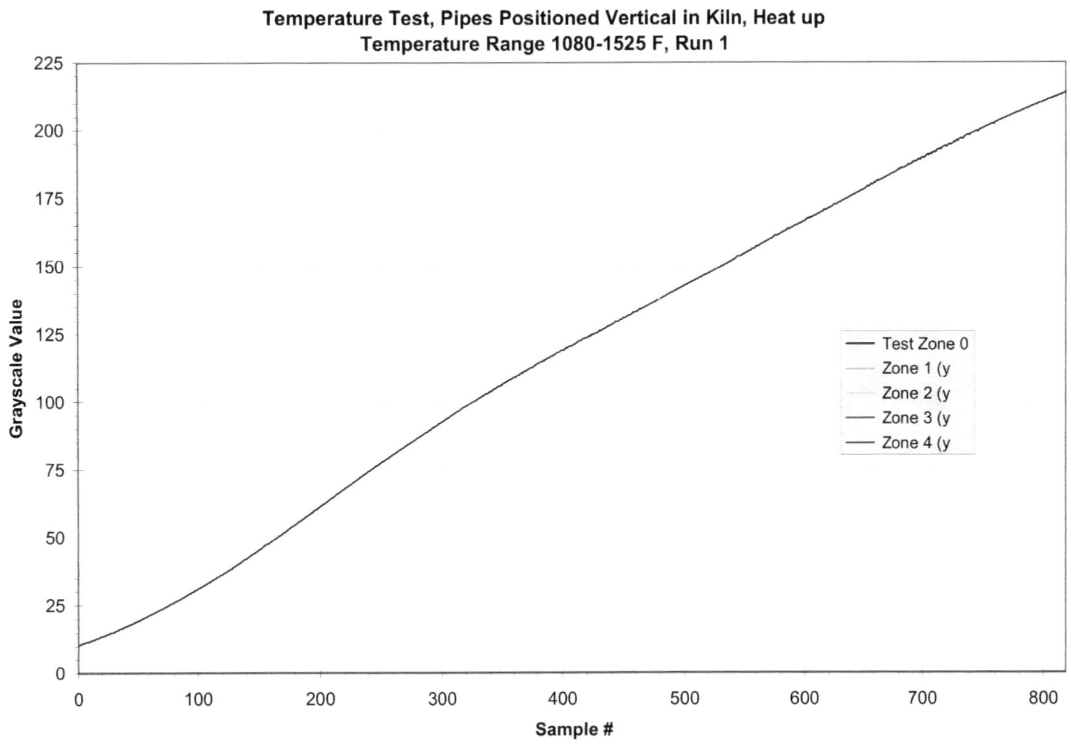

Figure 5. Heating Cycle

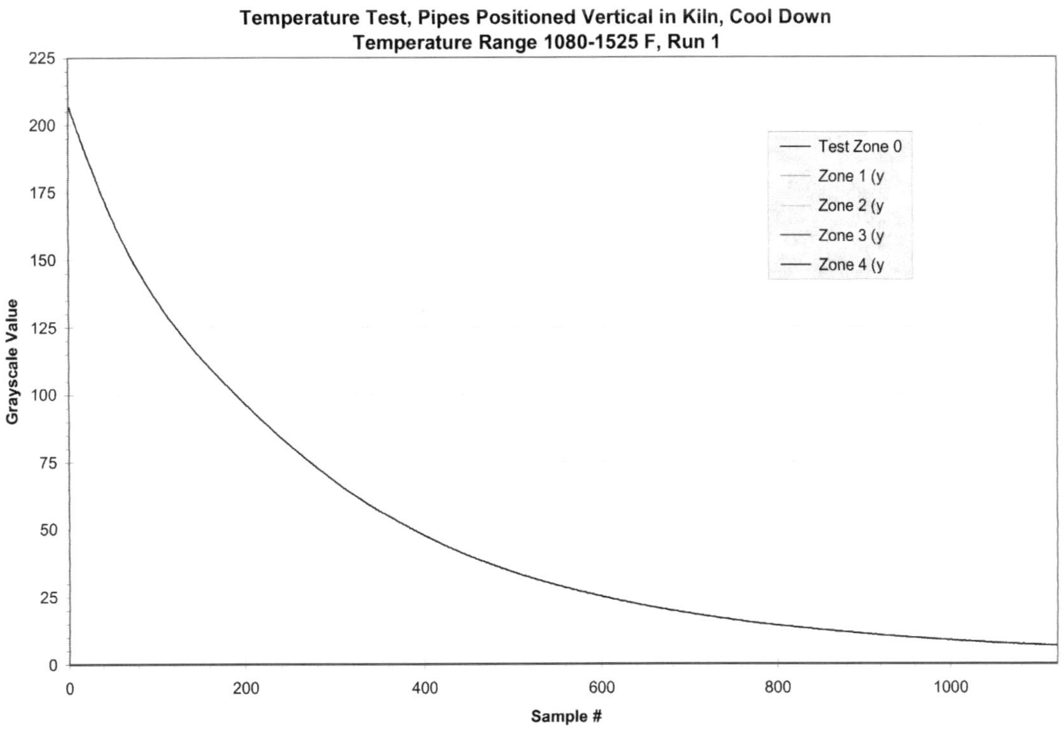

Figure 6. Cooling Cycle

Flame Measurement

Flame temperature can be monitored in the same way as described previously, however the flame size, or presence, can also be monitored. Figure 7 shows a live flame and Figure 8 shows a digital capture. Computing the flame area, or shape, can be done to track its changes over time.

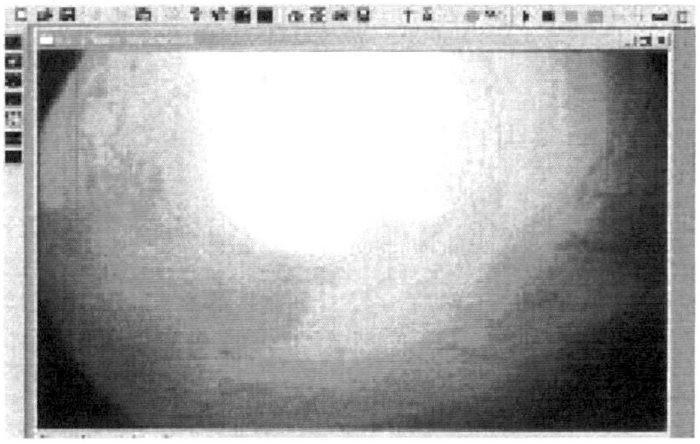

Figure 7. Kiln Flame (Live)

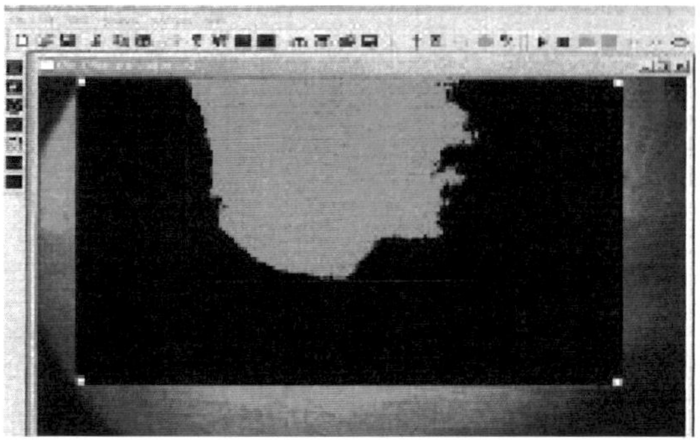

Figure 8. Kiln Flame (Digital)

Conclusion

Vision Technology presents an opportunity to increase effectiveness of furnace monitoring through hardware and software capabilities. The thermal design at the core of the hardware enables the system to provide the operator a view into the furnace and feed back on its thermal performance over time. This data can assist in detecting building upset conditions, but also provide a record for study to assist in improving the operational method of the furnace.

(45v) Straighten up! Non-Linear Compensation Using Feed Forward Control

Barbara Hamilton, Process Systems and Solutions, Emerson Process Management, Hampden , ME

Non Linear processes occur throughout the petrochemical plant. While the PID loop is surprisingly robust for most situations, there are cases where compensation of these non-linearity's is critical for stable control. Similarly, disturbances can cause major disruptions in control stability.

A method of feed forward control, using data based models, can provide robust control throughout the process range. Historical data are used to build models to drive the output toward the solution with PID control providing the final drive to zero error.

The model is also used to develop dynamic tuning. The result is a simple to configure, robust solution to manage non-linear loop and/or disturbance variations.

(45w) Adsorption of CO2 by Alkaline Activated Ca-Al2O3-SiO2 Material

Naim Faqir, King Fahd University of Petroleum & Minerals, Dhahran, Saudi Arabia and Reyad Shawabkeh, Chemical Engineering, King Fahd University of Petroleum & Minerals, Dhahran, Saudi Arabia

Natural gas contains a slight level of carbon dioxide which leads to several industrial and domestic heating problems. In addition, it is one of main greenhouse gases that contribute for global warming, where its pollution is a wide-spread problem arising from a wide range of industries, leading to complex contaminant in atmosphere. Given the increasing pressures worldwide to minimize the emission of greenhouse gases into atmosphere, CO_2 must be remediated efficiently and cost effectively.

In this short abstract a chemically treated limestone and alumino-silicate material were utilized for removal of CO_2 from natural gas using a combined adsorption-chemical reaction processes. This functionalized Ca-Al_2O_3-SiO_2 material was characterized according to its crystalline phase, surface area, pore size distribution and surface morphology. Then it was tested against different concentrations of CO_2 in CO_2-N_2 gas streams.

Pressure Gauges: Hidden Danger or Visible Safety

by Jeff Placek, WIKA Instrument, LP, Lawrenceville, GA

Introduction

In recent years, refining plants, petrochemical plants and chemical plants have been required to increase capacity, but have been provided with very few upgrades to existing equipment to accommodate the increased capacity requirements. This has caused plants to have increased stress and strain on their systems and operating units. For example, higher pulsating pressures and heavy mechanically-induced vibrations are translating to all areas of the processing equipment, especially pressure instrumentation.

Pressure gauges are one of the most important, yet most overlooked or ignored keys to ensuring the continued safety, productivity and profitability of process plants. Process pressure gauges are used to read pressure on transfer lines, discharge pumps, and other places where it is important to be able to read pressures.

Problems Associated with Misapplied, Faulty or Ignored Gauges

Incorrect, misapplied, damaged, faulty or ignored gauges can lead to a number of issues, including safety problems, downtime, and costly regulatory agency fines and other penalties.

Process Safety:

One group of problems relates to process safety issues, such as loss of containment that leads to fugitive emissions, fires and explosions. Bourdon tubes in malfunctioning gauges can rupture, causing the release of media, such as gasoline, oil or gas. Because this media is often volatile and corrosive, it can cause fires, explosions, or other serious safety issues, such as a toxic gas releases or acids spraying onto workers.

Media releases are common results of misapplied, faulty or ignored gauges. Fires and explosions are less common but they do occur. When reviewing plant instrumentation, findings have revealed instances where the wrong gauges have been installed in locations on pumps. In one instance, a gauge vibrated so severely when the pump came out of alignment that it caused the Bourdon tube in the gauge to crack. The media coming out of the tube leaked onto the pump, which was excessively hot because of age and misalignment, causing a fire.

Expensive Downtime:

A second group of problems relates to expensive downtime, which can result from the breakdown of process safety just mentioned, or simply from having to shut down a process in order to replace incorrect, misapplied, broken or otherwise damaged gauges.

1 - Safety-Related Downtime:

As a result of the problems identified above (emissions, fires, explosions, etc.), there can be any number of causes of subsequent downtime, including fire suppression activities, released media containment, line shutdown, personnel evacuation, and then eventually having to rebuild or repair the lines, replace equipment, etc. In addition, if there is a media release with no injuries or fatalities, the plant is required to notify the EPA. However, if there is a media release that does cause injuries or deaths, the plant is required to notify OSHA. These can lead to very long process shutdowns, anywhere from a day to a month or more. (More on OSHA and the EPA below.)

2 - Non-Safety-Related Downtime:

When a line fails, causing a gauge to fail, even if there is no release or safety incident, downtime can still occur as a result of the process needing to be shut down in order to replace the gauge. Gauge failure rarely causes other parts of the process to fail, although gauge failure is usually an alert that there is a problem. In most cases, a plant worker will simply replace the pressure gauge and move on. When another failure occurs, the worker will do the same, until the root cause of the failure, such as a misaligned pump, causes the plant to shut down the line, resulting in more costly, unplanned downtime and a major maintenance time.

Regulatory Agency Fines and Penalties:

A third group of problems relates to the potential for significant fines and other penalties from federal regulatory agencies (primarily OSHA and the EPA), as well as from state and local jurisdictions, that result from compromised process safety.

1 - OSHA:

OSHA addresses many process-related safety issues in 1910 Subpart H, "Hazardous Materials." More specific is 1910.119, "Process safety management of highly hazardous chemicals." In its Process Safety Management section, OSHA notes that, "Unexpected releases of toxic, reactive, or flammable liquids and gases in processes involving highly hazardous chemicals have been reported for many years in various industries that use chemicals with such properties. Regardless of the industry that uses these hazardous chemicals, there is a potential for an accidental release at times they are not properly controlled, creating the possibility of disaster."

OSHA goes on to note that: "To help ensure safe and healthful workforces, OSHA has issued the Process Safety Management of Highly Hazardous Chemicals standard (29 CFR 1910.119), which contains requirements for the management of hazards associated with processes using highly hazardous chemicals."

OSHA's standard emphasizes the management of hazards associated with highly hazardous chemicals and establishes a comprehensive management program that integrates technologies, procedures, and management practices.

While OSHA does not monitor gauges in specifically, they will enforce guidelines for worker safety. OSHA investigates injuries and deaths, and may levy fines or other penalties.

2 - EPA:

According to Greg Neneman, director of business development solutions for Valin Corporation (San Jose, Calif.), fugitive emission regulations from the Federal level have tightened up over the past several years. In many states, such as California, Oregon and Washington, the regulations are even tighter.

"One challenge companies face is ensuring compliance with emission regulations when they upgrade or expand their processes," points out Neneman. "They are not allowed to produce additional emissions beyond the original limit set by regulations."

From the standpoint of federal EPA regulations, each company that produces or emits volatile organic compounds (VOCs) is required to come up with a leak detection and repair program. "In some areas, county and local requirements are even tighter than federal requirements," continues Neneman.

In certain areas, companies that produce or emit VOCs are required by regulation to do audits on valves, flanges, gauges, and all other process connections on a quarterly basis. "And in some jurisdictions, the regulatory agencies can give plants 24 hours notice before they descend on that plant to check all of their records and conduct other investigations," adds Neneman.

In May 2012, the U.S. Environmental Protection Agency (EPA) levied a $1 million civil penalty on a plastics company for alleged violations of the Clean Air Act at two of its chemical manufacturing plants. In addition to paying the penalty, the company agreed to implement a comprehensive program to reduce fugitive emissions of hazardous air pollutants (HAPs) from leaking equipment such as valves and pumps. The settlement also required the company to implement enhanced work practices, including more frequent leak monitoring, better repair practices, and new efforts designed to prevent leaks. After being fined by the EPA, the company realized that instrumentation was one of the areas where they having fugitive emissions. Until that time, workers had been replacing gauges left and right, because they weren't aware of the problems that could occur. The gauges weren't a big expense to them, so they considered the replacement activity to be more of a nuisance than anything. It's when they had the rupture that they realized there was a real problem.

An audit of the plant, which involved reviewing at all of the instrumentation, found that in many applications, they were using the wrong instrumentation. To correct the issues, the right instrumentation was specified for them, and the correct gauges were ordered. The plant now has its instrumentation under control.

While conforming to EPA regulations has always been a challenge for process facilities, it is going to become even more challenging in the future as a result of EPA's recent introduction of new requirements related to fugitive emissions and leak detection and repair (LDAR) strategies. On April 17, 2012, the EPA Administrator signed a suite of final air quality regulations that affect both upstream and midstream oil and gas sectors - the production/processing and transmission/storage sectors of the oil and natural gas industry.

For the first time, EPA will regulate air emissions from natural gas wells that are hydraulically fractured, as well as other emission sources associated with exploration, production, processing, and transportation of oil and natural gas.

Promulgation of the final rules will result in new control equipment requirements, as well as increased requirements for monitoring, recordkeeping, and reporting for owners, operators, contractors and vendors.

The final rules require the use of emission controls, as well as work practices, through two different air regulatory programs. One is the New Source Performance Standards (NSPS). The other is the National Emissions Standards for Hazardous Air Pollutants (NESHAP).

EPA finalized a new NSPS at 40 CFR Part 60, Subpart OOOO, which regulates emissions of volatile organic compounds (VOC) and sulfur dioxide (SO2) from a variety of oil and gas exploration, production, processing, and transportation facilities, including natural gas wells, certain compressors and pneumatic controllers, storage vessels, and sweetening units at onshore natural gas processing plants. Subpart OOOO also includes startup, shutdown and malfunction provisions.

The EPA has also made it clear that oil and gas production and processing sources will be the focus of its National Enforcement Initiative (NEI) for fiscal years 2012 and 2013. With the new NSPS and revised NESHAPs in place, EPA now has the tools it needs to enforce in a comprehensive manner.

As a result of these new regulations, plants should look at everything more closely, including gauges.

How Problems Can Occur With Gauges

Technical problems can occur with process pressure gauges in a number of ways, which can be conveniently categorized into seven broad categories with specialties or subsets. The most common reason for gauge problems is that workers often install gauges without knowing how they should be installed or accessorized. As a result, the gauges can fail because they were never designed to function in those specific situations. They can also fail because they were installed incorrectly, or because they were not properly maintained.

One way to eliminate or minimize these problems is to consider specific pressure gauge accessories. These can prove to be the extra step needed to ensure a successful, proactive preventive maintenance program, and for extending the life of the gauges. Accessories can provide additional protection for the gauges during their lifetimes.

1 - Mechanical Vibration.

This is oscillating, reciprocating or other periodic motion of a rigid or elastic body or medium forced from a position or state of equilibrium. Typically, vibration is the main cause of pressure gauge failure in process plants. Pump applications produce most of this vibration.

Gauges that aren't designed to withstand vibration may be installed in applications where vibration occurs, causing the gauges to come out of accuracy. Even more serious, a gauge can vibrate so much that the Bourdon tube inside it can crack, releasing media."

One way to prevent vibration-related problems is to install liquid-filled gauges, which help to dampen the vibration and also protect the internal components. The liquid case filling also lubricates the internals and prevents corrosion.

2 - Pulsation.

When media rapidly cycles through the pressure system of the gauge, pulsation occurs. If the pointer pulsation increments are greater than 5 percent of the full scale value, you must intervene to prevent damage to the gauge.

A restrictor or throttle screw can solve most pulsation problems. It will slow down the media by reducing the size of the orifice and averaging out pressure fluctuations. For extreme pulsation, such as spiking and surging, a snubber works best. It allows more control over the amount of snubbing and the flow of media into the pressure system.

3 - Temperature:

If a gauge is exposed to a media with very high temperature, the metal alloy in the gauge may not be able to handle the temperature, causing the Bourdon tube to fail and release media.

Long pipes that separate the pressure gauge from the process, or small cooling elements with radiator fins, protect the pressure instrument from high process temperatures by allowing ambient air to flow over the surface area. In environments with steam, using a siphon with the pressure gauge cools the steam down and drives it back to its liquid state before it gets to the gauge. With very high-temperature media, a diaphragm seal with a cooling element will isolate and prevent the hot media from reaching the gauge.

4 - Spikes or Overpressure.

In some cases, process media may be under extreme pressure. Pumps also can cause spikes and over-pressure, both of which shorten gauge life. Gauges can fail as a result of these pressure spikes. When workers turn pumps on or off, or open or close valves, a surge of media or oil goes through the pipe and hits the pressure gauge, which causes a huge spike. For example, if a line is pumping 100 psi of oil, someone may install a 100 psi gauge, not realizing that this leaves no allowance for any overpressure, surging, or spiking. When pressure does increase, it causes overpressure on the gauge, sometimes to the point of where the Bourdon tube ruptures and leaks media.

One way to prevent this problem is to use a different type of gauge, such as 200 psi gauge for a 100 psi line. Another option is to accessorize the gauge with an over-pressure protector, which would shut off the pressure if there were a surge or spike. Overpressure protectors are designed to avert spikes and the resulting overpressure damage. These protectors provide protection for pressure gauges against the effects of pressures exceeding their maximum pressure ratings.

5 - Corrosion.

Corrosion problems can occur with gauges if the process media is not compatible with the pressure gauges' wetted parts material. Media can sometimes be so corrosive that it corrodes through the Bourdon tube and leaks onto the ground.

When the media is not compatible with the wetted parts material of the pressure gauge, a diaphragm seal is required to isolate the gauge from the media. The benefit of a diaphragm seal is that it can be made of special alloys that are designed to withstand corrosion. It is essential to select a diaphragm material that is compatible to the process media.

6 - Clogging.

In some processes, the media may eventually crystallize or may otherwise harden. If a standard gauge is used in such an application, it will eventually get clogged and will not be able to read the pressure accurately. For example, if a gauge gets clogged and ends up reading zero pressure, when in fact it is under tremendous pressure, there will be an unexpected and dangerous release of media if and when the piping system is opened. Diaphragm seals protect the pressure gauge from solids, particles and viscous media by creating a barrier between the process and the pressure inlet of the gauge. There is also an option to equip the diaphragm seal with flushing ports to constantly flush material out from the diaphragm surface.

7 - Mishandling and Misapplication.

Gauges can also fail prematurely if they are misapplied. For example, a gauge may have been ideal for the application when it was installed. However, if there are equipment modifications or more components subsequently added, this can change process parameters, resulting in the gauge not longer being right for that application. Misapplication can lead to serious problems associated with failure and subsequent media release.

It is also possible to accidentally damage gauges during installation. Gauges should be installed using the proper sized wrench on the flats. Gauges should never be installed by turning the case. After installation, gauges can also fail if they are mishandled or used improperly.

Why Problems Can Occur With Gauges

In a typical process plant, about 25 percent of pressure gauges have failed or are about to fail.[1]
A lack of management focus and worker education and experience are key reasons plants experience issues with gauges.

Management Focus.

Gauges are often viewed as inexpensive commodities, with management failing to realize that they are very critical parts of the processes. Often, plants will replace failed gauges with the same replacement gauges, which will also fail, only to have to be replaced again. Eventually, their luck may run out and a serious failure may occur because the incorrect gauge has been installed.

Why do gauges tend to receive such little attention from management? Management tends to focus on the expensive equipment - $100,000 or million dollars items. They often don't understand the importance of focusing attention on $100 gauges. This is not the fault of management. The problem is that no one has explained to them just how important gauges are.

As noted earlier, though, ignoring instrumentation can cause serious safety and financial problems. Plant process gauges can be thought of in the same way as vehicle gauges. If the temperature gauge on an automobile is defective, it is possible to destroy the engine as a result of overheating that is not registered by the defective temperature gauge. Similarly, if the oil pressure gauge on an automobile is defective, it is again possible to destroy the engine as a result of seized components resulting from loss of oil that is not indicated by the defective oil pressure gauge.

Worker Education and Experience.

A lack of worker education and experience also contributes to gauge issues. In many instances, when maintenance technicians look at gauges, they don't pay much attention to them unless they leaking or look as if they are going to explode. They don't realize all of the information that that gauge can provide. In addition, technicians may not realize that a gauge has failed or is about to fail. Again, as long as it is not leaking media, they will often just keep walking. This is rarely the fault of the technicians. Many have not been properly trained to understand process pressure gauge performance.

Along the same lines, the industry has been experiencing a "brain drain" for several years, with large numbers of experienced technical workers and engineers retiring and being replaced by younger people who lack the education and experience to understand what gauges are telling them, or, worse, not being replaced at all.

According to research conducted by Cambridge Energy Research Associates in 2007, more than half of all oil-field professionals will reach retirement age during the next decade. The Interstate Oil & Gas Compact Commission found that the average age of production and service company employees is 46 to 49 years, and the average retirement age in the oil and gas industry is 55 years. [2]

Historically, plants had specialists whose jobs focused exclusively on instrumentation. They spec'd the gauges, maintained them, and made sure they were operating correctly. However, as plants got leaner in recent years, they started letting these people go. Plants first shifted gauge responsibility to the pipefitter whose main responsibility was to make sure the pipes were put together correctly and not leaking.

[1]WIKA data aggregated from more than 250 instrument audits.

[2]Peter Peek, Jocelyn Fenard, Patrick Gantes, Christine Theiler. December 2008, "Skills Shortage in the Global Oil and Gas Industry: How to close the gap" Part I. Pages 14-15.

Then, as plants began letting pipefitters go, they shifted the responsibility to the unit operators or technicians, who now install the gauges. These people usually don't have any experience or knowledge of how to install instrumentation. They usually just replace old gauges with similar gauges. The problem is that the gauges are usually the wrong ones to begin with. The result is that, these days, few plants have people who have much if any education and experience with instrumentation.

Taking Stock of Process Pressure Gauges

When it comes to improving the safety, productivity and profitability of process plants through an effective gauge management program, companies can take a couple of steps. First, an understanding of the vital importance of gauges must be established. Second, workers must understand what kind of gauges will be effective.

Focus on Gauges:

World-class process plants maintain their assets proactively. In such plants, pressure gauges receive as much attention as pumps and control valves. Management understands the importance of proactively maintaining their assets, and they realize that one of the best ways to maintain their most expensive assets is to properly source and maintain the gauges that protect these assets. That is, they understand the value of spending $100 for a gauge that can subsequently prevent damage or destruction to a $100,000 or $500,000 piece of equipment and the resulting negative safety, downtime and regulatory implications.

There is an interest in predictive maintenance these days in process plants, which is a good thing. A lot of this, though, involves expensive software. In addition, when something is out of kilter, the system may be set up such that technicians may have to wait for it to get far enough out of the normal range before it will generate an alert on the computer. However, properly-maintained pressure gauges, studied by people who know what to look for, can be very inexpensive but reliable and useful mechanical predictive maintenance tools.

Gauge Procurement Considerations:

Over time, as noted earlier, the status of pressure gauges has unfortunately diminished to that of commodities in many process plants. In order to ensure process plant safety, productivity and profitability, it is essential that plant professionals understand the importance of selecting the right gauges for applications.

It is also important that these be high-quality gauges, not "commodity" gauges. That is, not all process gauges are created equal. The so-called "lowest price" gauges appear attractive at first, but are rarely the "lowest cost" in the long run, because they usually have to be replaced more often, leading to excessive downtime and inconvenience. Worse, they may fail unexpectedly, leading to catastrophic results.

Consider these 10 steps to select the right gauges that will help to ensure safe, productive and profitable plant processes.

1 - Know what is in the pipe. This way, you can select a gauge alloy that will be compatible with the process media.

2 - Next, know the temperature of the media. If it is very hot, for example, it will be unsafe to use a standard pressure gauge. A long pipe, siphon or diaphragm seal assembly may be needed to cool the media sufficiently before it reaches the gauge internals.

3 – Third, know the ambient conditions surrounding the gauge. For example, if it is corrosive around the gauge, you need to make sure that the external components of the gauge such as the case and window can handle the environmental conditions.

4 - Next, know the pressure in the piping that the gauge is meant to measure. A best practice is using a gauge that is designed for two times the normal operating pressure. For example, if normal pressure is 100 psi, use a 200 psi gauge.

5 – Know the conditions that affect gauge wear. These could include mechanical vibration, pulsation, pressure spikes, etc. For example, if vibration exists, a gauge with a liquid-filled case is recommended. If there is pulsation, use a throttle plug or restrictor accessory.

6 - Select gauges that are designed for a particular service. Process type gauges are designed with strong Bourdon tubes and other heavy duty components. If a gauge is misapplied, the internals of the gauge may not be rugged enough to resist failure and subsequent media release in these applications.

7 - The gauge case must be rugged and have the correct safety features. In some case designs, if there is a rupture of the Bourdon tube, media will be released out the front of the gauge, which can spray workers. Select a gauge where media would release out from the back of the gauge.

8 - Accuracy is important. In most process plants, +/-0.5% accuracy is the norm, and +/-1% is acceptable.

9 - Size is also important. The gauge dial needs to be large enough that operators can read it from a distance – four inches or larger.

10 - Finally, the gauge should be mounted properly. A gauge is usually mounted by its stem, but it can also be mounted in a panel or against a flat surface.

Looking Outside for Assistance

As noted earlier, it can be difficult for plant management to devote enough time and attention to an effective process pressure gauge program. Similarly, it can be difficult for technical personnel to get up-to-speed in terms of education and experience related to process pressure gauges.

There has been a growing trend over the last 30 years in virtually all industries for companies to outsource services that are not considered "core competencies." That is, companies have come to realize that it makes sense to focus their attention on their missions and what they do best, and leave secondary and peripheral activities to third parties that specialize in these other activities and services. Becoming experts in pressure gauges is certainly not a core competency for refining plants, petrochemical, and chemical plants, so this is an area that can easily be outsourced to an experienced third-party audit team.

If a plant elects not to gain the experience that it needs on its own, free assistance is available. To ensure that gauges are working properly and are able to protect the valuable assets that they were designed to protect, a five-phase instrument review is recommended. A best practices review identifies gauges with unwanted conditions and recommends corrective actions to ensure optimum performance. The review identifies issues that affect reliability, longevity and safety. Such a review can uncover evidence of pulsation, vibration, pressure spikes, excess process or ambient temperature, installation damage, clogging, chemical incompatibility, and other problems that can lead to serious problems.

Phase 1 - Interview. This involves discussions between the third-party audit team and plant personnel. During the interview, plant personnel learn the purpose of the instrument audit, how it will be conducted, and how they will benefit from the audit. The interview also helps determine the best way to match the audit to the needs of that specific plant.

Phase 2 - Pre-Meeting. This involves determining the units or sections of the plant to audit, as well as outline roles and responsibilities. It also identifies prohibited and exempt areas of the plant. The third-party team reviews problematic instrument installations and discusses safety procedures with plant personnel.

Phase 3 - Instrument Application Audit. This involves having key plant personnel lead the third-party team on an actual plant walk-through. The audit team reviews the target gauge installations, identifies gauges that demonstrate unwanted conditions, as well as those that demonstrate installation best practices.

Phase 4 - Post-Meeting. This involves explaining the instrument audit results to key plant personnel, specifically noting gauge installations that pose immediate safety concerns. Together, the audit team and plant personnel determine which installations to investigate immediately and which to block out. The audit team and plant personnel then set schedules for the follow-up meeting.

Phase 5 - Follow-up Meeting and Presentation. The third-party team presents its general findings and the executive summary resulting from the instrument audit. The team also presents a spreadsheet that lists details about the gauges from the instrument review, color-coded by severity. The team also provides recommendations for improvements based on experience and best practices. Together, the audit team and plant personnel develop a strategy for moving forward.

The plant should also consider leveraging the audit team in other ways. For example, over time a plant's storeroom inventory can get out of hand. A qualified third-party audit team can help plant personnel rationalize, consolidate and manage their inventory of pressure gauges. For example, if a plant stocks spares for 10 or more ranges, the audit team may be able to help reduce that number to five or six ranges.

Essentially, when used properly, gauges play a significant role in plant safety, productivity and profitability. They act like silent alarms, indicating when something is seriously wrong or helping to diagnose deeper issues within systems and processes that can eventually lead to serious safety issues. To ensure safety and reliability at your plant, make certain the proper gauges are in place by reviewing your processes, gauges and your inventory. Everything is at stake.

REFERENCES

1. WIKA data aggregated from more than 250 instrument audits.
2. Peter Peek, Jocelyn Fenard, Patrick Gantes, Christine Theiler. December 2008, "Skills Shortage in the Global Oil and Gas Industry: How to close the gap" Part I. Pages 14-15.

ExxonMobil
Research and Engineering

Best Practices in Pilot Plant Piping

Richard Palluzi
April, 2013

The ideal piping for a pilot plant would be:

- Leak resistant
- Easily modified
- Able to accommodate varying requirements
- Easy to install
- Low cost
- Available in many different materials of construction
- Available in all sizes
- Have fittings available in numerous configurations

ExxonMobil
Research and Engineering

Leak Resistance

- Pilot Plants are more subject to leakage than process plants
  - – Smaller size
  - – Larger number of joints
  - – Constant modifications
  - – Need to accommodate accessibility, flexibility and change
- Common types of piping used in pilot plants
  - – Piping
    - + Threaded
    - + Welded and flanged
    - + Groved
  - – Tubing
    - + Compression fittings
    - + Vacuum fittings
    - + Mechanically joined

Best Practice: Tubing

- Available in numerous materials
- Fittings are readily available up to 1" OD
- Easier to install and modify
- Requires less skill, tools and training for installation
- Smaller and lighter than pipe
- Lower cost than pipe
- Ability to bend
  - – Reduces need for fittings (lower cost and less potential for leakage)
  - – Easier to install in tighter locations
  - – Allows for easier removal of components for cleaning, modification or inspection

Best Practice: Compression Fittings

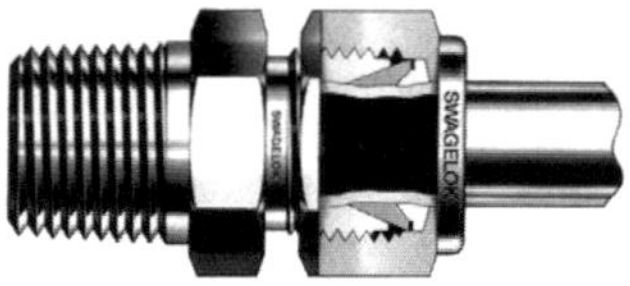

- Better leak resistance than threaded fittings
- Requires few tools and less training
- More expensive than pipe fittings
  - Significantly faster to install so labor savings usually make then cheaper overall
- Each fitting is a potential disassembly point
  - Do require clearance although much less than piping
- Significantly better make/break life than pipe fittings
  - But not infinite!
- Numerous specialty fittings available
- Many small valves and associated components available
- Most compression fittings have limitations on tubing hardness
  - Confirm the tubing you ordered is within specification

5

ExxonMobil
Research and Engineering

Best Practice: Compression Fittings

- Recognize that compression fittings larger than ¾" may require mechanical assistance
- Compression fittings have both minimum and maximum wall thicknesses
  - Too thin and the tube may collapse without swaging
  - Too thick and the tube will not deform enough to prevent blowout
- Drilled through compression fittings, used for dip tubes, thermocouples and similar uses, lose significant pressure ratings

OD(inches)	Typical Pressure Derating
½ and less	25%
5/8-3/4	50%
7/8 - 1	75%

6

ExxonMobil
Research and Engineering

Best Practice: Vacuum Fittings

- Significantly better leak performance than threaded fittings
  - Usually requires some welding to avoid additional joints
- More expensive than compression and pipe fittings
- Each fitting is a potential disassembly point
- Zero clearance fittings
- Long term make/break life
- Gasket surface is more prone to damage
  - Awareness can reduce frequency
- Decision to use O-ring style or metal gasket is usually based on temperature and pressure

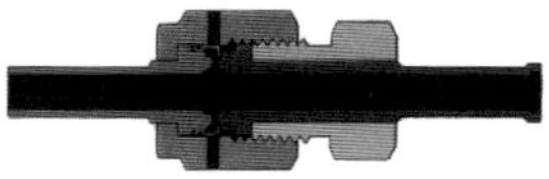

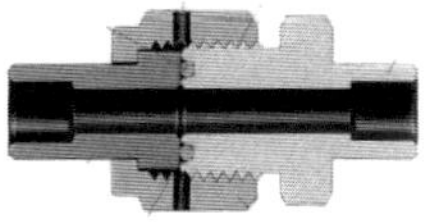

7

ExxonMobil
Research and Engineering

Best Practice: Threaded Piping

- Make sure the die is sharp and in good condition or the threads are likely to be rounded and more prone to leakage
- Make sure the threading is the right length
  - Longer or shorter lengths will leak
- Use good quality fittings
  - Particularly unions
- Avoid threaded connections in cyclic service
  - Temperature
  - Pressure
- Lightly coat unions with lubricating oil before each assembly
  - The coating greatly helps it to seal in gas service
- Recognize that threads over 1" NPS are increasingly difficult to seal
  - Particularly in gas service

8

ExxonMobil
Research and Engineering

Best Practices: Flanged Pipe

- Use raised face flanges whenever possible as they require less force to seal
- Avoid threaded flanges
  - Adds another potential leak point
  - Often makes bolt alignment difficult if retightened
- Tubing is usually a better approach for pipe sizes ½" and smaller

Best Practice: Specialty Piping

- Understand when specialty piping may be a viable or more effective option
- Grooved piping allows for fast and easy disassembly
  - Lower pressure (1,000 psig is typical)
  - CS or SS only
  - High materials costs
  - Some specialty equipment (grooving tool) required
  - Sizes below 1" NPS very limited

- Sanitary fittings allow for ease of cleaning
  - Available for piping or tubing
  - 304 or 316 SS only
  - Lower pressure (300 psig is typical)
  - High materials costs
  - Sizes below 1/2" NPS very limited

Best Practice: Specialty Piping

- Flare compression fittings are excellent for thermal cycling but are generally only available in CS and brass
  - Good for refrigeration systems
- Longer ferrule designs and cone and collar type compression fittings extend the range of compression fittings to 15,000 or 60,000 psig respectively
  - Much more expensive

Best Practice: Mounting

- Use vacuum fittings where frequent removal or cleaning is required
  - Zero clearance makes layout easier
  - Ability to replace gaskets maximizes service life

- Mount the valves not the tubing wherever possible
  - Minimizes the stress on the fittings created when operating the valves
  - Will reduce long term leakage
- Avoid unsupported elements
  - Long spans
  - Heavier components
  - Flimsy supports
  - All lead to increased leakage

- Use bends to allow for thermal expansion
  - Even small temperature swings promote leaks over time

Best Practice: Reduce Joints

- Fewer joints always mean fewer potential leak points
- Use specialty fittings to minimize joints
  - Long term savings outweigh costs over time
  - Requires some planning or stocking
- Weld joints that are:
  - In difficult to reach locations
  - Not likely to need to be opened
  - Are on systems subject to routine or frequent temperature swings
  - Under insulation

Best Practice: Remember Maintenance

- Almost everything eventually needs maintenance
- Make sure easy removal is provided for:
  - Instruments
  - Rotating equipment
  - Filters and strainers that need to be removed to be changed
  - Reactors
  - Regulators
  - Relief devices
  - Control valves
- Decide if you have clearance or need zero clearance fittings

Best Practice: Leave Room

- Pilot plants are always modified
- Building a unit in too small a space:
  - Adds to construction costs
  - Makes future changes difficult or impossible
  - Greatly increases maintenance costs
- The Golden Rule: Plan on at least twice the largest space you think you'll need
- Piping runs eat up space
- Insulation easts up space

15

ExxonMobil
Research and Engineering

Best Practice: Training

- Train your operators and craft support regarding best piping practices
- Always use a holdback whenever making or breaking any fitting
  - Otherwise the fittings o neither side will start to leak
- Make up the fitting the proper amount
  - Not too much, not too little
  - Consider gauge blocks for compression fittings
- Never remove a support
- Make sure any changes or modifications are properly supported

16

ExxonMobil
Research and Engineering

Best Practice: You Get What You Pay For

- Buy good quality pipe, tubing and fittings
- Use high quality gaskets and sealants
  - Teflon tape and pastes generally work better than dopes
  - Pastes require a minimum time for hardening
  - Silver plated vacuum gaskets seal better than plain stainless
  - Spiral re-enforced gaskets seal easier and last longer than flat gaskets
  - Paying more for a good quality seal always ends up being more cost effective!

17 ExxonMobil Research and Engineering

Best Practice: Heat Tracing

- Avoid joints wherever possible
- Use bends to accommodate thermal expansion
- Use standoffs to avoid having supports becoming heat sinks
- Use vacuum fittings when joints are necessary
- Mark insulation for joint location for ease of maintenance

18 ExxonMobil Research and Engineering

(54b) Successful Operation of a 1 BPD Fischer-Tropsch Pilot Plant

Juan Inga and **Lindsay Hernandez**, SGC Energia, Houston, TX

In 2012 SGC Energia commissioned, started up, and operated a pilot scale Fischer-Tropsch process for the demonstration step of the XTLH solution in Pasadena, TX. This plant was used to confirm design plans for future larger scale projects as well as to gather additional data. Running a Fischer-Tropsch reactor at a nominal capacity of 1 BPD, SGC Energia gained valuable information from lessons learned and the challenges of the start-up process. Upon the conclusion of the successful run, this moveable skid is being relocated to provide future operator training for a commercial unit. This commercial plant will be located in the Gulf area.

SGC Energia has operated different FT reactor technologies at a 1BPD level, plus others on a smaller scale. Commercial sized samples have been produced from these operations. Currently the SGCE Fischer-Tropsch reactor choice greatly reduces the risk involved, making the system more robust; and therefore more equipped for commercialization.

(54c) Continuous Improvement in Pilot Plants for Safer Operation

Kiran Gupte, BP Refining & Logistics Technology, Naperville, IL

Continuous Improvement in Pilot Plants

for Safer Operation

Kiran Gupte

BP Refining & Logistics Technology

150 W. Warrenville Rd

Naperville, IL 60563

BP's Refining & Logistics Technology group has successfully designed, commissioned and operated pilot plants in support of refining processes over the last 20 years. Of course, over the years the management practices have evolved with newer methodologies. The "Continuous Improvement Process" is important to critically review all operations to further strengthen the safety in design and operation. In addition to the old and reliable "Process Safety Review (PSR)" process, the newer "Management of Change (MOC)" procedures play the most critical role. An "Inherently Safer Design" is not only for a new pilot plant during the conceptual design phase but it should also be considered in every improvement. The "Continuous Improvement" process helps to bring out and consider smaller steps in improving operation of the pilot plants. The cumulative effect of every improvement step however can never be over emphasized. This presentation will highlight the challenges of operating small scale pilot plant and methods to continually improve these units for safer operation. As the pilot units get used for ever expanding types of projects, the attention to "Operating within the Design Envelope" is important for safe operation with zero incidents. All these aspects are an integral part of the overarching Operations Management System. The safe practices developed are applicable to all types of equipment including small-scale catalyst screening reactors, small-medium autoclave reactors, distillation apparatus, and large pilot plant and demonstration reactors handling a broad range of hydrocarbon feedstocks.

(54d) Using Dimensional Analysis to Design Pilot Plant Experimental Programs

Jonathan H. Worstell, Consultant, Richmond, TX

Pilot plants are capital intensive and expensive to operate. The most expensive item of a pilot plant budget is its experimental program. A poorly designed pilot plant experimental program requires more operating campaigns than originally budgeted. Thus, such a pilot plant will "go beyond" its budget --- a non-optimal situation.

This presentation discusses using Dimensional Analysis as a method for designing pilot plant experimental programs. First, Dimensional Analysis helps identify the major parameters impacting the chemical process. Second, Dimensional Analysis shows how to minimize the experimental program for a given pilot plant, thereby reducing its total operating expense.

Dimensional Analysis is not new: it has just fell out of favor with chemical engineers. It is time to rethink its use.

Design of Batch Scheduling

Aaron Herrick, Chemstations, Inc.

Chemstations™ www.chemstations.com

Abstract

Design and scale-up of new batch processes presents a unique set of challenges versus design of continuous processes. One key difference is optimization of the time required for each process step. Rules of thumb are often used due to lack of data or tools.

Improvements in dynamic simulation software have allowed for more rigorous modeling of increasingly complex systems and scenarios. Modeling of batch process steps in the design phase can highlight potential problem areas such as heat balance limitations, short residence times, reversal of flows, and pressure/flow dependencies. This allows for efficient optimization of step scheduling using the latest dynamic simulation techniques. Examples of batch designs are presented using CHEMCAD simulation software. A new method is included for simplification of handling any number of process steps and control elements.

Chemstations™ 1

Agenda

- Current practices in design of batch scheduling
- Improvements made possible with
  - Rigorous modeling methodology
  - Microsoft Excel™ to drive a schedule in a simulator
- Example using CITRON
- Example using WINTEK
- Summary & suggested procedure

Chemstations™ 2

Current practices in design of batch scheduling

- Methods
  - Best guess & experience
  - Lab scale & pilot scale testing
  - Use a simulation
- Pros
  - Fast
  - Fairly easy
  - Gives a basic time and heat duty analysis
- Cons
  - Suboptimal schedule times
  - Might not account for equipment/process limitations
  - Difficult to do safety analysis until after process is running

Chemstations™ 3

CITRON (current practice)

Recover 120 kgs of 99.9999% pure Citronellal from 170 kgs essential oil using a 500L pot still with a 10ft packed column.

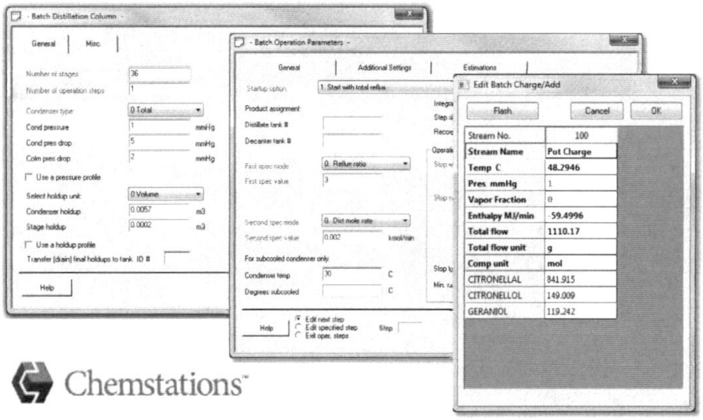

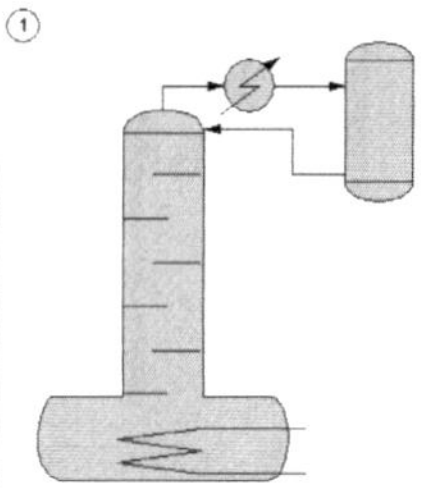

Chemstations™

4

CITRON (current practice)

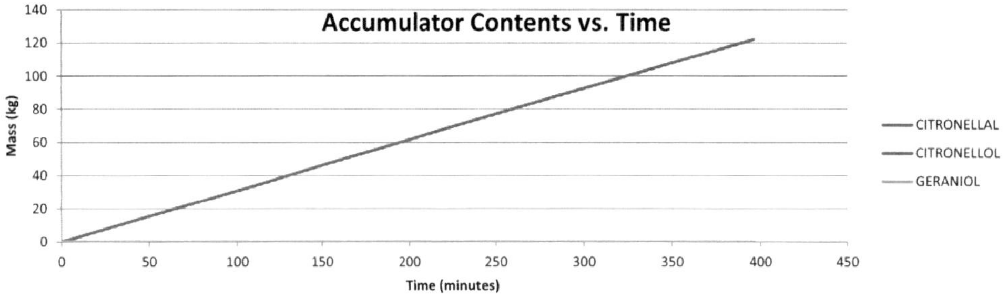

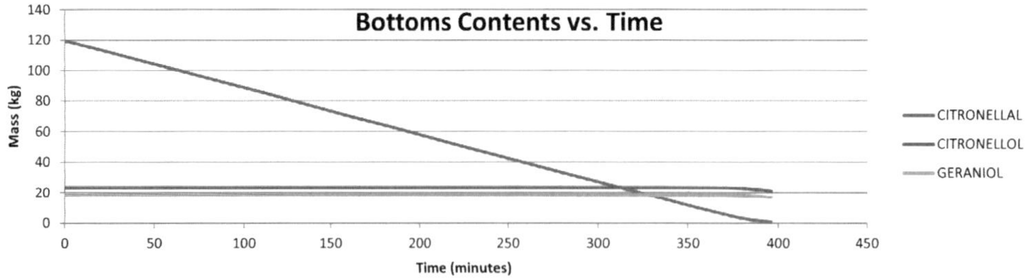

Chemstations™

5

CITRON (current practice)

What have we learned from the model?

- Heat duty/flow/time relationship
- Basic operation steps/timing
- Limited equipment sizes and specifications:
  - Heating requirements
  - Condenser requirements
  - Column dimensions

Chemstations™ 6

CITRON Dynamic (advanced practice)

- Expand our previous example to include
  - Rigorous heat exchanger geometry and performance
  - Event sequencing using DATAMAP to Microsoft Excel
  - Utilities modeling
  - Column metal heat transfer
  - Heat duty/ cooling water on control
  - Dry column startup to total reflux
  - Detailed engineering, e.g., nitrogen sweep on vessel, insulation, etc.

Chemstations™ 7

CITRON Dynamic

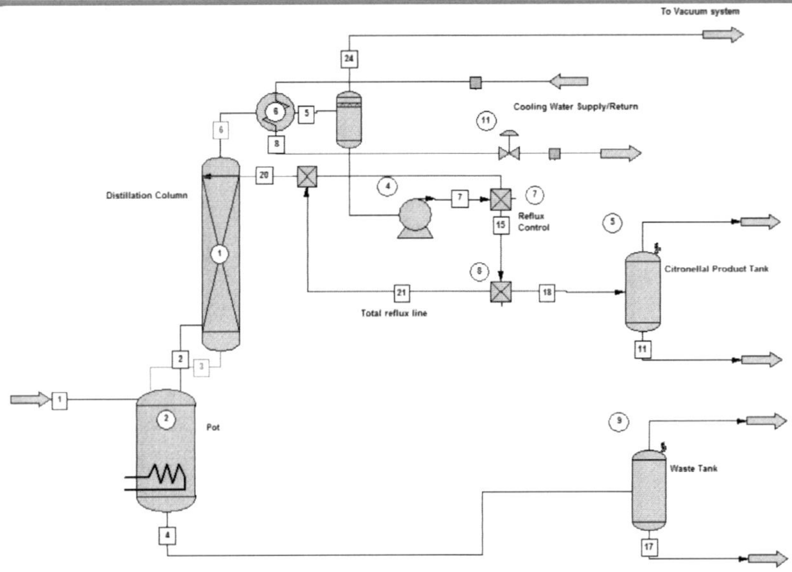

Chemstations

8

Schedule of Events

- A schedule of events is made in Excel and must be connected to CHEMCAD:

Schedule of Events

Time(min)	CW Valve mode	CW Valve Opening	Total reflux	Product Tank	Waste Tank			Notes
0	4	0	1	0	0	36	0	Startup
21	4	0	1	0	0	29	0	Lower heat duty
25	4	5	1	0	0	29	0	Open CW Valve
50	4	5	0	1	0	29	0	Switch off Total reflux
473	4	5	[illegible]	[illegible]	[illegible]	0		Heater off, drain to tank
540	4	0	[illegible]	[illegible]	[illegible]	0		CW off

Schedule of heat duty to pot

Chemstations

9

CHEMCAD to Excel connection

- CHEMCAD sends time information to Excel cell A2 via datamap
- Excel uses time, schedule, and VLOOKUP() function to determine values of process variables
- CHEMCAD collects the current time step data from row 2 of the Excel sheet and enters it into the flowsheet via datamap

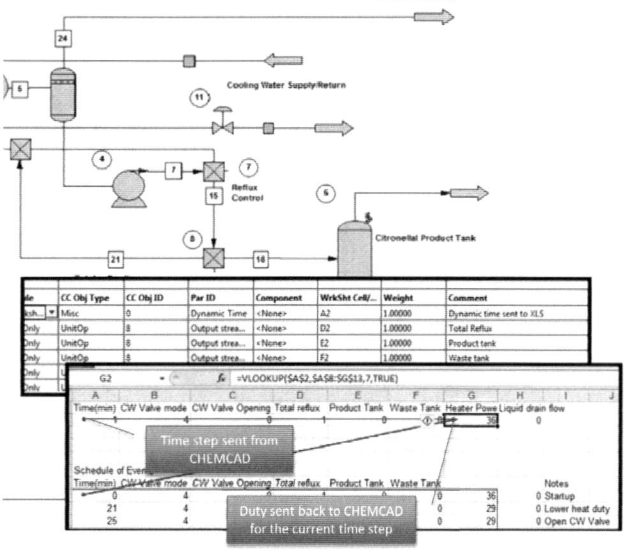

Chemstations™ 10

CITRON Dynamic Operations

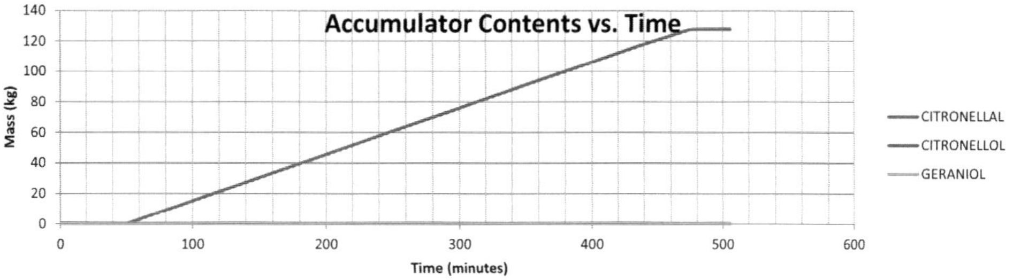

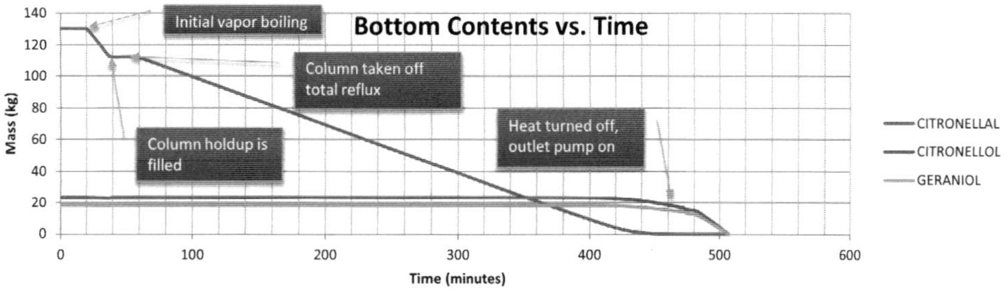

Chemstations™ 11

Comparison of CC-BATCH and CC-Dynamics Column Bottom Flow

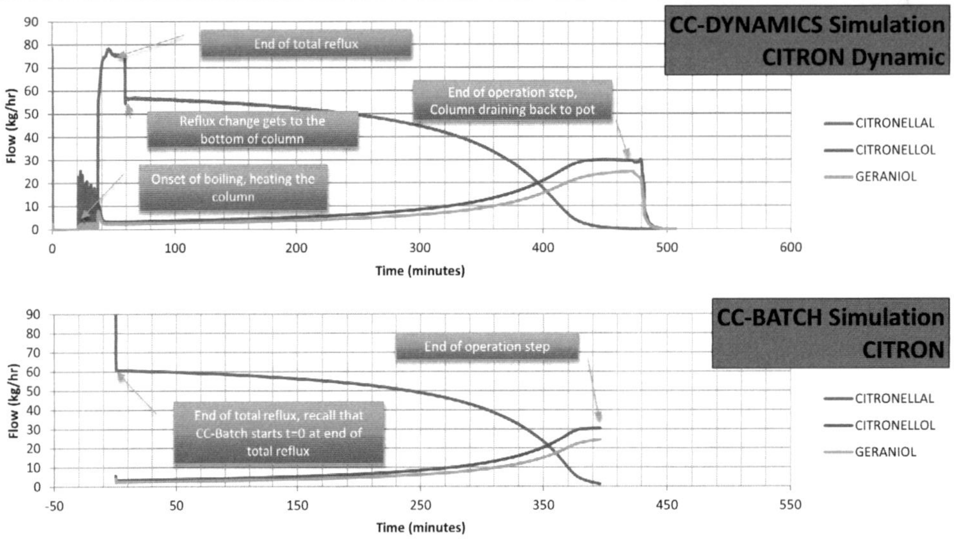

Chemstations™ 12

CITRON Dynamic Conclusions

We learned from using dynamics:

- Detailed startup procedure
- Detailed operation steps/Sequence details
- Equipment performance limitations
- Vacuum load
- Utility demands
- Equipment optimization is now possible: checking condenser capacity, vacuum system capacity, column flood %, column insulation requirements, etc.
- Higher fidelity simulation provides higher fidelity economic calculation (campaign time and costs)

Chemstations™ 13

WINTEK Batch Dehydration

Skid mounted solvent dehydration plant

Stripping water from a solvent stream in a two bed adsorber (mol sieve) system. One bed is active and one bed is regenerating under vacuum. Process was already built and operational before modeling analysis.

Process is scheduled with a defined sequence for opening and closing valves to allow one bed to regenerate while the other adsorbs. Rigorous equipment and piping models and pressure/flow calculation included (allows for reversible flow).

Rigorous simulation allows us to simulate effect of malfunction (RB3 blowdown valve malfunction)

Chemstations™ 14

WINTEK Process Flowsheet

2 Adsorption beds

2 PID controls

10 Actuated ball valves

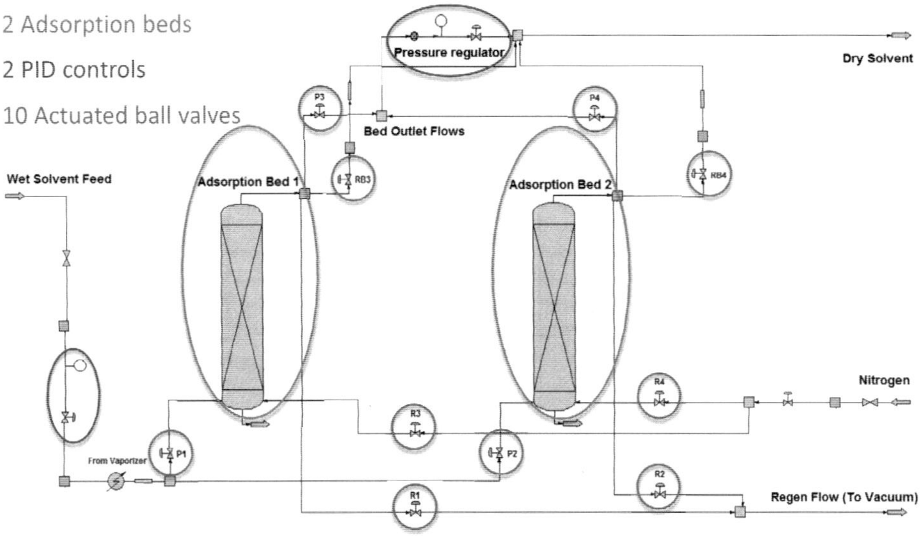

Chemstations™ 15

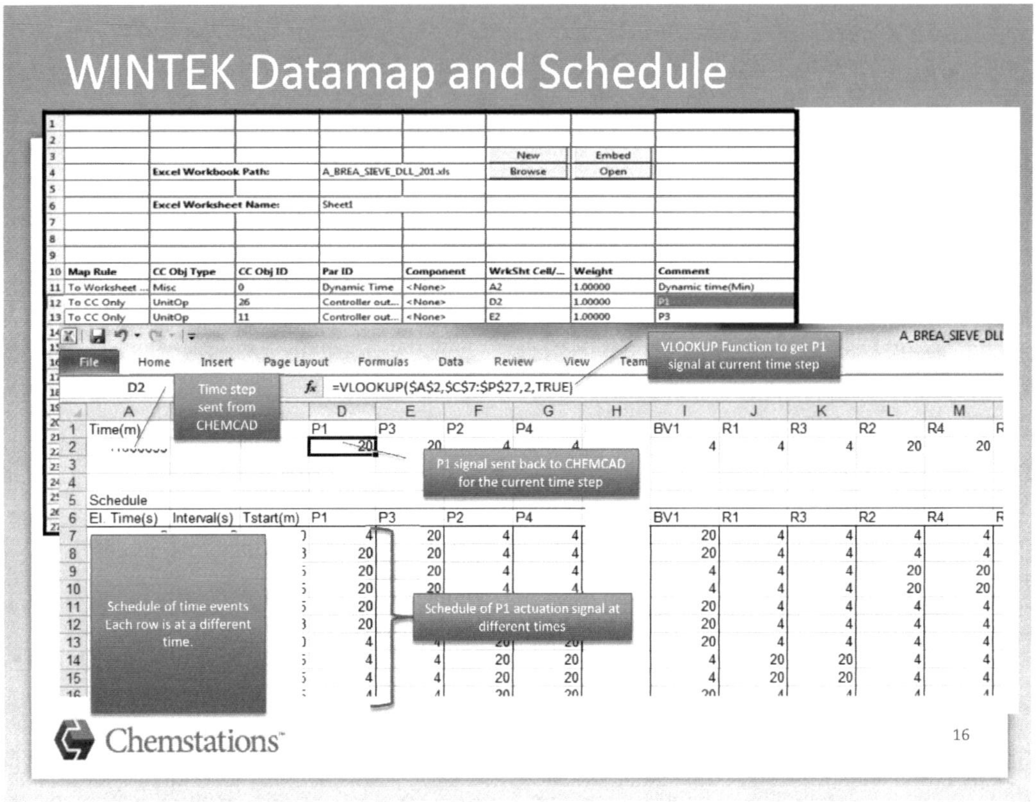
WINTEK Datamap and Schedule
Excel Workbook Path:
A_BREA_SIEVE_DLL_201.xls
New
Embed
Browse
Open
Excel Worksheet Name:
Sheet1
Map Rule
CC Obj Type
CC Obj ID
Par ID
Component
WrkSht Cell/...
Weight
Comment
To Worksheet ...
Misc
0
Dynamic Time
<None>
A2
1.00000
Dynamic time(Min)
To CC Only
UnitOp
26
Controller out...
<None>
D2
1.00000
P1
To CC Only
UnitOp
11
Controller out...
<None>
E2
1.00000
P3
A_BREA_SIEVE_DLL
File
Home
Insert
Page Layout
Formulas
Data
Review
View
Team
VLOOKUP Function to get P1 signal at current time step
D2
=VLOOKUP($A$2,$C$7:$P$27,2,TRUE)
Time step sent from CHEMCAD
Time(m)
P1
P3
P2
P4
BV1
R1
R3
R2
R4
P1 signal sent back to CHEMCAD for the current time step
Schedule
El. Time(s)
Interval(s)
Tstart(m)
Schedule of time events Each row is at a different time.
Schedule of P1 actuation signal at different times
Chemstations
16

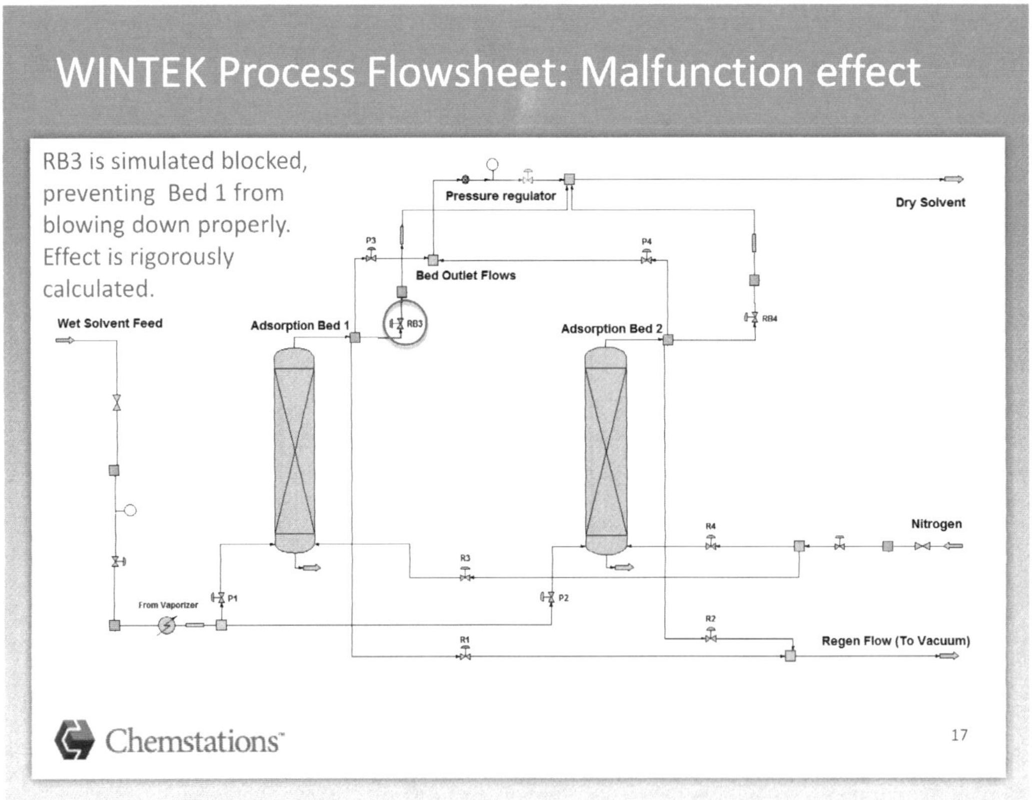
WINTEK Process Flowsheet: Malfunction effect
RB3 is simulated blocked, preventing Bed 1 from blowing down properly. Effect is rigorously calculated.
Pressure regulator
Dry Solvent
P3
P4
Bed Outlet Flows
Wet Solvent Feed
Adsorption Bed 1
RB3
Adsorption Bed 2
RB4
Nitrogen
R4
R3
From Vaporizer
P1
P2
R2
R1
Regen Flow (To Vacuum)
Chemstations
17

WINTEK Adsorber Beds

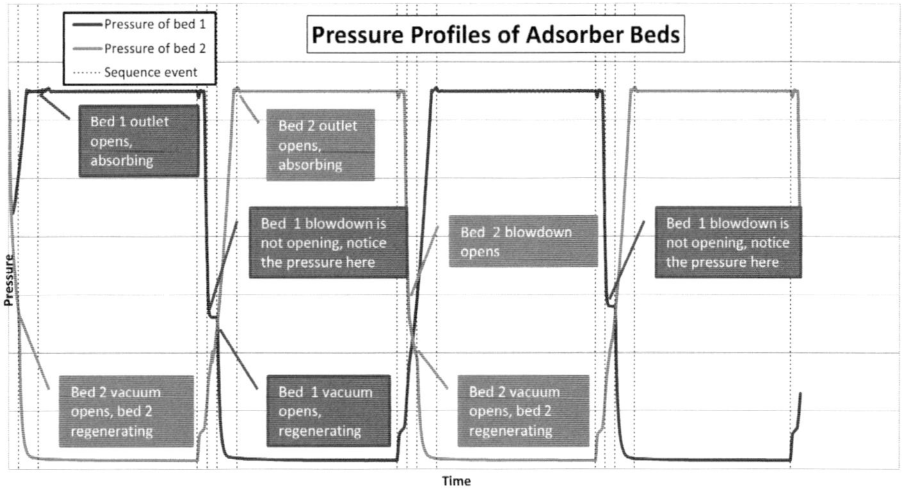

Chemstations™ 18

WINTEK Anhydrous Solvent Flowrate

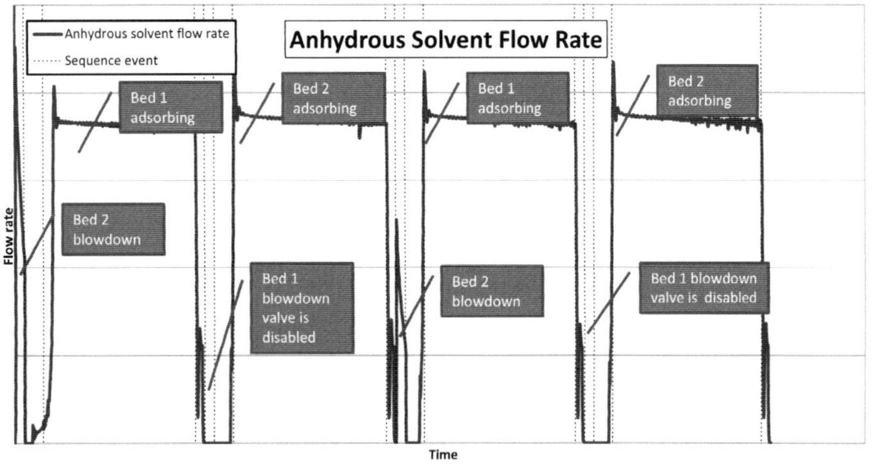

Chemstations™ 19

WINTEK Process Flowsheet: reversing flow

P3 and P4 open, flow is forward through one valve, reverse through the other

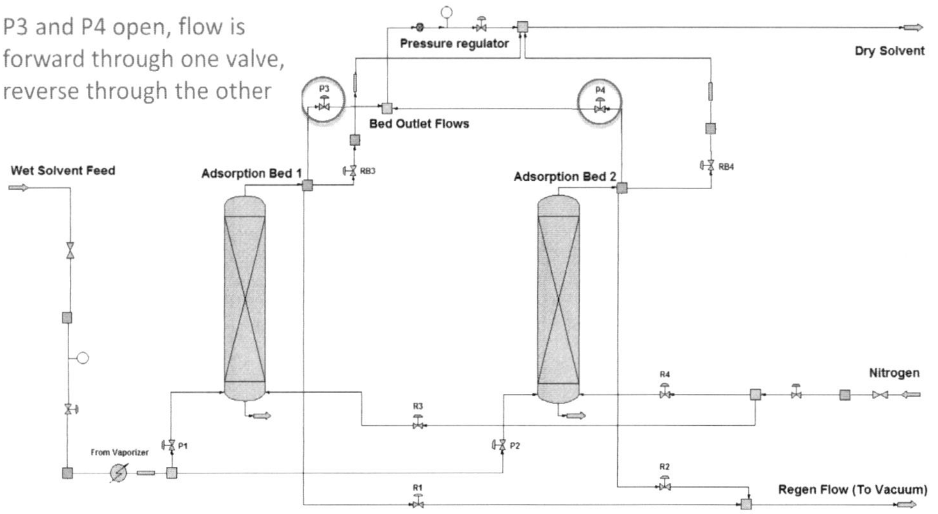

Chemstations™ 20

WINTEK Sequence of outlet and regeneration flows out of adsorber beds

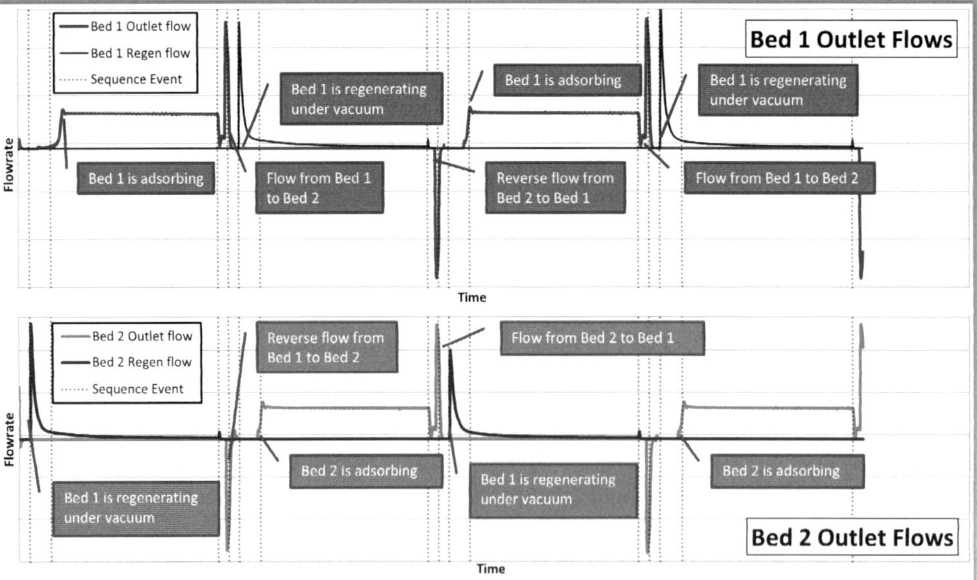

Chemstations™ 21

Optimization of schedule (future)

- By using an optimization engine (either the built in engine in CHEMCAD, or an external one connected to the Excel spreadsheet) the schedule of valve events can be optimized to maximize on-stream performance (blue area of chart).

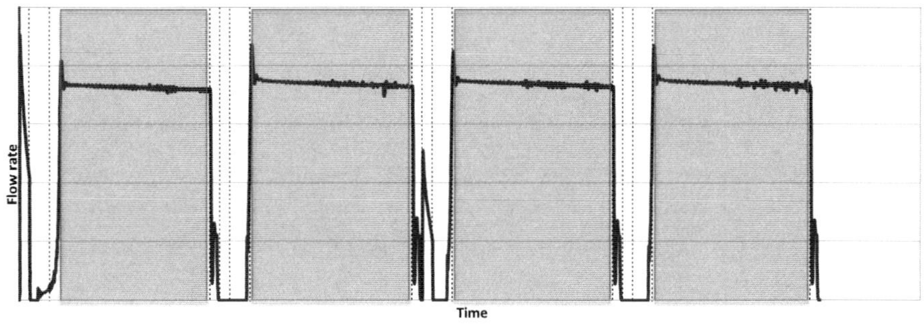

Chemstations™ 22

WINTEK Conclusions

What did we learn from our model?

- Discrete event scheduler with pressure and flow calculations (including reversible flow) allowed optimization of the process schedule
- Ability to identify and quantify
  - Bottlenecks
  - Sequence timing issues
  - Equipment limitations
  - Malfunction effects (bed 1 blowdown valve sticking)
  - Control loop tuning issues
- Ability to use the model for process scaling: process can be scaled up or down to meet customer requirements using the model before building/assembly
- Ability to optimize schedule of events

Chemstations™ 23

Suggested Procedure

1. Start with a simplified model (like CC-BATCH) to get heat duties, initial equipment sizes/specifications, and an initial timing/schedule
2. Build a rigorous dynamic model with as much detail as required (but no more than required) to solve the engineering problem (e.g., only add reversible flow if it is a concern)
3. Build an event sequence control scheme using DATAMAP and Excel starting with the information gathered in Step 1 above. Progressively improve the sequence by running the dynamic model with the scheduler

Chemstations™ 24

Special Thanks

Paul Winter, President and Owner

Wintek Corporation

http://www.wintek-corp.com

230 U.S. Highway 206 STE 401
Flanders, New Jersey 07836

Wintek has been providing equipment solutions to process applications operating at vacuum conditions since 1986, and serves the biodiesel, chemical, environmental, food, pharmaceutical, and plastics industries. Our focus is on process and equipment reliability for your application at a reasonable cost.

Chemstations™ 25

INLINE SYSTEM TO DETECT HYDRATES, WAX AND SAND

IMAGING Thomas M. Canty, P.E.
President, J.M. Canty Inc
Buffalo, NY 14094

ABSTRACT

Inline imaging systems are capable of viewing and digitally analyzing hydrates, waxes and sand in crude oil flow lines at pressure through 10,000 psi and temperatures up through 300 deg C. The technology relies on lenses that are made of fused glass which enable a quality, high speed image that detects and quantifies the size, shape, concentration and color of the suspended components in a crude oil stream. This allows for research and testing in addition to inline measurement and control to assure proper flow

INTRODUCTION

Vision systems are developed to the point where they can provide analysis and control in the oil and gas industry on a reliable and accurate basis. Design of hardware and software, and advancements in Ethernet communication technology have allowed vision based systems to see into process environments such as crude oil to detect and analyze particle constituents. Applications involving the monitoring and control of Produced Water concentrations, Water in Oil analysis, Multi-cut Sand, Oil and Water, Froth (Flotation Control) and Interface can also be effectively handled by vision with the benefit of visual optimization by the operator.

Vision analysis has been proven for many years in applications such as polystyrene bead manufacturing, sugar crystal growth, cell growth, abrasive manufacturing and many others, providing an extra dimension to process analytics that enables the process engineer to make evaluations based on his own expertise and then to use the information to better set software parameters that enable the system to more accurately analyze and control the process.

VISION TECHNOLOGY - HARDWARE

The conditions of the application require a vision system to have the ability to survive in an extreme environment, maintain explosion proof electrical ratings and be easily used by operators in order to be effective. The first point of discussion in this regard deals with the hardware mounted to the process.

Fused Glass Technology

The viewing mechanism of the vision system is most important and must be fitted with rugged, high pressure windows through which illumination and imagery can be conducted. The best option is the fused, glass to metal seal because it is rugged and can withstand pressure, temperature, shock and vibration. In addition, because it is fused directly to the metal probe it can be polished flush with the metal leaving no sites for particle adherence and contamination build up (see Figure 1). The fused glass interface (Fig. 2) is ground and polished to a flat, crevice free surface which remains clean during operation.

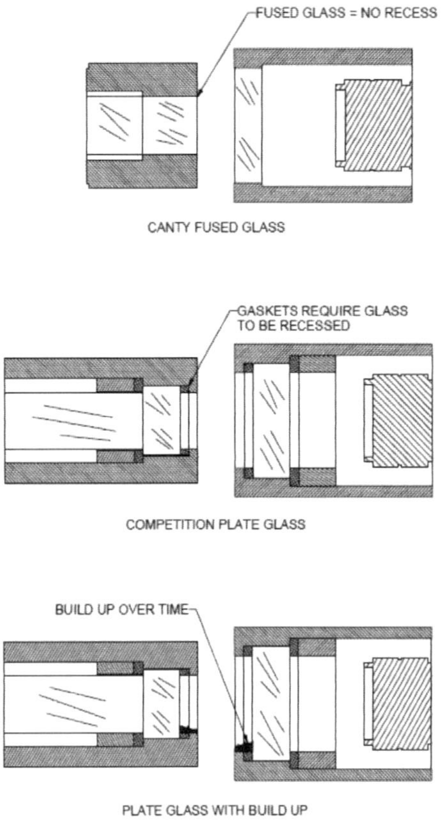

Fig. 1 – Fused Glass vs. Plate Glass Design and Cleanliness

Illumination

Illumination is critical to the success of vision based technology, and a consistent, high intensity source is the building block that allows software to define particles consistently and accurately. Another important feature of a complete system is the ability to control the measurement zone where the particles are illuminated and seen by the camera. The light pipe component shown in Figure 2 has adjustment capability in order that the gap between the camera and light pipe (measurement zone) can be adjusted to optimize the sensitivity of the system. Fluids can range from relatively clear (low turbidity) to quite heavy in particulate matter (high turbidity) and the control of the measurement zone enables the illumination to be controlled by means other than adjustment of intensity from the bulb. A typical light system is shown in Figure 3.

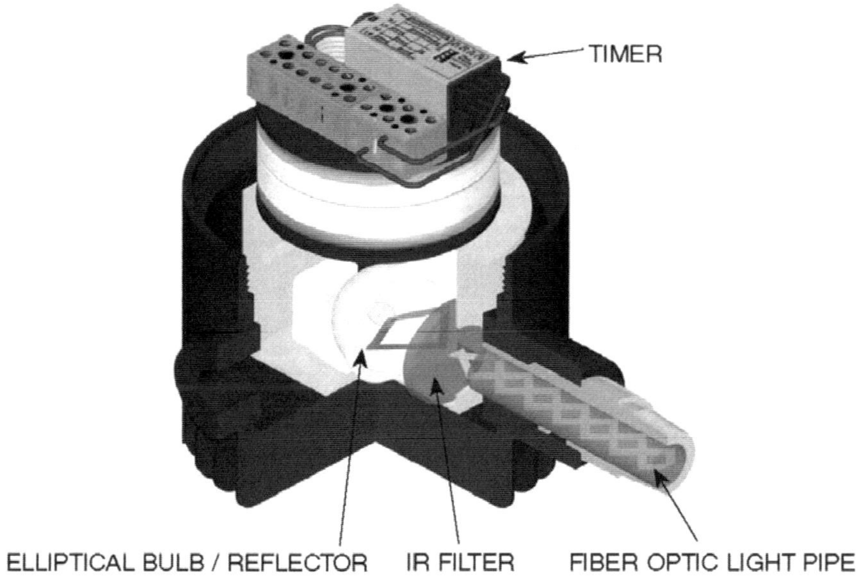

Figure 3. High Intensity Light Section

Cleanliness / Jet Spray Ring Technology

Development of a Jet Spray Ring generates a high pressure vortex rinsing action to remove tough deposits from Vision Systems. Originally designed for the U.S. Space Shuttle program, the Jet Spray Ring accepts a rinsing/cleaning solution and distributes the pressurized material through the specially designed rinse ports. These rinse ports have been engineered to provide maximum cleaning power in a compact package. The Jet cleaning action expels all debris from the Vision System glass, providing an exceptional view into the process. For petroleum applications the spray ports are designed into the instrument probe and maintain the configuration on the measurement zone.

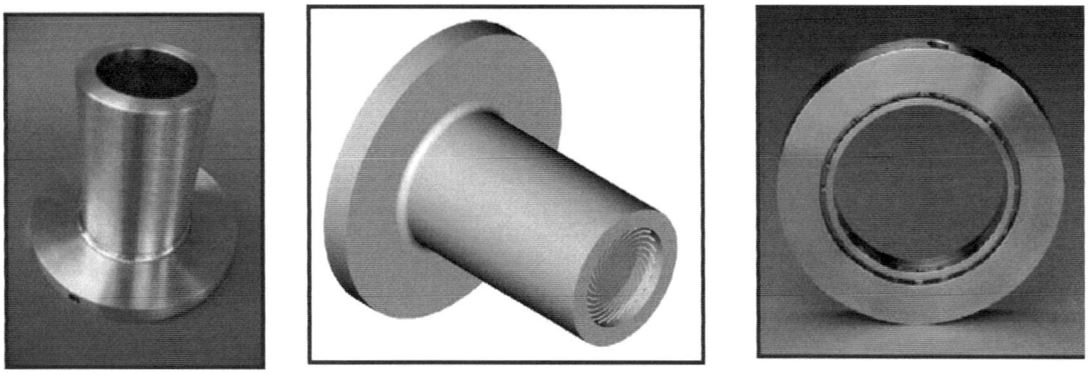

Fig. 4 – Typical Jet Spray Rings

Fig. 5 – Jet Spray Ring Cleaning Action Sequence

Typical Instrument Configurations

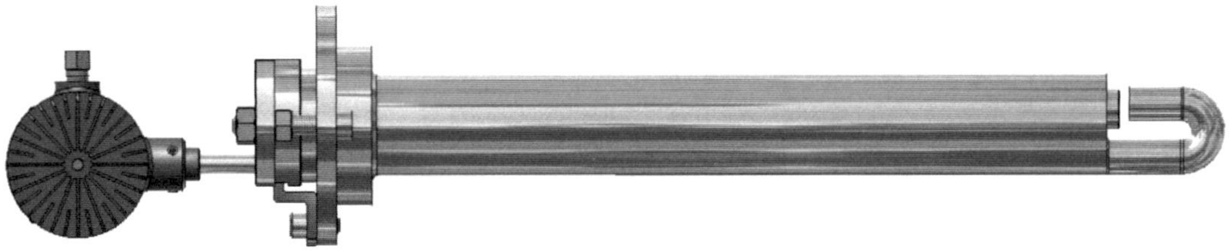

Fig. 6 - Inline System for 6" Flow Line and Larger

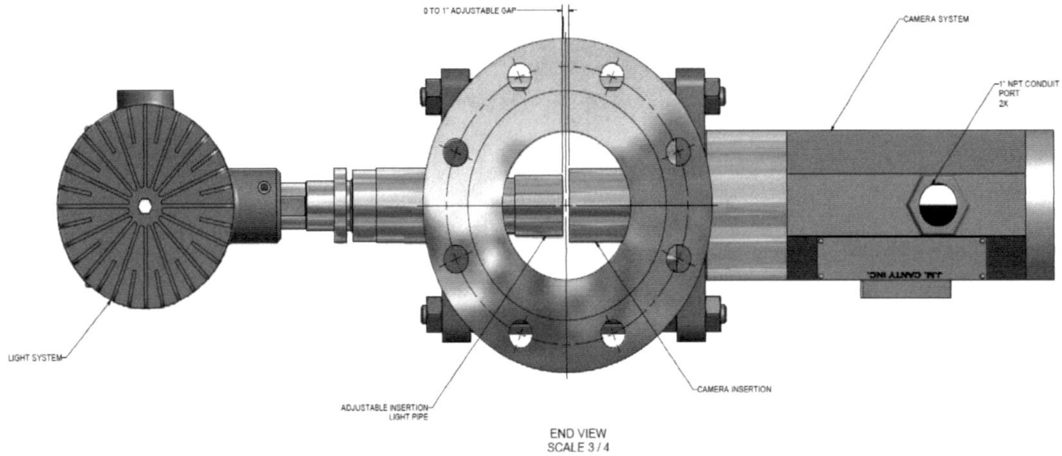

Fig. 7 - Inflow Systems ½" – 4" Pipeline

VISION TECHNOLOGY - SOFTWARE

The second point of discussion deals with the software. Vision systems are generally based on the same concept as the human eye – brain interaction. Objects are seen in our field of vision because they stand out from the background. This delineation is created by differences in light intensity around the object caused by shadows or colors. The vision system detects particles in much the same way. Intensity differences from the background field occur at particle boundaries. The vision system detects these and is able to form the particle based on defined threshold information which sets the level of intensity change that defines a particle boundary. The threshold information is input by the operator based on what he sees on screen. This is the powerful advantage of vision over other types of instrumentation. Reliance on indirect calibration is not required. The actual particles are used to define what is a proper size detection and what is not. This is depicted in the Fig. 8-9.

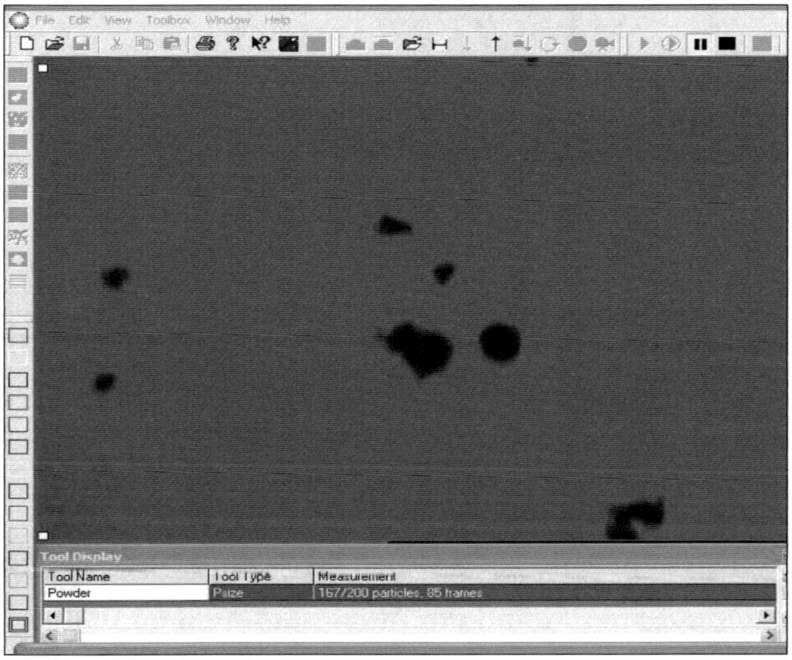

Figure 8. Live Stream View

Figure 8 shows an image of oil particles in the stream which can be used to define to the software what the particles actually are. The next step is to digitize the image and test thresholds to properly define what a particle is based on the actual view of the particle. Figure 9 shows a digitized image of the particle capture which yields particles as they are actually seen. This ability to select the threshold is a way the operator can confirm that the software is capturing particles correctly for analysis.

Fig. 9 – Proper Threshold Selected

The discussion to this point will serve as a basis for the remainder of this paper. The hardware and software principles are largely the same regardless of applications. The discussion to follow will focus on the data that can be extracted from the process image to enhance the production and control of the aforementioned applications.

DETECTION OF WAX and HYDRATES

The presence of precipitated wax in a flow line or pipeline can lead to clogging issues. Detection of wax precipitation can alert the operator to the pending problem and prevent a stoppage in production at best, and a loss of a flow line at worst. As process temperature cools, the wax takes on a solid form which tends to adhere to metal surfaces. Build up over time will choke the line. Figure 10 shows results of the system's ability to track wax precipitation in field trials as a function of temperature. Figure 11 shows a typical image of a crude oil which contains water and precipitated wax. The circular ring shapes are water droplets and the semi transparent, large formations are wax particles.

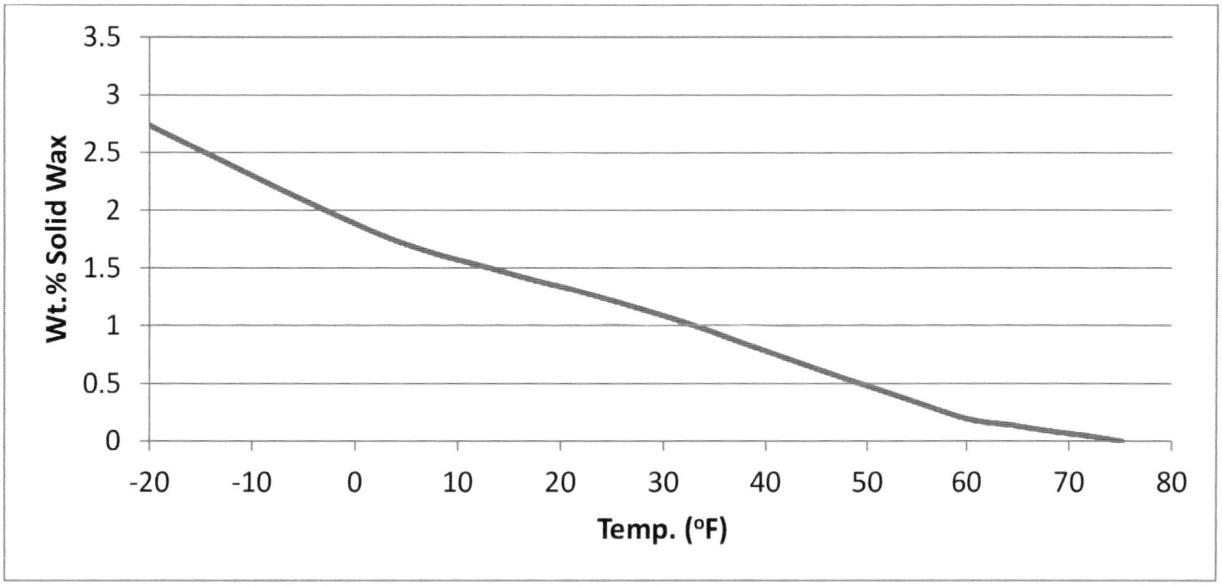

Figure 10. Wax vs Temperature

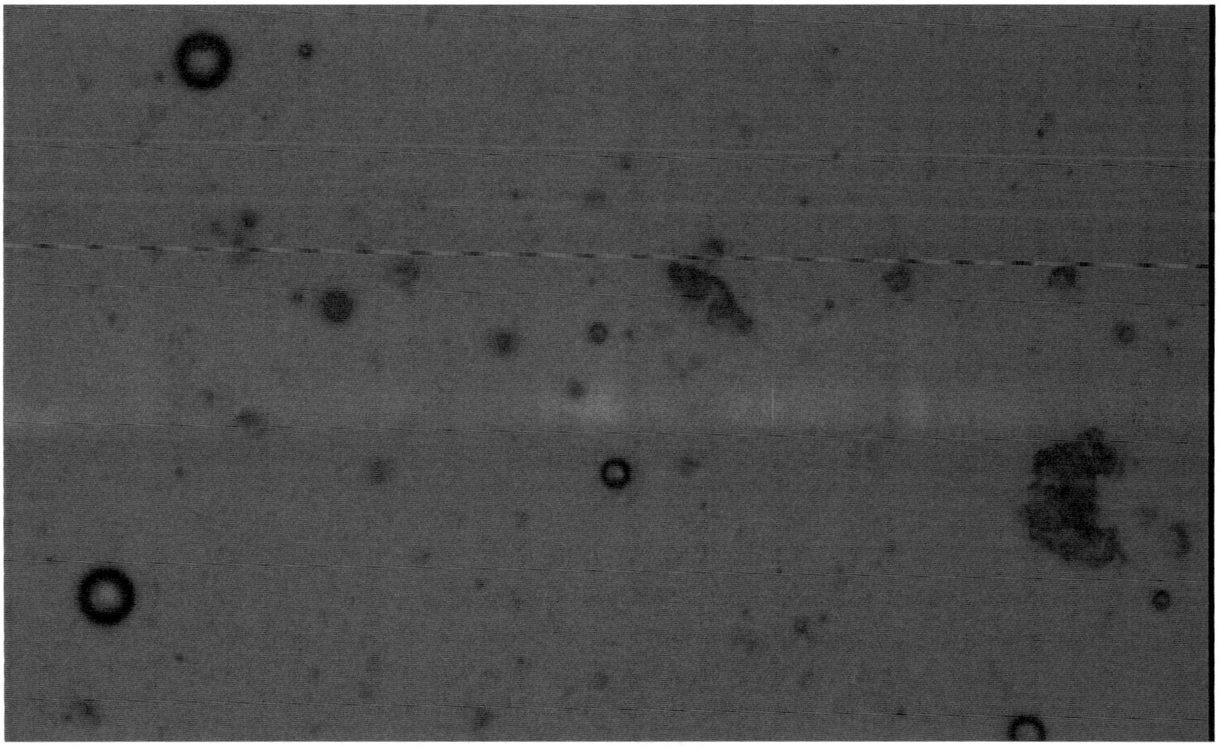

Figure 11. Typical Image of Crude with Water and Wax

Hydrates are detected in a similar fashion as they crystallize out of solution. Methane droplets will start to freeze, then agglomerate and finally plug the line if left untreated long enough by either temperature control or chemical injection. As their formation proceeds they will obtain an opaque shell which is indicative that the process is well underway and action must be taken to avoid disaster.

DETECTION OF SAND

The detection of sand is a critical function in the separation process. Once removed from the oil it must be cleaned, injected or dumped if present in small enough quantities. Detection is critical in any case. Viewing a produced water image one will see oil droplets, sand and other contaminants. General acceptance levels to dump to the environment are 30 ppm which is not a lot. When sand is detected in the outflow of the separator it is an indicator that separation must be adjusted to handle the present characteristics of the crude. Unlike oil droplets, sand, and other solids will almost always be non circular in the view. Vision software allows the categorization of these shapes for identification. In the image following a sand particle is seen in the center and upper left corner. These can be identified by aspect ratio or circularity as an example. Concentration of each constituent can be determined.

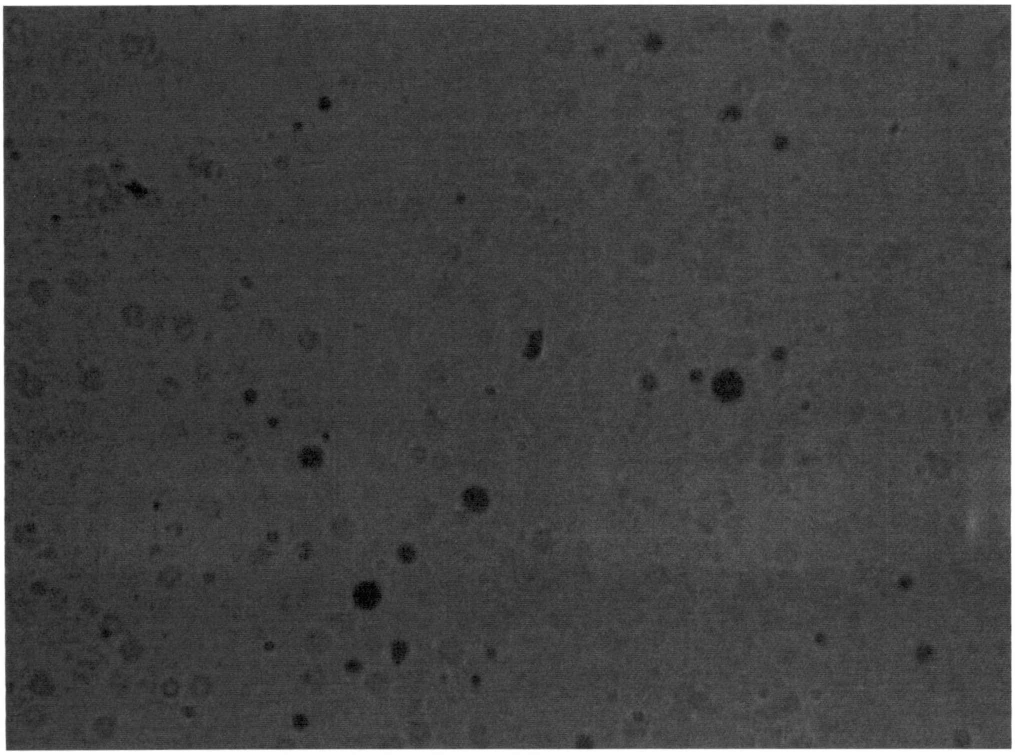

Figure 12. Sand Particles in Produced Water

CONCLUSION

Vision technology has the ability to maintain its viability in the harsh oil environment through design and manufacturing inherent in its components. Good visual optics, lighting and control of the measurement zone are all critical elements to success. Software has the advantage over other technologies due to its capability to analyze shape as well as size. This makes the difference between counting particles and classifying particles for a full understanding of how the process is behaving.

An Introduction to the America Invents Act

David W. Holt
256-517-5191 dholt@babc.com

Bradley Arant Boult Cummings LLP

babc.com ALABAMA | DISTRICT OF COLUMBIA | MISSISSIPPI | NORTH CAROLINA | TENNESSEE

Agenda

- Basics of Patent Law
- The America Invents Act
  - First Inventor to File
  - Other Changes
    - Prior Use Defense
    - Best Mode
    - Post-Grant Proceeding
    - False Marking
    - Joinder

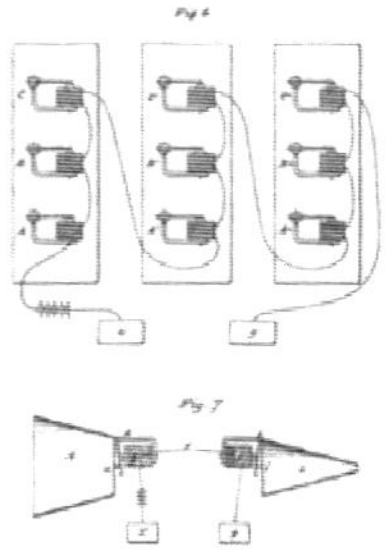

Pat. 174,465 (1876)
Alexander Graham Bell

Bradley Arant Boult Cummings

What is a Patent?

- A patent grants the owner the right to exclude others from making, using, selling, offering for sale, and importing the subject matter bounded by the claims of the patent.
- Period of exclusivity is 20 years from the filing date.*

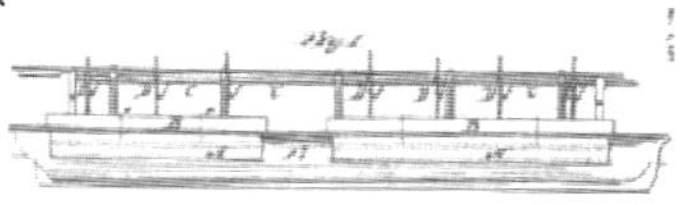

Pat. No. 6,469 (1849): Manner of Buoying Vessels
Abraham Lincoln

Bradley Arant Boult Cummings

Requirements for Obtaining a Patent

- "Process, machine, manufacture, or composition of matter" (Sec. 101)
- Novel (Sec. 102)
- Not obvious (Sec. 103)
- Useful (Sec. 101)

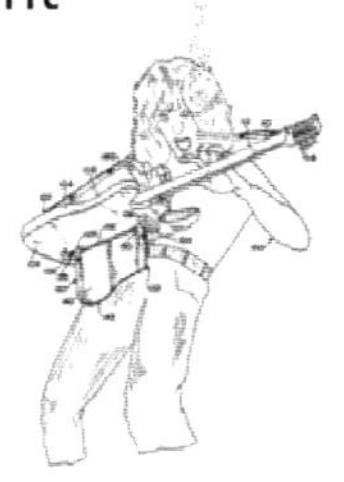

Pat. No. 4,656,917 (1987) for "Musical Instrument Support"
Edward L. Van Halen

Bradley Arant Boult Cummings

Patenting Flowchart and Approximate Timeline

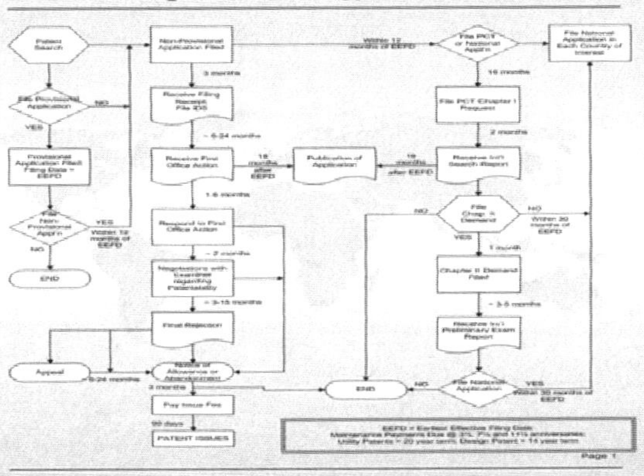

Parts of Patent Application

- Abstract (brief explanation)
- Specification
  - Field of invention
  - Background (problems solved)
  - Summary
  - Detailed description
  - Drawings
  - Claims (defines scope of monopoly)

Bradley Arant Boult Cummings

The America Invents Act

First Inventor to File

- "First to Invent"
  - US system from 1789 to Mar. 15, 2013
  - Patent issued to party who could prove prior invention
- "First to File"
  - Most of the rest of the world
  - Patent issued to first party to file, if not anticipated by published prior art (i.e., absolute novelty).
- "First Inventor to File" – **New under the AIA**
  - US system beginning Mar. 16, 2013
  - Patent issued to first party to file, with exceptions
  - No more swearing behind prior art to show earlier "invention"
  - But 1 year grace period after inventor public disclosure

Public Disclosure Grace Period

- The AIA retains an inventor's **one year personal** grace period for filing an application after disclosure, if the public disclosure :
  - Was made by the inventor;
  - Was made by a third party that obtained the information from the inventor
- BUT, no grace period for a third party disclosure of an independently-created invention

Derivation Proceeding

- The AIA replaces the interference system with a derivation proceeding.
  - Derivation proceedings will determine whether an inventor named in an earlier-filed application derived the claimed subject matter from the inventor of a later-filed application.
  - Derivation proceedings will apply to applications having priority claims after March 15, 2013.

Scenario A

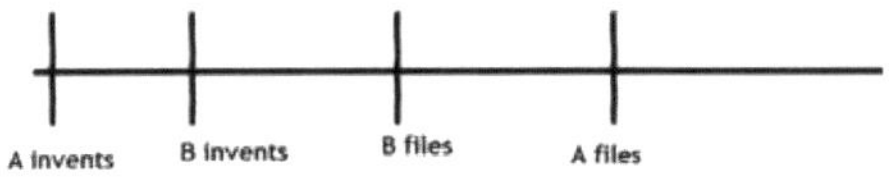

First to Invent: ***A*** obtains patent (so long as he wasn't dilatory in filing)

First to File: ***B*** obtains patent.

First Inventor to File: ***B*** obtains patent.

Scenario B

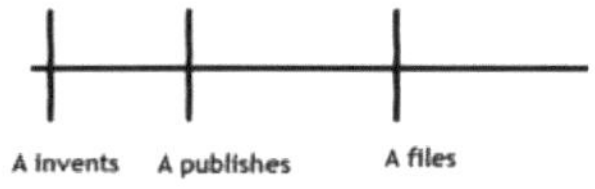

First to Invent: ***A*** obtains patent if filed within year of publication.

First to File: A does ***NOT*** obtain patent, invention in the public domain.

First Inventor to File: ***A*** obtains patent if filed within year of publication.

WHY? USA has a 1-year grace period for inventors.

Scenario C

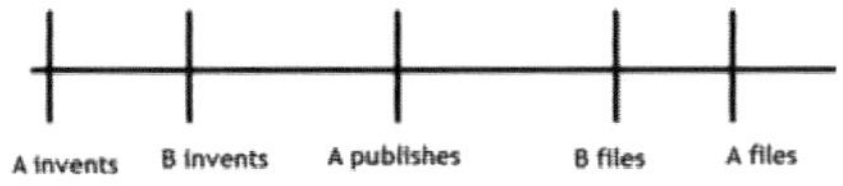

First to Invent: ***A*** obtains patent if filed within year of publication.

First to File: ***No one*** obtains patent, given to public domain.

First Inventor to File: ***A*** obtains patent if filed within year of publication.

Scenario D

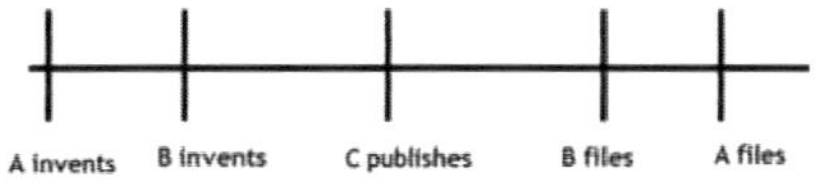

First to Invent: ***A*** obtains patent if filed within year of C's publication.

First to File: ***No one*** obtains patent, given to public domain.

First Inventor to File: ***No one*** obtains patent, given to public domain.

WHY? In new USA system, 1 year grace period limited only to publications by or derived from inventor.

Scenario E

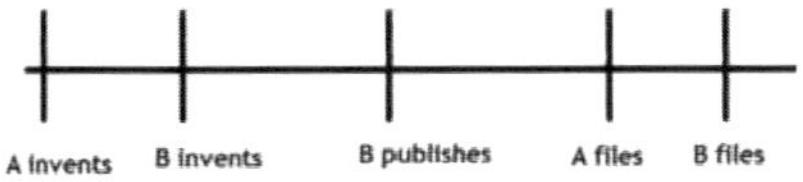

First to Invent: ***A*** obtains patent if filed within year of B's publication.

First to File: ***No one*** obtains patent, given to public domain.

First Inventor to File: ***B*** obtains the patent.

NOTE – In this scenario, because B publishes first, he receives a 1-year grace period to file, which knocks out A's filing as prior art. Thus, B is both second to invent and second to file but gets the patent anyway.

Who Benefits from the AIA?

- Companies with efficient "bench-to-office" procedures
- Companies willing to use Provisional Patent Applications or Preemptive Publication
- Universities/ research institutions and micro entities get lower fees
- Entities that may publish before filing (consider foreign effects)
- Entities willing to pay $4800 for accelerated prosecution for 6 month examination and 1 year decision

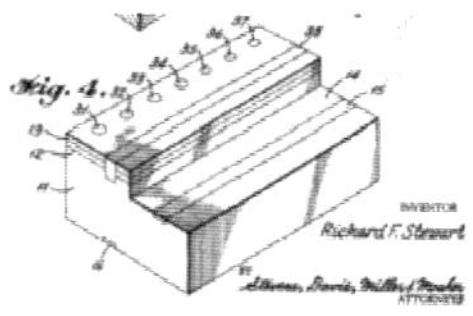

Pat. 3,138,747 (1959)
Integrated Semiconductor Chip

Owned by Texas Instruments, Inc.

Who Loses in the AIA?

- Garage Inventors
- Entities that delay filings
- Patent Applicant Pocketbooks (fees ↑)

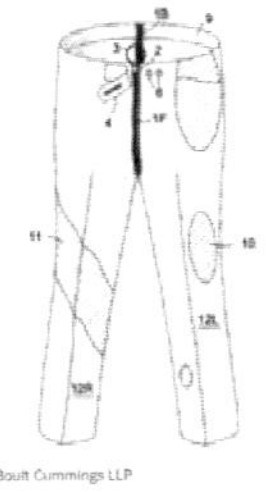

Pat. 6,161,223 (2000)

"Pants Separable at Crotch for Style Mixing"

Prior Use Defense

- Common in first-to-file countries
- Historically not available at all; added in 1999 as defense for parties making secret commercial use of certain business methods prior to patent being applied for

Pat. 223,898 (1880) for Electric Lamp
Thomas Edison

Prior User Rights

- The AIA has expanded the prior use defense to apply to *any* patent issued after Sept. 16, 2011, "consisting of a process, or consisting of a machine, manufacture, or composition of matter used in manufacturing or other commercial processes."
- Cannot be asserted against university-owned patents
- Lots of limitations

Best Mode Defense

- The AIA eliminates the best mode defense, i.e., accused infringers can no longer claim that an asserted patent is invalid due to the patent holder's failure to include the best mode.
- However, the AIA does not eliminate the best mode requirement under 35 U.S.C. § 112 for patent applicants when submitting an application.

New Post-Grant Proceedings – Effective 9/16/12

- Post-Grant Review
  - Challenge patent on any grounds within 12 months of issue
- Inter Partes Review
  - Challenge patent on novelty or obviousness grounds
  - Replaces Inter Partes Reexamination
- Supplemental Reexamination
  - Ex parte proceeding to "scrub" new prior art or inequitable conduct
  - Patent Owner initiated

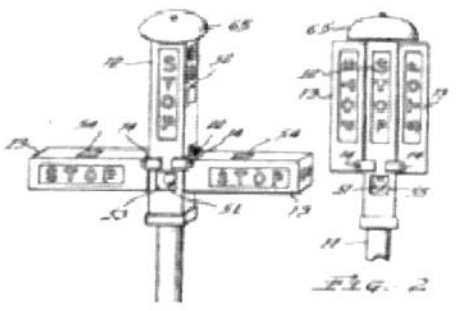

Fig. 2

Pat. No. 1,475,024, "Traffic Signal" (1924) G.A. Morgan

False Patent Marking Provisions

- The AIA eliminated qui tam false marking suits, but the U.S. may still sue for civil penalties.
- The provision applies to all cases filed on or after Sept. 16, 2011 **and** all pending cases.
- The AIA created a new cause of action for false marking for private parties that have suffered a competitive injury due to the false marking.
- Virtual Marking

Joinder of Multiple Defendants

- For multiple accused infringers to be joined in the same action:
  - The right to relief must be asserted against the infringers jointly and severally, or arise out of the same transaction(s) or occurrence(s) relating to the making, using, importing, offer to sell, or selling of the same accused product or process, **and**
  - Questions of fact common to all defendants will arise in the action.

Any questions?

David W. Holt
dholt@babc.com
256-517-5191

(63b) Effective IP Strategies Under the AIA

Mark Rygiel, Sterne Kessler, Washington, DC

The recently enacted America Invents Act (AIA) represents the most significant change to the U.S. Patent Law System in sixty years. The AIA effects all aspects of the U.S. patent system from obtaining patents, enforcing patents in litigation, and attacking invalid or overly broad competitor patents. This presentation will look at the AIA and how successful companies in the chemical, energy, process and product development industries can utilize effective intellectual property strategies under the AIA to protect their technology and thwart competitors. Particularly, this presentation will examine how companies can use new procedures under the AIA to quickly build a patent portfolio, strengthen issued patents, and enforce issued patents in litigation. Additionally, this presentation will address strategies for navigating the existing competitor patent landscape by using the new post-grant procedures created by the AIA.

(63c) Protecting Your Inventions Before It Is Too Late

Anthony Venturino, Novak Druce and Quigg, LLP, Washington, DC

The presentation will cover patent and trade secret basics, how to identify an invention, how to protect it by patenting or keeping it as a trade secret, how to share inventions during joint development, and how to manage employees to protect inventions and trade secrets.

(63d) Patents 201: A User-Guide to Patents in the Energy Industry

Jeffrey J. Phillips, Winston & Strawn LLP, Houston, TX

Many past AIChE presentations relating to patent law have focused on the basics of patentability. The purpose of this presentation is to help chemical engineers and their management understand how to use their patents, and how to avoid common pitfalls associated with competitors' patent portfolios. Specific topics that will be covered include:

- Patent licenses (should I grant one, do I need one from someone else);
- Enforcement of patents (how to stop an infringer);
- Obtaining defensive patents for R&D work (why the costs are usually justified);
- Freedom-to-operate opinions (do I need one, yes);
- Merger and acquisition due diligence (caveat emptor); and
- Anatomy of a "strong" patent portfolio.

This presentation is targeted for a mixed audience with technical, business, and legal backgrounds, and will include real-life examples from my experiences counseling and litigating in the petrochemical and biofuels industries.

(64a) An Ethylene Plant Start-up with Total Recycles: Dynamic Simulation of De-Ethanizer Transient Behaviors

Ha Dinh, Shujing Zhang, Jian Zhang and Qiang Xu, Dan F. Smith Department of Chemical Engineering, Lamar University, Beaumont, TX

Flaring emissions during plant turnarounds have always been a major environmental and economical concern in the chemical process industry. A significant case is during plant start-ups. Traditionally, off-spec products are directed to flare system as wastes until operating units are in specifications. One of the recent flare minimization practices is the start-up with total recycles, in which off-spec streams from downstream are directed to system input until the process operates normally. Dynamic simulation is employed as a virtual tool to generate a new strategy and provide guiding procedure to operators.

In this study, an operation procedure of an ethylene plant start-up is investigated via dynamic simulations. Three case studies are presented to demonstrate the efficacy of dynamic simulations. The base case is the conventional method of flaring off-spec products during the first three and a half hours to fire and stabilize three cracking furnaces before starting charged-gas compressor (CGC) system. Off-spec C_2-hydrocarbons are withdrawn from DeC_2 overhead and recycled back to feed CGC as an alternate to either one (Case One) or two furnaces (Case Two). This recycled stream is also mixed with another off-spec one from C3 splitter and additional natural gas to make up to achieve an equivalent molecular weight as CGC normal feed. Dynamic simulation contributes a great deal in evaluating new strategies as well as predicting process dynamic behaviors. It also helps achieving more precise timing to control the recycle loop; hence, flaring is potentially minimized while the start-up time is shortened. One observation is the difference of the recycle stream temperature and the expected value entering the first-stage compressor, a solution of which is confirmed by dynamic simulations. Following this representative, utilizing dynamic simulation as a driving force for operation during start-ups, shutdowns and malfunctions (SSM) conditions effectively provides an in-depth understanding and provides more detailed planning and expectations for all related personnel before the real action is proceeded.

(64b) A Process-Integration Framework for Abnormal Situation Management (ASM): A Systematic Approach with Application to Qatar's Industrial Needs and Opportunities

Fadwa T. Eljack, Department of Chemical Engineering, Qatar University, Doha, Qatar, Mahmoud El-Halwagi, Chemical Engineering, Texas A&M University, College Station, TX and Qiang Xu, Dan F. Smith Department of Chemical Engineering, Lamar University, Beaumont, TX

Qatar is a leading producer of a variety of gas products including LNG, GTL and a wider spectrum of petrochemical intermediates and specialty chemicals. The country is blessed with the third largest natural gas reserve in the world and its booming industry includes the largest LNG train, the world's biggest ethylene cracker, the largest GTL plant and the list continue. The industrial fortune however comes at a cost in terms of impact on the environment. And although the importance of flare reduction is a worldwide struggle, the State of Qatar has a vested interest in reducing its flaring rates to enhance the quality of life for it's citizens. In 2009, the State of Qatar officially partnered with the World Bank in it's program for Global Gas Flaring Reduction, aiming to be a global leader in environmental protection. Flare reduction will directly contribute to limiting the emissions of environmentally harmful waste of gas and the reduction of CO2 and other GHGs emissions. A concentrated effort to reduce flaring in Qatar's industrial cities can have a dramatic impact on the sustainability of gas industrial industries, quality of life of local populations in addition to reducing Qatar's CO2 footprint. Initial efforts from 2007-2010 lowered flaring rates by 14%. Nonetheless, Qatar still flares 1.9 billion cubic meters annually, indicating reduction efforts must continue. Evolving environmental regulations and Qatar's commitment to flare gas reduction makes better management of industrial processes a key topic and research priority area for the State of Qatar.

The literature indicates that the reasons for high flaring sources is a combination of lack of end of use options for the unwanted gases during operation, and weak environmental regulations. Companies have access to relatively inexpensive natural gas and often feel that managing associated gases is too much of a hassle. With the rising prices of energy and oil, industry now has motivation to better manage their natural resources and they have started to investigate means of recovering these vented and associated gases. It is here that there is still great potential for research work.

In this paper, the authors propose a project that focuses at development of a novel approach for optimal management of process facilities. It is based on simultaneous design and operational optimization by identifying key flaring and venting sources, causes and consequences of process upsets that result in flaring and by applying recent process design and optimization tools. Tradeoffs between cost, production, mass utilization, energy efficiency, safety, and environment impact will be established through multi-objective optimization models (Pareto curves), which can be used by decision makers to determine optimal flaring policies. The developed design and operational methodologies and associated models will introduce for the first time a comprehensive, systematic, and generally applicable approach that combines abnormal situation management, systems dynamics, and process.

(64c) Effect of Crosswind On Flaring Combustion Efficiency and Subsequent Increase in 8-Hr Ozone Concentration in the Houston-Galveston Ozone-Nonattainment Area

Ziyuan Wang, Jian Zhang, Qiang Xu and Thomas C. Ho, Dan F. Smith Department of Chemical Engineering, Lamar University, Beaumont, TX

Emissions from industrial flares have been identified as a significant VOC emission source in a region heavily concentrated with chemical plants, e.g., the Houston-Galveston (HG) Region in Texas. The amount of flaring emissions from each flare depends directly on a flare's combustion efficiency (CE). Several studies have shown that high crosswinds may reduce the flaring CE to a level well below the currently assumed CE of 98% and increase the amount of flaring emissions. A decrease of CE from 98% to 96 % would effectively double the currently estimated flaring emissions based on the assumed CE of 98%. The impact of crosswind, therefore, can be significant in the increase of regional ozone concentration due to the increase of flaring emissions. The objective of this study was to perform CAMx simulations to quantify the effect of crosswind in the increase of regional 8-hr ozone in the HG Region during the 2006 ozone episode from May 31 through June 15, 2006.

In this study, a correlation between the crosswind speed and the flaring CE was first established based on the experimental obtained CEs available in the literature and the computer simulated CEs predicted from a computational fluid dynamics (CFD) model. The derived correlation equation was then used to generate dynamic flaring emissions for each CAMx grid with wind speeds provided from the CAMx meteorological file. The generated wind speed-induced flaring emissions were then incorporated into the total emissions for CAMx simulations. The CAMx simulation results with and without the wind speed-induced flaring emissions were then compared to quantify the effect of crosswind on the increase of 8-hr ozone concentration. The results have indicated that the flaring CE may range from 98% to 91% during the episode depending on the specific wind speed at the location and time of simulation and the observed 8-hr ozone concentration increase may range from 0.3 to 2.2 ppb depending on the overall meteorological conditions and the dynamic distribution of NOx and VOCs in the region. It is worth reporting that the ozone increase has been observed to occur predominately in a VOC-controlled area during the simulation. More detailed simulation results will be reported at the meeting.

(64d) CFD Simulations and Simple Correlations to Predict Flare Performance Under Various Operating Conditions

Kanwar Devesh Singh[1], Preeti Gangadharan[1], Daniel H. Chen[1], Helen H. Lou[1], Tanaji Dabade[2] and Xianchang Li[2], (1)Dan F. Smith Department of Chemical Engineering, Lamar University, Beaumont, TX, (2)Mechanical Engineering, Lamar University, Beaumont, TX

CDF Simulations and Simple Correlations To Predict Flare Performance Under Various Operating Conditions

K. Singh, P. Gangadharan, D. Chen, H. Lou, D. F. Smith Chemical Engineering Dept.,

X. Li, T. Dabade, Mechanical Engineering Dept. Lamar University, Beaumont, TX

With high uncertainty in accounting for the VOC emission inventories in the Houston-Galveston-Brazoria (HGB) area, flares have been identified as an important underreported or unreported emission source and the evaluation of their performances are currently under-way[1]. The assumption of a very high DRE (Destruction and Removal Efficiency >98%), if operated in compliance with 40 CFR § 60.18, has been under scrutiny and recent flare tests performed on industrial-scale flare systems have indicated otherwise[2]. Most of these tests indicated that, even if operating under the guidelines, flare efficiency can go far below 98% depending on factors like air-to-fuel ratio, steam-to-fuel ratio, flare-tip velocity, vent gas heat content, and crosswind[2]. In this work, computational fluid dynamics (CFD) simulations of industrial ethylene flares are performed to predict the effect of these parameters on flare DRE, CE (Combustion Efficiency), and VOC (Volatile Organic Compound) emissions. Easy to use correlations were developed in this work to predict the flare performance and CO/formaldehyde emissions under various operating conditions. The CFD simulations employ a validated reaction mechanism suitable for the combustion of C_1 to C_3 light hydrocarbons coupled with Eddy Dissipation Concept (EDC) turbulent models.

---

[1] http://www.tceq.state.tx.us/assets/public/implementation/air/sip/hgb/hgb_sip_2009/09017SIP_Ch6_ado.pdf

[2] TCEQ 2010 Flare Study Project Final Report; PGA No. 582-8-862-45-FY09-04

Overview of P.E. Licensing Process in United States

William R. Parrish, P.E. and Cory D. Jensen, P.E.

Professional engineering licensure was established in the United States to protect the health, safety and welfare of the public from unscrupulous or incompetent decisions of engineers. Licensure provides a proven and accepted assessment of experience, education and ethical standards.

This presentation gives an overview of the benefits of being licensed, the key players involved in becoming licensed and the path to licensure. The overview presents the background information to help put the other session presentations in perspective. More details about licensure are available in the March issue of *Chemical Engineering Progress (CEP).*

Value of being licensed

Being licensed shows other that you have a strong commitment to high technical and ethical standards. It can provide you with better career opportunities and provide a competitive edge when seeking employment. Many employers state that they would choose a licensed engineer over an unlicensed engineer, given both are otherwise equally qualified. Also, companies often offer incentives to encourage their employees to become licensed.

Some engineers think that licensure is important only in consulting or expert testimony. However, licensure is often required in firms doing design, construction and startup of processes. State laws also may include those teaching "advanced" engineering subjects be licensed.

Some states allow an "industrial exemption" for large companies who have the financial resources to cover liabilities arising from employee decisions. Most companies encourage licensure but many oppose eliminating the industrial exemption if they have operations in multiple states because it requires their engineers to be licensed in multiple states. There is a push to eliminate the industrial exemption. This important topic is discussed in another presentation.

No one can predict the exact future course of their career. However, being licensed may provide unexpected rewards and opportunities.

Key players in licensure

Becoming a P.E. involves education, experience and examination. Besides your coworkers, who can help guide your job experience, there are two important players in becoming licensed. The first is your local engineering board that establishes the standards for education and experience and ability to take the examinations. You will work with the board throughout the licensure process.

The second important player is the National Council of Examiners for Engineering and Surveying (NCEES) (www.ncees.org) . This is a nonprofit organization whose members are the various engineering licensing boards. The NCEES was created to provide some uniformity of standards of education, experience and examination. The membership establishes Model Law to offer uniformity. However, Model Law is a set of suggested best practices and no board follows Model Law exactly. All exams are administered and graded by the NCEES.

The licensure process

Obtaining a P.E. involves the following steps:

1. Having the appropriate formal education
2. Passing the Fundamentals of Engineering (FE) exam
3. Obtaining the appropriate experience (engineering internship)
4. Providing required documentation to board
5. Passing the Chemical Professional Engineering (PE) exam

While the general requirements are uniform across all jurisdictions, exact details vary. It is critical to work with your local board so that you understand and meet the exact requirements. A brief synopsis of each step is given below. More details are provided in the *CEP* article and in other presentations in this session.

1. Formal education: All jurisdictions require BS engineering degree, or equivalent, from an EAC/ABET-accredited four-year engineering program. Foreign degrees and other degrees are considered but must be approved by the board.
   Current Model Law states that as of 2020, engineers taking the PE exam must have a Masters or equivalent (MOE). This requirement is strongly opposed by AIChE and many other organizations. More details are presented in another presentation in this session and in the *CEP* article.

2. FE Exam: This exam tests your fundamental knowledge of engineering. It is easiest to pass when you are first eligible to take it (usually senior year or shortly after graduation). Another presentation and the *CEP* article provide more details of the exam and the upcoming changes in the exam.

3. Practical experience: Your work experience must be appropriate engineering practice. Typically four years are required although some boards permit fewer. In some jurisdictions advanced degrees can reduce the number of years required.

4. Documentation: The exact format of documentation and what is required varies between jurisdictions so it critical to work with your local board on this. However, all will require proof of education, passing the FE exam and work experience. The work experience will be supported by several references, who will also vouch for your ethical behavior.

5. PE Exam: This is the final step in becoming a licensed professional engineer. The exam is designed to be passed by a "minimally competent engineer with four years of practical experience". This exam tests the practical application of your engineering knowledge. Both a session presentation and the *CEP* article provide more details on the exam.

Summary

The AIChE recommends that its members become licensed to help ensure a competent and ethical chemical engineering workforce. While the process may seem daunting at first, working with your fellow employees and your local board will make the effort much easier. Each year about 400 chemical engineers become P.E's. They have made a significant investment in themselves and possibly in their future career.

(78b) Licensure--An Academic Perspective

S. Ranil Wickramasinghe, Department of Chemical Engineering, University of Arkansas, Fayetteville, AR

Most Chemical Engineering faculty members are not licensed. Why? Is it important for faculty to be licensed? Most undergraduates take the Fundamentals of Engineering exam but do not take the Professional Engineering exam. Why? Should faculty be licensed? Is licensure important for faculty who are conducting research? Should we encourage faculty to obtain their PE license? In this presentation these questions will be discussed.

TESTING FOR COMPETENCY - THE PROFESSIONAL ENGINEERING EXAMINATION

Anne Bertelsmann, PhD, PE, Galveston Bay Refinery, Marathon Petroleum Company LP, Texas City, TX
And
Denise Chastain-Knight, PE, CFSE, CH2M HILL, Augusta, GA

Introduction

The decision to pursue engineering licensure is a personal one, and the road includes many hurdles to overcome. Engineers take this road for a variety of reasons, with different motivations and experiences along the way, and finish the trek with the PE Exam. The Principals and Practice of Engineering Examination (PE Exam) is a nationwide exam used to ensure that licensed PE's are "minimally competent". The exam must include test content applicable to the competency being measured, effectively differentiate between competent and not competent candidates, and provide a consistent competency measure from administration to administration. This paper will define minimal competency, discuss the purpose of the Professional Engineering Exam, exam format using the Chemical Engineering Exam as the example, development methodology, item quality metrics and offer suggestions for test taking preparation. It will also describe the function of the NCEES Exam Development Committee and their role in developing a fair, effective exam, and explain why it takes weeks to get a result for a computer scored exam.

Path to Licensure

Whether an engineer undertakes the path to licensure early or late in their career, each must successfully negotiate the same hurdles. Each step in the process has a specific goal to either develop or to measure the engineer's capabilities toward professional competency. Figure 1 on the next page illustrates the path to licensure.

The quest to licensure begins with obtaining a Chemical Engineering degree. The next step is to pass an exam that measures competency on engineering basics. This first exam is the Fundamentals of Engineering Exam (FE) – also sometimes called the EIT (Engineer in Training) Exam. The FE exam is typically taken either during the senior year in college or shortly after graduation. The types of items resemble "textbook" questions and are designed to test understanding of the engineering fundamentals as taught in a typical engineering degree program. Currently, the FE examination consists of a general morning session (common to all disciplines) and a choice of seven discipline specific afternoon modules. Starting in the beginning of 2014, the FE examination will convert to a computer based test, which is discipline specific (i.e. no general module). For more information on the FE exam content and the conversion to computer based testing, refer to the NCEES website (http://ncees.org).

Figure 1: Path to PE Licensure

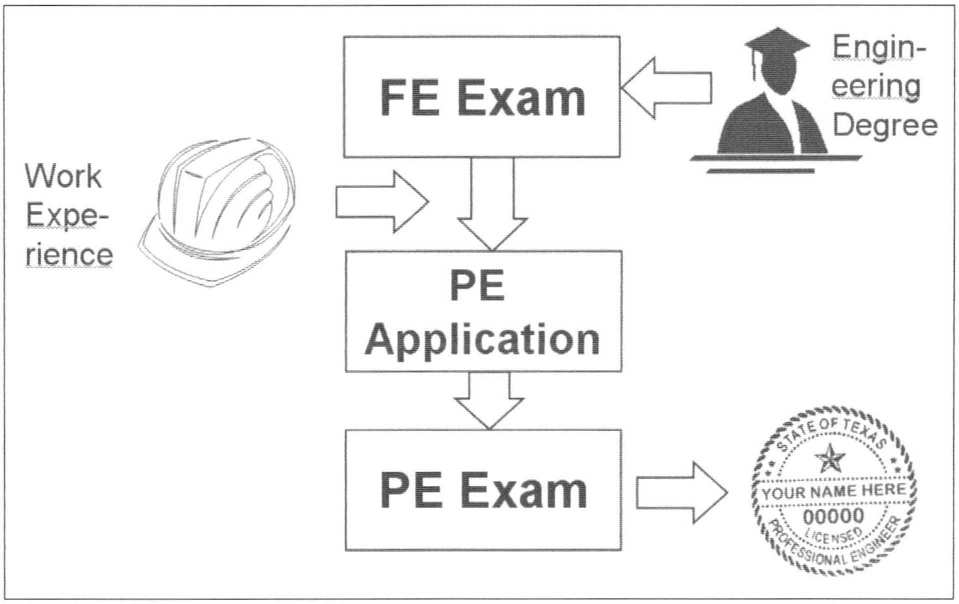

Following graduation, engineers enter the practice phase of their careers where fundamental knowledge is applied to real world application. In this phase of their careers, engineers use the theoretical principals learned from their education to solve problems that ultimately deliver physical products to market. Most states require a minimum of 4 years engineering related work experience before allowing candidates to sit for the PE exam. In order to become licensed, the candidate needs to submit an application to the state licensing board, describing education and experience, documenting passing of FE exam, and providing character and skills performance references. Format of the application and requirements for references, transcripts etc. vary from state to state. Once the application is accepted by the board, the candidate can register for the PE exam. After passing the PE examination, the engineer is licensed and can obtain the PE stamp. It is through this process that the state licensing boards ensure that only qualified candidates become licensed.

PE Exam Purpose

The PE exam is the final step in the road to licensure and its purpose is to ensure that only competent engineers become licensed. This helps to uphold the high standards of licensure and protects the health and welfare of the public. Since the exam is the same nationwide, it ensures consistency between states. Thus, one of the main requirements for a PE examination is that it can distinguish between competence and incompetence and that it can demonstrate this reliably and consistently. It is important to remember that "competent" should not be equated with "expert". The purpose of licensure is to identify individuals who have at least the minimum knowledge at which a person can practice professional engineering in such a manner that will safeguard life, health and property and promote the public welfare. That is why the term "minimally competent" is used.

A minimally competent engineer:

- should be able to perform identified tasks
- should meet minimum education and experience requirements
- is beginning to specialize in a sub-discipline
- may safely practice in another sub-discipline even though he/she does not practice in this area on a daily basis

The definition of "minimally competent" has implications on how the questions on the exam are written. The PE exam differs from the FE exam in that it addresses both the principles and the practice of engineering. So, while the FE questions are typically textbook related, questions on the PE exam are predominantly practice related. The questions build upon the experience that an engineer has gained during his or her first few years of working and need to recognize that chemical engineering experience can be gained in a variety of areas. The difficulty level of the questions is geared towards an engineer with about four years of experience.

PE Exam Content

The content of the examination is determined through a survey of practicing chemical engineers (PAKS study = Professional Activities, Knowledge and Skills), which is repeated about every 6-8 years to ensure that the examination stays current. The PAKS process includes conducting a technical content survey with input from practicing professionals. This survey is conducted over a period of weeks to ensure that the respondent pool size is adequate and is sufficiently diverse to be representative of the overall profession. Survey results are analyzed by a team of licensed Chemical Engineers, most with no prior exposure to PE exam development work, under the direction of a psychometrician who ensures that proper procedures are followed. The PAKS team will determine both the domain (subject area) content and item weight of the subdomains. PE exam content is established based on the PAKS study, and is published in the format of the PE exam specification available from the NCEES website (http://ncees.org/exams/pe-exam/) and is summarized in Table 1 on the next page.

The ChE exam specification was recently updated and became effective for the April 2011 exam administration. It includes six major domains and fifteen sub-domains that investigate theoretical and practical knowledge. Exam content is distributed within the major and subdomains according to the percentages shown in Table 1 to ensure a consistent content basis for every exam. The specification also includes guidelines for topical content within each sub-domain. For example the exam will include 4% of the items from sub-domain VI. C. Operation. Figure 2 includes example content from the 2011 exam specification.

Table 1: April 2011 ChE PE Exam Specification Summary

Knowledge Area	**Approximate Percentage of Examination**
I. Mass/Energy Balances and Thermodynamics	**23%**
A. Mass Balances	10%
B. Energy Balances and Thermodynamics	13%
II. Heat Transfer	**16%**
A. Mechanisms	9%
B. Applications	7%
III. Kinetics	**11%**
A. Reaction Parameters	6%
B. Reactors	5%
IV. Fluids	**16%**
A. Mechanical-Energy Balance	12%
B. Flow and Pressure Measurement Techniques	4%
V. Mass Transfer	**14%**
A. Phase Equilibria	5%
B. Continuous Vapor-Liquid Contactors	8%
C. Miscellaneous Mass Transfer Processes	1%
VI. Plant Design and Operation	**20%**
A. Economic Considerations	1%
B. Design	10%
C. Operation	4%
D. Safety, Health, and Environment	5%

Figure 2: Sample Exam Specification Sub-domain Content

C. Operation 4%

1. Process and equipment reliability (e.g., testing, preventive maintenance, startup/shutdown procedures, robustness)
2. Process improvement and troubleshooting (e.g., debottlenecking, experimental design and evaluation, optimization)

PE Exam Format:

The PE examination is given twice a year - once in April and once in October. It is an eight hour exam consisting of a four hour morning session and a four hour afternoon session with a lunch break in between. There are a total of 80 multiple choice questions (40 in the morning, 40 in the afternoon). Each question has four answer choices of which one and only one is the correct answer. Each question is stand-alone, i.e. it is independent of the other questions on the exam and can be solved with the information given in the problem, sometimes with the help of standard references. Some questions involve calculations while others are qualitative. All questions count equally and there is no penalty for guessing.

Each of the 80 multiple choice questions on the exam is of the same general format. In exam development, questions are referred to as "items" rather than questions because not every item has a question mark (like the one in the example below). Figure 3 illustrates the anatomy of an item, i.e. its general format and identifies the different parts of the item.

Figure 3: Anatomy of an Item

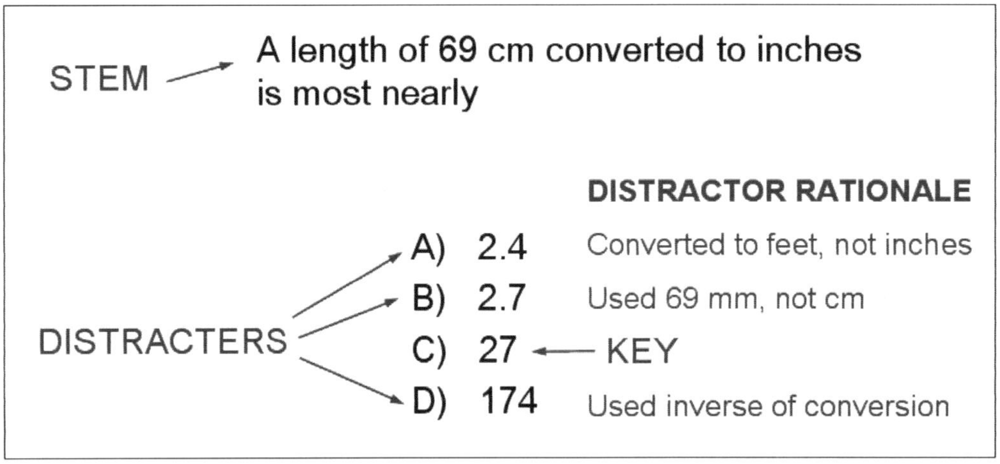

Each item consists of a stem and four answer choices. The stem (or problem statement) identifies what calculations are needed. It includes the information that is provided to solve the problem (sometimes even in table format) and states what is required for an answer. Beneath the stem are the four answer choices. Of these four answer choices, one (and only one) is the correct answer – the key. The other three are so-called distracters. They are incorrect answers – but they are not random numbers.

Each of the wrong choices represents a typical error that a less than minimally competent engineer (or someone not familiar with this specific topic) may make. Each distracter has to have a "rationale", i.e. an explanation which error was made. So, when a candidate is taking the exam and calculating a number that is exactly one of the given answer choices it does not necessarily mean that the problem was solved correctly. Maybe the candidate just made one of the "expected likely errors"...

The distracters can have a significant impact on the statistics of an item. If one of the distracters is obviously wrong (such as a negative absolute pressure), then even less than competent candidates will disregard it. That reduces the feasible choices, improves the "guessing" performance, and increases the chance that a not minimally competent candidate gets the question right. The Exam Development Committee strives to develop items that include the key, and three equally likely errors as distracters.

Exam Development

The PE Examination (as well as the FE, and the Surveying Exams) is a service that NCEES provides to the licensing boards. As such, NCEES provides the framework and the oversight for the exams, but the responsibility for the content rests with each exam committee. The committee writes the items, assembles the exams and reviews the statistics. NCEES has established rules and procedures that the exam committees follow to ensure consistency. Committees report back to NCEES with their exam quality measures. NCEES also provides the logistical support (psychometric consultant, publishing and printing, administrative support, maintaining records and the item bank, administering the exam, scoring, etc) and provides all security functions. For the larger exams, NCEES also pays and arranges for volunteer travel. For the smaller exams, this is the responsibility of the sponsoring society.

The Chemical Exam Committee is the group that actually writes and reviews the Chemical PE Exam. To be on the committee, one must be a licensed PE and express interest through the NCEES website or through a committee member. The committee has about 50 active members and meets twice a year in Clemson, South Carolina at the NCEES headquarters. Not everyone comes to every meeting, of course, so typically 30 – 35 committee members present at any given meeting. During the meeting items are written and reviewed by committee members who work with one another to critique and improve the items. Task teams also work to assemble, review and pretest exams.

Before an exam is administered, each item will have no less than six documented formal reviews and likely have been assessed by more than 20 people. Committee members are asked to complete detail oriented and often challenging engineering work and can earn PDH (professional development hours) for this activity. Developing the PE exam helps committee members to keep their engineering skills current and to give back to the profession. Without the work of this committee, the PE exam and with it the licensing wouldn't happen.

Preparing an exam that effectively and consistently measures minimal competency is a challenging task that succeeds or fails based on the ability of the committee to predict item performance. Each committee relies on a structured process and employees a variety of quality tools and techniques to improve the opportunity for success. Figure 4 is a workflow of the exam development activities. Items are written by a PE and submitted for review. Many items are prepared by the committee members at each meeting, while others are submitted by PEs who do not attend the meetings. The individuals who write and review items are referred to as Subject Matter Experts (SME). The author and two reviewers will independently work the item and develop the rationale for each distracter. Through this process the item is often revised and edited to improve its presentation and quality. This step is considered one of the most critical activities of the committee because items are the building blocks of the exam and each must meet high quality standards. Each committee member is responsible for item quality and is granted authority to modify draft items as needed to meet quality objectives, or reject items that fail criteria and cannot be edited to usable items. Finally the item is given a final check by the committee chair (or authorized designee) to verify that documentation is complete, both reviewers agree to the item content, and all quality checks have been completed, and approves the item for acceptance into the item bank. The item bank is a repository for items that are ready to be placed on exams.

Figure 4: Exam Development Process

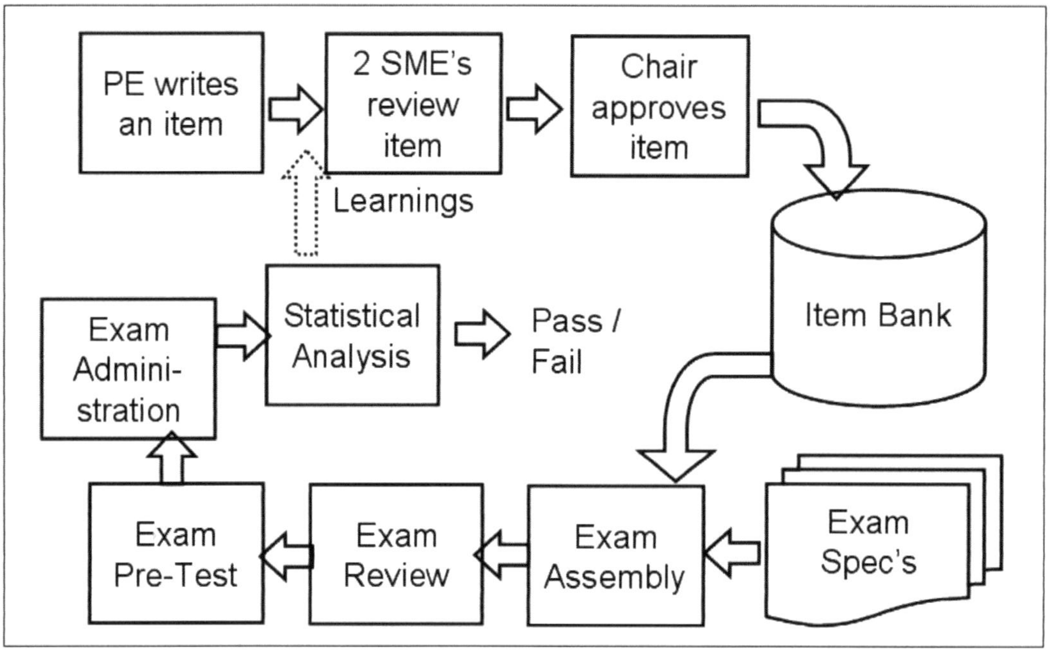

Exams are built, reviewed, revised, reviewed, quality tested and reviewed over a series of meetings. Each exam will have three independent team reviews, a pretest and a quality assessment. There are specific checks required at each workflow gate with task teams assigned to maximize the number of eyes that evaluate any given exam. An exam is compiled using items from the bank following the content distribution of the specification and sequenced in the specification order. The build team will then conduct a complete review of the exam, looking at each exam item for individual quality and checking for overall content flow and replacing items if necessary. The exam is then reviewed in two more meetings by independent teams who provide a "fresh set of eyes" and continue to improve the content to ensure that each item on the exam and the overall exam meet quality criteria. They also verify that the documentation is complete in accordance with the NCEES exam development requirements. Each exam must also have a pre-test where a group of committee members take the exam in as if they were actual test takers. After working all the items on the exam, they answer a series of questions about the items designed to predict item performance, then perform a distracter evaluation. Comments from the pre-testers are incorporated into the exam. In the final meeting before administration, a quality team conducts one last review of the exam verifying that all preceding checks (and associated documentation) are complete, all comments have been addressed, specification distribution is correct, and that author diversity limits and other NCEES criteria are met.

When the exam is administered and scored, data is compiled that can be used to evaluate the quality of the exam as a whole, and individual items. The data is analyzed and used to help the committee to learn what makes items perform effectively, and improve our performance prediction skills. Authors are provided a 'report card' so they know how items they wrote performed and have the opportunity to see how those items evolved in the review process. At each meeting, the committee will review items that had great statistics and try to find out what made them perform well. Similarly, we will review items that didn't perform well and identify traits that are undesirable in future items.

Item Statistics

Item quality is ultimately measured by the statistics generated from the candidate responses after an exam administration. Before discussing the statistical analysis of the examinee responses, one must first consider what makes an item a "good" item. The definition of a good item is one that is capable of distinguishing between "minimally competent" and not "minimally competent" candidates, does not discriminate against any particular group of test takers, and performs consistently in multiple exam administrations. To achieve those goals, it must be relevant, not confusing, and measure competency.

Most importantly, the content of the item must be relevant. Relevance is defined as the appropriateness of the topic and the difficulty of the item. The topic of the item must test knowledge or skills that are directly related to the exam specifications. For example: a question on high voltage power distribution may be relevant for an electrical engineer, but will not provide information on engineering competency of a *chemical* engineering candidate – no matter how well written it is. If the item does not relate to the exam specifications, it does not belong on the exam. Second, the difficulty level and complexity of the item must be such that a "minimally competent" engineer with 4 years experience can solve it in about 6 minutes. When writing items, authors often think about the interesting and challenging problems that are encountered in practice; however, if an engineer with 20 years experience requires several hours or days to solve the problem, what are the chances that a junior engineer will get the answer in 6 minutes? These type of problems, as interesting as they may be, do not make good exam items.

Another important principle to consider is that items should not be confusing to people who know the answer. When an item is confusing, a competent engineer who understands the concept being tested may not select the key, thus the question does not help to distinguish between competent and not competent. The so called "trick" question does not evaluate a candidate's competence and is considered a "bad" item to be weeded out in the exam development process.

It is equally undesirable to use items that give the answer away to people who don't know the answer. If the item includes clues to the key, then an incompetent engineer can choose the right answer without understanding the subject matter; thus the item does not measure competency. Clues can take many different forms:

- If some of the answer choices can be ruled out as impossible (like negative absolute pressures), the chance of guessing the right answer goes up.
- Grammar can give clues when one of the answer choices "stands out", i.e. is more detailed or does / does not match the grammatical structure of the item statement

For example, an item that asks about the 19th president of the United States, where three of the four answer choices are well known pop stars, does not require the candidate to know anything about US history to pick the correct answer.

Following an exam administration, a psychometric analysis is performed (psycho – metric = measuring the mind) on examinee responses. The analysis is solely based on information provided on the marked answer sheets without regard to the actual content of each item, but quite a bit of information about an item can be observed just from examinee responses. The psychometric analysis will evaluate a number of things. Of course, the simplest analysis is to determine how many candidates selected the key. Statistically, there is a 25% chance of guessing the correct answer because there are four choices. An item where less than 25% of the candidates selected the key may be flawed because it may have directed candidates towards preferentially selecting a distracter over the key. The next question is to ask *who* got the question right. Ideally, the candidates who got the question right were the same people who performed well on the exam overall. The analysis looks at the distribution of answers between the distracters and who selected them. This analysis also considers if all distracters were selected equally or if there was one distracter that was particularly attractive; maybe it even got selected more frequently than the key.

The objective of this analysis is to confirm the validity of each item as a tool capable of measuring engineering competence. An item that follows the basic three criteria (relevance, no confusion and no clues) can be expected to distinguish between competence and incompetence; however, sometimes clues, confusion or complexity can creep into an item unintentionally and be missed during the review before the exam is given. Statistics are then used to identify potential flaws. One such flaw is a mis-key. Mis-key is when the committee identified for example option A as the correct right answer, but option B is actually correct. Another flaw is an ambiguous item where more than one of the options could be correct. Despite the committee's efforts, such flawed items occasionally make it through the review process and are placed on an exam. The rigor of review process prior to exam administration minimizes the likelihood of a flawed item; however, it cannot totally eliminate them. The statistics after the exam will reflect item imperfections and provide an opportunity to compensate for item flaws in exam scoring so that candidates are not negatively impacted by flawed items.

Figure 5 is an example of the statistical analysis report for an item prepared by the psychometric consultant. One such sheet is prepared for each of the 80 items on an exam. The top left corner shows the "number of cases", i.e. the number of examinees who took the exam in this administration. In this example, there were 2910 candidates. So, this example is not from any Chemical PE exam where typically about 300 candidates sit per exam. The number below the case number (highlighted in yellow) is the Mean (or the P+ value). This number shows what percentage of the candidates selected the key; in this case, 65%. P+ alone indicates that the question is of appropriate difficulty for the average test taker; not too easy and not too hard. Statistically, the higher the P+, the easier the item.

Figure 5: Item Statistics

Number of Cases	2910
Mean (P+)	0.65
Delta	11.44
Pearson	0.33
Biserial	0.43
Mean of Passers	0.78
Mean of Failers	0.53
Mean Near Cut	0.68
Phi	0.26
DIF Grade	

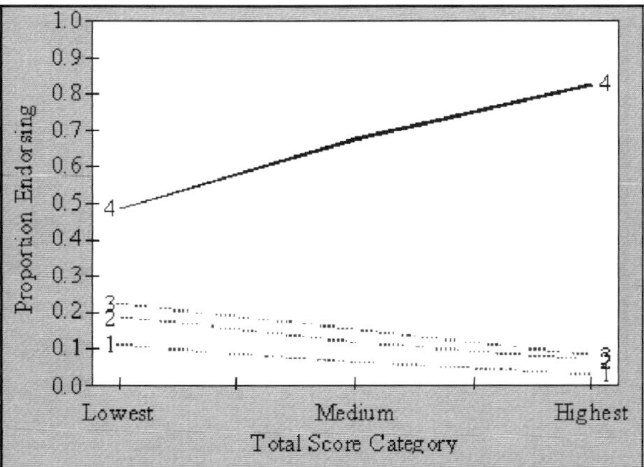

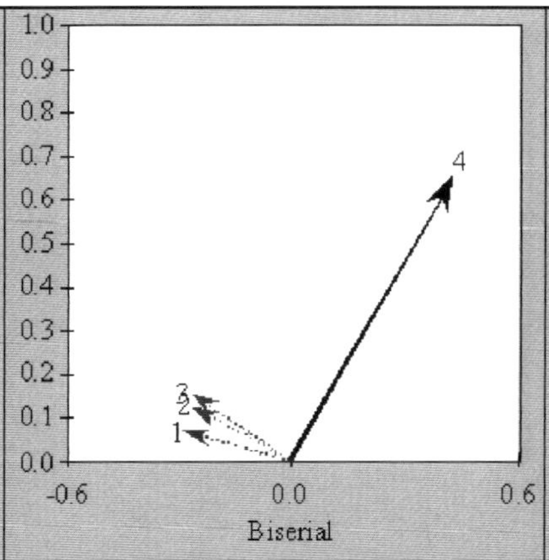

Alt.	Lowest	Low	Medium	High	Highest	%	N	Mean Scr	Biserial
1	0.11		0.06		0.03	6.7%	195	43.0	-0.29
2	0.19		0.12		0.07	12.5%	365	44.1	-0.27
3	0.22		0.15		0.08	15.4%	449	44.3	-0.27
4*	0.48		0.67		0.83	65.2%	1896	50.1	0.43
OM	0.00		0.00		0.00	0.2%	5	42.4	-0.20
Mn Scr	38.68		48.50		58.27				
Grp N	1025		1022		863				

The statistics not only evaluate how many candidates selected the key, but how selecting the key on this item correlates to overall exam scores. Test takers are divided into three groups of roughly equal size based on overall score on the exam; the low scoring third, the medium scoring third and the high scoring third. For each of these groups, the report in the table under the left graph shows how many selected which answer choice. For example on this question, "4" (or "D") was the key as indicated by the asterisk. Of the low scoring third, only 48% got the right answer, while 83% of the high performers selected the key. Thus, someone who performs well on the exam overall (which is an indication of competence) is significantly more likely to select the key than someone who scored low (may indicate lack of minimal competence). This spread shows that the question is capable of testing competency.

Figure 5 also illustrates that the incorrect answers were relatively evenly distributed between the distracters, indicating that none of the distracters appears to be overly attractive or easily ruled out. Another way of looking at the same data is presented on the right of the page. The table shows average overall scores for the people who selected each of the answer choices. In this case, the people who selected the key had an average overall score of 50, while the people who selected the distracters had overall scores in the range of 43 to 44. Five candidates with an average score of 42 did not answer the item. This clearly demonstrates that the people who selected the key were more likely to do well on the exam overall, and indicates that the item can effectively measure competency.

The data is evaluated in yet another way by the cumulative statistics in the lower left: Mean of Passers indicates that of those candidates who passed the exam, 78% selected the key. Mean of Failers shows that only 53% of the people who failed the exam selected the key. The larger the spread between these two numbers, the better the item is at measuring competency.

A handy way of summarizing the "who got the question right" analysis is the R-Biserial number. Basically, this number is related to the slope of the lines in the graph on the left. For a good item, this slope for the key should be positive and the larger the better because it expresses the difference in performance of the low and the high scorers. Theoretical range for the R-Biserial number is from -1 to +1, positive values indicate that the candidates who selected the key also scored higher on the overall exam. A negative R-Biserial number indicates that higher scoring candidates selected one of the distracters, which could be an indication of a mis-key or ambiguous item. The R-Biserial number is highlighted in yellow on the left and shown in the picture on the right. The R-Biserial number is important because it evaluates how well an item meets the three basic criteria. We previously observed that the sample item is of appropriate difficulty based on the percentage of correct answers (P+ value). The R-Biserial number shows that we did not confuse the competent candidates (a high percentage of high performers got it right) and apparently it did not give any clues to the less competent candidates (. It also appears to be relevant with regard to the topic since the high performers (those with better understanding of chemical engineering) did better on this than the low performers.

To further illustrate the concept of the R-Biserial number, consider an example of an item that might produce undesirable statistics. The item is asked on the final exam for a graduating class of chemical engineers: "Which bar in town has the most beers on tap?" One would expect a certain percentage of students to select the correct answer, but are those candidates the most knowledgeable students who spend a lot of time in the lab, or the "party crowd"? This question would likely have a negative R-Biserial number, indicating the overall least knowledgeable students are more likely to select the key, which means this item is not an effective measure of a graduating chemical engineer competency, nor is it relevant to chemical engineering knowledge and skills.

Determining Competency

The individual item statistics are the building blocks for determining the overall exam performance. To make the move from the individual item statistics to determining competency of the candidate, a passing score, also called cut score, needs to be established. Figure 6 on the next page illustrates the philosophy behind relating the score distribution to the cut point, or passing score. Total score is simply the number of items answered correctly. For the Chemical Engineering PE Exam, the total score has a theoretical range of 0 to 80. Most candidates have at least some knowledge and get lucky guessing on some questions, so scores on the very low end are uncommon.

If it were known upfront which candidates are competent and which ones are not, the scores of the two groups could be plotted as shown in Figure 6. The at least minimally competent engineers have higher scores, the not minimally competent engineers have lower scores. However, there will be some overlap between the two groups. The better individual items differentiate between competency and incompetency, the smaller the overlap. However, because there is a range of competency and because a person's performance can vary, there will always be some overlap and thus the cut point decision is not clear cut.

Figure 6: Score Distribution

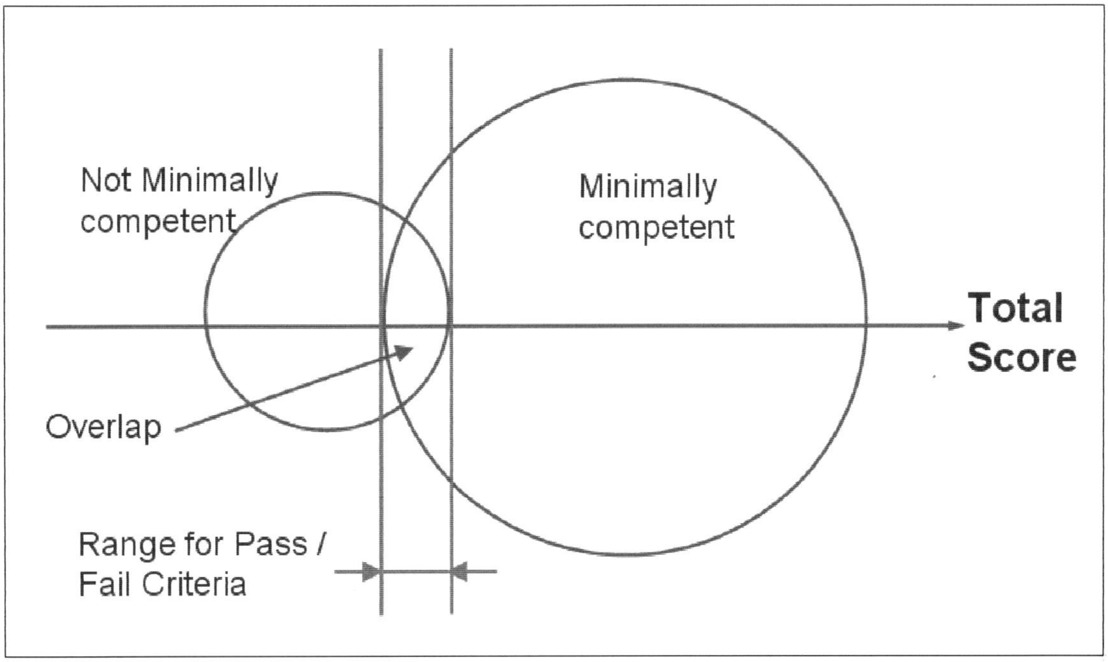

The overlap represents the range where the pass/fail decision should be made. All the at least minimally competent candidates on the right side should pass, all the not minimally competent candidates on the left should fail. With any decision point within that overlap range, there is a potential for competent engineers to fail and/or for less than minimally competent engineers to pass. Failing a competent engineer means that they will need to re-take the exam and includes the risk that they may give up on licensure. Passing a less than minimally competent engineer may allow an under-qualified person into public practice. Since the second type of error is more consequential, there is a tendency towards raising the bar when determining the pass/fail criteria.

Without knowing upfront which candidates are "minimally competent' and which ones are not (after all, it is the purpose of this exam to establish this), statistical measure are used to determine the pass/fail criteria, i.e. establish the "cut score". The cut score is the number of correct answers that a candidate needs to have in order to be considered "minimally competent". One way of determining the cut score is through panel of licensed PE's. This process is fairly involved, since it requires about 15-20 PE's who can volunteer 2-3 days of their time and travel to attend to cut score session. The panel must have good diversity (age, gender, area of practice, geographic location, demographics, etc…) and the cut score study must be completed before the results of the exam can be reported back to the state boards and ultimately to the candidates who are eagerly waiting for the scores. For these reasons, a cut score panel is usually only done when there was a change in exam specifications where the exam results cannot be directly compared to recent prior exams. The first exam where the format or specification is changed is called the "anchor exam" and cut point is established by a formal study.

For a cut score study the diverse panel of volunteers (all licensed PE's) meet and spend about a day to work the exam in a timed environment just like the actual administration. They can use any references, cannot consult with one another and they don't have access to the exam solution. The next day, they receive their score, and review and discuss each item as a group. They look at the statistics similar to those described above, but lacking the pass/fail metrics, review the solution, the distracters and discuss the topic and difficulty of each item. On the basis of that discussion, the panel members make a judgment call on how many "minimally competent" engineers out of 100 would answer this question correctly. A psychometrician guides the session and collects these estimates from each panel member. This data is then used in statistical analysis, generating a proposed passing score. The proposed passing score can then be adjusted (for example, to err on the "safe side"). This then becomes the basis for the pass/fail decision for the anchor exam.

Such a cut score study is not practical for every exam at every administration. So as long as there is no change in the exam specification, it is possible to "equate" the current exam to a previous exam. The equating process, heavily supported by a psychometrician, compares examinee performance on the current exam with examinee performance on previous exams. From this comparison it can be determined if the current exam is harder or easier than the previous exams, and cut point adjusted accordingly. If the data indicates that the exam is easier, the cut score is adjusted upward and a candidate will need to answer more questions correctly in order to pass. Conversely, candidates are not penalized if they take a harder exam because the cut point is revised downward and fewer questions must be answered correctly to pass. Equating is NOT the same as grading on a curve. When grading on a curve, the passing percentage would be fixed and the cut score adjusted to achieve that percentage. With equating, the competency of a candidate is measured on personal performance, and not dependent on the individual candidate's position within the test taking population, nor on the difficulty of the exam.

The psychometricians calculate other measures from the data. Coefficient Alpha measures the internal consistency or statistical homogeneity of the data, thus basically expresses the reliability of the exam. A measure called "Subkoviak" provides an estimate of the consistency of the pass/fail decision or an estimate of the probability that an individual would receive the same pass/fail decision on multiple administrations of the exam. All these measures are regularly reviewed to ensure that the exams can consistently measure competency.

After an exam is administered and all the answer sheets are scored, the statistical item analysis is performed. Exam Committees need to review potentially flawed items in case the key needs to be adjusted and the exam rescored. Then the cut point is determined, either through a cut score study with a panel or through equating. Only then can the pass / fail decision be made. State boards are notified with the results, perform state specific pass/fail assessments and then notify candidates of the results. NCEES recommends that only pass/fail should be reported to the candidates, though some states provide the actual score (or a "scaled percentage" based on a fixed passing percentage) to candidates. Candidates who fail get a diagnostics report so that they know how they performed in the various subject areas. NCEES recommends to only report the pass/fail decision for two reasons: The cut score can vary from exam to exam, so a candidate may pass one year and another candidate taking the exam another year may fail with the same score. The other reason is that the exam is designed to make a pass/fail decision on minimum competency. In that regard, the PE exam is different from an exam like the SAT, which is designed to measure "aptitude" over a wide range of abilities. The PE exam is not designed to measure ability accurately at the far ends of the scale. The PE exam does not distinguish between the "outstandingly competent" and the merely "extraordinarily competent engineer". Similarly, it cannot distinguish between the "totally incompetent" and the "absolutely clueless". Thus, a candidate who received a score of 75 out of 80 is not necessarily a better engineer than a candidate with a score of 72 out of 80 – both are "minimally competent" and pass.

The skeptical reader asks; "Does this really work?" Can an exam such as this really measure competency as the statistical "number magic" appears to indicate? The answer is found in the historical exam pass rates. NCEES does not report an overall pass rate for the exam, but breaks it up into a pass rate for first time takers and for repeat takers. Table 2 includes pass rates for exams given under the current Chemical PE Specification. The most recent pass rates for other exams can be found on the NCEES website.

One specific item to note about the data in Table 2 is that there is a gap between the pass rate for first time takers and for repeat takers. The pass rate for first time takers is typically rather high; in the 70 – 80% range for the Chemical PE exam. This high pass rate is because the examinee pool is already a pre-selected group. In order to be admitted to the exam, a candidate had to pass the FE exam, had to prepare an application with documentation of engineering experience, and had to be approved by the state board, a process that weeds out some of the less qualified potential candidates. Also, especially in chemical engineering where there are many jobs that come under the industrial exemption and do not require licensure, the people who decide to pursue licensure are already a non-random subset of all engineering graduates. This contributes to the relatively high pass rates for first time takers. The portion of the candidate population that passes on a repeat attempt includes in candidates that are minimally competent but may have been inadequately prepare in their first exam sitting.

Table 2: Chemical PE Exam Pass Rates

Chemical PE Exam Pass Rates	**April 2011**	**October 2011**	**April 2012**	**October 2012**
1st Time Pass Rate	78%	75%	74%	71%
Repeat Pass Rate	31%	29%	31%	20%

How to Prepare for the Exam

All the fascinating information presented in this paper aside, chances are the reader really wants to know how to prepare for the exam. Preparation philosophy will differ for a young candidate near the 4 year experience mark from that for a veteran engineer. One theory is that any minimally competent engineer can pass the examination without any preparation if he or she had unlimited time - time to look up references, read up on concepts and topics outside one's familiar area of practice, find the right equations and conversion factors, etc. But there is a time limit, and 6 minutes per question is not a lot of time.

Exam preparation should focus on building speed and efficiency for solving items. Through study and preparation, a candidate should become familiar with reference content, placing tabs to help find frequently used tables or compile a personal reference of frequently used information. Common equations and frequently used conversion factors should be memorized. Candidates should reactivate the habit of using math: solving equations, working with variables and using a scientific calculator instead of a computer. Speaking of calculators, it is strongly recommended to prepare for the exam using one of the approved models. Precious time can be wasted searching for the logarithm button on an unfamiliar calculator. Approved calculator models can be found on the NCEES website.

Exam readiness is achieved through lots and lots of practice. Preparation classes are useful for refreshing material basics, but are no substitute for rehearsal. A candidate should prepare by solving as many sample problems as one can find. By solving many, many problems, one will automatically remember the frequently used equations and conversion factors, enhance math skills and remember where to find information. The process also develops test day confidence because the items encountered on the exam have been encountered in similar form during exam preparation. Candidates should conduct one or more mock exam day(s) to practice pacing and practice maintaining focus for 8 hours. NCEES publishes a sample exam for each specification that has been developed by the exam committees in the same manner as a "real" exam. One author of this paper frequently recommends to potential exam candidates that they practice with readily available materials, complete a timed trial with the NCEES exam, then continue practicing with more materials.

Finally each candidate should prepare for the test environment and stage to be at peak performance on test day. Exams are administered in finite locations within a state and often require candidates to travel. Plan a travel itinerary that will allow arrival at the test site early, rested and unstressed. Have a satisfying but not heavy breakfast and make a plan for a relaxing lunch. Consider the test room conditions, dress comfortably and in layers. Be aware of the materials that are allowed in the test hall, and prepare accordingly. Candidates will be turned away without a valid government photo ID. Cell phones and other forbidden materials may be confiscated or admission refused. The exam is worked in a printed booklet, which contains sufficient space for working the questions and pencils are provided. Candidates need to bring an approved calculator, unapproved calculators will be rejected. The exam is open book, so candidates also need to bring any references that they want to use during the exam. References must be bound (no loose paper), but can be marked with flags. Review the test day policies on the NCEES website and plan accordingly (http://ncees.org/exams/exam-day-policies/).

Closing Comments

Professional Engineer Licensure is a personal decision and requires one to undertake a dedicated and challenging journey. Candidate should embark on this journey knowing that the exam development process is designed to ensure each exam measures for minimal competency in a consistent manner, and items are statistically validated. Finally, the factors influencing exam score processing timelines are understood. The authors of this paper encourage engineers to pursue licensure. We speak for ourselves and our fellow PEs when we say that nobody ever regrets the resources and effort spent obtaining a PE, and we appreciate the industry approval and instant expert acceptance granted by those two little letters following our names. P.E. Professional Engineer.

(78d) The Current Hot Button Issues in Domestic Professional Licensing

Amos Holt, Southwest Research Institute, San Antonio, TX

Amos Holt, a past president of ASME and current VP for Environmental, Safety and Quality Systems will review the issues currently receiving the greatest attention by the PE community which includes all of the various state and territorial licensing boards, the National Council for Examiners of Engineers and Surveyors as well as the numerous professional society stakeholders and educational organizations and societies. The discussuion will touch on the Model Law and its currect sections on the Masters or Equivalent (MOE) requirements as well as Industrial Exemptions. The various positions of key players covered including the position of a consortium of professional societies called "Licensing that Works" or LTW will be summarized. This group includes both ASME and AIChE and represents about 240,000 engineers thtroughout the world. Amos will tell you what you will likely be hearing more about in the next few years and how it might affect both current PE licensees and prospective PE licensees.

The California Divide – And Why You Need To Engage Your Legislators

Diane Spencer, PE, Lawrence Livermore National Laboratory, CA
and Emmett Miller, PE, consultant and AIChE Fellow, CA

Abstract

California is crippled by a great divide that has nothing to do with tectonic plates. California's professional engineering statute gives civil engineers free reign, allows electrical and mechanical engineers to work in their disciplines so long as that does not conflict with civil engineering, and acknowledges by title only that other engineering disciplines exist.

This division presents problems for engineers whether they have a professional engineering license or not. For example, Spill Prevention Control and Countermeasure (SPCC) plans are mandated by federal regulations. Such plans are prepared by chemical engineers all over the country, but NOT in California. SPCC plans are part of the exclusive domain of civil engineers.

A fire protection engineer was once told he could not design a fire suppression system because that was the exclusive domain of mechanical engineers. A chemical engineer was once told he could not submit a Risk Management Plan (RMP) (required by Cal EPA) without a civil engineer stamp because even the hazard analysis on a facility that uses chemicals is the civil engineer's domain. These are just a few examples where engineers are not allowed to work on projects that similarly trained engineers are expected to do in the rest of the country and the rest of the world.

So how do engineers in California navigate this regulatory nightmare? And how do we influence the legislators, engineering board, and others to nudge the statute in the right direction?

California's answer included forming a coalition of professional engineering societies that work together to present a united voice. The California Legislative Council of Professional Engineers (CLCPE) not only represents the interests of engineering professions to the legislature, but serves as a pro bono resource for the legislature on technical issues as well.

This paper will discuss: (1) our lessons learned from working with legislators; (2) California's professional engineering statute and why it is written as it is; and (3) why one needs to be engaged in the legislative process. Professional engineers are entrusted with protecting the health, safety, and welfare of the public. Getting engaged has the benefit of letting legislators understand the impact of their decisions and allows engineers to serve the communities where they live, and provide a safer, better community for everyone.

California's Problem

While most other states utilize some form of a generic registration model for engineers, California's model is discipline-specific, with civil engineering held on a pedestal. The definition of "civil engineering" in the Professional Engineers Act (California Business and Professions Code Sections 6700-6799.) is so broad that it encompasses nearly the entire spectrum of engineering. This overly broad definition, coupled with a mandate that only civil engineers can do civil engineering, means there is little room for those in other engineering disciplines to legally work in their own professions. The law allows overlap from civil engineering into all other disciplines, but other disciplines cannot overlap into civil engineering. So, not only are chemical engineers impacted, the other engineering disciplines are impacted as well.

Section 6731 states the following, with parts underlined for emphasis:

> "Civil engineering embraces the following studies or activities in connection with fixed works for irrigation, drainage, waterpower, water supply, flood control, inland waterways, harbors, municipal improvements, railroads, highways, tunnels, airports and airways, purification of water, sewerage, refuse disposal, foundations, grading, framed and homogeneous structures, buildings, or bridges:
>
> (a) The economics of, the use and design of, materials of construction and the determination of their physical qualities.
>
> (b) The supervision of the construction of engineering structures.
>
> (c) The investigation of the laws, phenomena and forces of nature.
>
> (d) Appraisals or valuations.
>
> (e) The preparation or submission of designs, plans and specifications and engineering reports.
>
> (f) Coordination of the work of professional, technical, or special consultants.
>
> (g) Creation, preparation, or modification of electronic or computerized data in the performance of the activities described in subdivisions (a) through (f)."

For many years the conventional wisdom has been that the term "fixed works," as used in the statute, (see Section 6731 above) somehow limits the scope of civil engineering, leaving room for other engineers to practice their craft. In order to clarify where that boundary lies, Senator Mimi Walters asked the California Legislative Counsel Bureau for a legal opinion (2011). The findings were shocking. "Fixed works" includes not only the structures, but also all the attachments to the structure. So, if a mechanical engineer designs a pumping system or a chemical engineer designs a fractionating column, and that hardware and the hardware is secured to a foundation, those engineers have crossed into the exclusive domain of the civil engineer. Other engineers are similarly impacted.

When a reasonable person reviews this list of studies or activities, it becomes apparent that the list precludes even scientists from legally studying the phenomena and forces of nature. And in most jobs, engineers are expected to produce engineering reports or to coordinate the work of other professionals.

How Did California Get Into This PE Mess?

The perverse nature of the current Professional Engineers Act in California is a result of good intentions gone awry.

In March 1928, a dam northeast of Castaic, CA, broke and flooded the valley below, wiping out the town of Santa Paula and flooding other towns downstream. Hundreds of people died and millions of dollars' worth of property was damaged.

The Professional Engineers Act was then created to ensure that only qualified people (civil engineer PEs) could work on dams and other fixed works. The legislators made the law as inclusive as possible to make sure there were no loopholes. Later, the Legislature expanded the Act to include Chemical, Electrical, Mechanical, Petroleum, Industrial and Metallurgical Engineering as Title Act Disciplines. With the passage of time, the Board for Professional Engineers, Land Surveyors, and Geologists (BPELSG, aka "Engineering Board") was empowered by the Legislature to create Title Disciplines on its own authority, which it did, eventually increasing the count to 15 title protected disciplines (Title Act disciplines), These are defined by Board Rules (16CCR 400-476) and have minimal stature in law. Along the line, Mechanical and Electrical Engineering were converted by statute to practice protection (Practice Act discipline)[1].

This patch job to add mechanical engineering and electrical engineering to the Professional Engineers Act has been ineffective (per Legislative Counsel's review). Inclusion of Title Act disciplines constitutes an even lower tier where those engineers are not able to take responsible charge of work within their field of expertise. In summary, the field is far from level.

The problem with the Professional Engineers Act has been known for many years. The difference now is the Legislative Counsel's legal opinion that was issued in July 2011 (www.CLCPE.org). According to that opinion, companies that rely on engineering services are on notice that the only engineering discipline allowed to conduct engineering work (not just supervise) in California is civil engineering. If a project requires a chemical engineer, a dual-licensed engineer will be needed to meet the competency requirements of both a chemical engineer and the legal requirements of a civil engineer. A dual license is needed for the other engineering disciplines as well.

---

[1] Four of the title disciplines (Corrosion, Manufacturing, Quality and Safety) were dropped ~1999 for lack of an NCEES exam.

The Industrial Exemption

Common to most states, California has an Industrial Exemption that allows companies engaged in manufacturing to use unlicensed engineers in their businesses— except for civil engineering work. The justification for the Industrial Exemption has been that such employers are capable of and competent in evaluating their own engineering talent, that they do not need interference from the state in doing so, and that failure that will impact the public is subject to strict liability.

There are exemptions for people who perform some specific and narrow civil engineering services. Architects performing architecture and building contractors following building codes are examples. But no general exemption exists for engineers who are not civil engineers if they perform "civil engineering." The Industrial Exemption even states that it does not apply to civil engineering.

With the recent legal opinion, however, all engineering practice in connection with facilities that are geographically fixed is now regulated as civil engineering, and the Industrial Exemption is therefore meaningless.

Engaging the Legislature

In the late 1940's, engineering societies in California organized the California Legislative Council for Professional Engineers (CLCPE) in order to work together as a resource to assist the California legislature, support laws that benefit the public and profession, and help keep society members informed about legislation that might affect them. Application of the Professional Engineers Act was not a problem to engineers in the state so long as the Engineering Board and regulators maintained a liberal interpretation. Likewise, CLCPE did not intensify its effort to change the law until the literal interpretation ultimately reinforced by the Legislative Counsel ruling came into focus.

CLCPE's efforts to change the law began in the 1980s. Our members helped draft a bill and found a senator to carry it. In California, the legislator who carries the bill is the "author" and organizations that support the bill, such as CLCPE, are "sponsors" regardless of who actually drafts the bill. Drafting a bill and finding "authors" is a process CLCPE repeated about a dozen times.

Let's reframe the process we used in California in terms that may help others take advantage of our lessons learned.

1) Coalition. We had a consortium of engineering societies working together as a coalition. This provided more credibility in that we were able to speak with one voice instead of individual organizations promoting separate proposals. A coalition also allows a sponsoring organization such as CLCPE to represent a larger portion of a legislator's constituency. AIChE local section leaders met with a senator to promote the one bill that chemical engineers sponsored without going through CLCPE. The senator listened intently about the impact of current law, asked a few clarifying questions, and then asked about how many chemical engineers were in the state. She was not satisfied with our answer, stood up and thanked us for our time.

Creating a coalition of several engineering societies enabled us to afford a lobbyist. AIChE and local sections are a 501(c)(3) nonprofit organization, which restricts how much money can be spent on lobbying. CLCPE however, was organized as a 501(c)(6) nonprofit, which allows for lobbying, but since the societies and CLCPE are all volunteer organizations, we do not have an intense political presence, and that restricts campaigning for or donating to candidates or political parties. Unfortunately, the opposition we face in California has major campaign funding and they openly support candidates that vote as the opposition wants them to vote.

Another advantage of working within a coalition of engineering societies goes beyond promoting legislation. The conversations we had when we met allowed us to gain insight into the problems each of the different disciplines were facing and we heard personal testimony about those various issues. This perspective has been invaluable.

2) *Hiring a lobbyist*. Over its lifetime, CLCPE has successively employed three lobbyists who helped with efforts to change the Professional Engineers Act, in addition to supporting the overall mission. A good lobbyist helps focus the discussion and facilitate consensus on language for a given bill. They also know how to navigate through the legislature, explain the bill or other proposals to legislators, legislative staff, and to others, and to find legislators who will author a bill. Typically engineers have full-time jobs, and working a bill is very time-consuming. The first of CLCPE's lobbyists was an engineer who later became Executive Officer of the Engineering Board. The other two were professional lobbyists with legal backgrounds.

3) *Language for the bill*. Earlier this paper discussed drafting language for a bill and then entering that bill into the legislative process (i.e., finding an author). "Entering" is the key word here. In the late 1990s, our bill attracted attention from the Professional and Technical Consultants Association (PATCA). In our quest to expand the law to allow engineers to practice their own profession, we needed to be careful not to adversely impact scientists (e.g., chemists) and other technical professionals. After several meetings with PATCA, we were able to work out compatible language. Going forward on subsequent bills, we deliberately engaged with the chemists (PATCA) early in the process.

4) *Bill's "author"*. Over the decades we have been pursuing change, our bills have been authored by different senators and assemblymen. In several cases, the legislator had been an engineer in his/her former career. One such senator, who had been an aeronautical engineer, had commented that it took multiple disciplines to design an aircraft. A legislator with a technical or science background readily grasped the issue we were dealing with. One of the first committees to hear a bill dealing with engineering (i.e., the bills we were promoting) was either the Business and

Professions Committee (B&PC) in the Senate or the B&PC in the Assembly. When seeking an author we often approached members of these committees as well. The intent was to get early support and move the bill through the legislative process.

5) *Finding support*. In addition to sponsors, it helps to find other organizations that will support a bill through (1) letters, (2) testimonies, and (3) lobbying efforts. We collected binders of letters from individuals, municipalities, regulators, and corporations. These included Chevron, the Chemical Industry Council of California, and Genentech. The letters provided a lot of anecdotes on how this perverse law affected people. The volume of the letters attracted the attention of legislators.

Once a bill is scheduled for a vote, testimonies at the hearing can be very compelling. The city of San Diego, California Fire Prevention Officers Association, a division of the California Fire Chief Association, along with various individual department fire chiefs and fire marshals, individuals impacted by current law, and others have testified at legislative hearings over the years. The Dean of Engineering at University of California—Davis testified on behalf of the University of California at the hearing in January 2012, stating that the state is educating engineers who cannot work in their own professions. The dean was shocked to find out that the legislators still would not vote to support changing the current law.[2]

The third level of support is when companies/ organizations involve their lobbyists. However, we found it very difficult to convince companies that had lobbyists in Sacramento to get engaged at this higher level.

6) *Other curve balls*. We experienced numerous unexpected challenges. One such curve ball happened in 1997, when chemical engineers tried independently (without CLCPE) to change the law. The Engineering Board required that we provide a Sunrise Package[3]. Essentially converting engineers from Title Act (defined in by BPELSG Board Rules) to Practice Act (defined in the Professional Engineers Act) is viewed as creating a new engineering discipline. We were given one week to prepare this. Fortunately, CLCPE had prepared something similar 10 to 15 years earlier, which we were able to mine and then create the required Sunrise Package. The Engineering Board still did not support our bill. Note: Keeping a good archive of material produced in supporting legislation could have unexpected value in the future. The key is being able to find the needed archived information when it is needed. Another advantage is that many of the same people remained engaged in CLCPE for long periods.

---

[2] Incidentally, the western region of ASCE sent their lobbyist to testify in opposition at the same hearing where the UCD dean testified.

[3] Sunrise is a process under which an occupation or profession wishing to receive state certification or licensure must propose the components of the legislation, along with cost and benefit estimates of the proposed regulation

What We Proposed and Why

Our approach to changing legislation evolved over the years. As mentioned earlier, chemical engineers made an attempt in 1997 to raise chemical engineering to a Practice Act (defined by statute). Our opposition protested that our attempt was part of a turf battle among the engineering disciplines. Legislators do not want to be drawn into controversies (e.g., infighting). Our next attempt involved CLCPE, with the intent to define all the Title Act disciplines as Practice Act disciplines. Because we had a coalition involving more than 15 engineering societies, we were able to combat this accusation. Still we did not succeed.

We also shifted our approach and proposed legislation that was closer in alignment with generic engineering. Specifically, we proposed that engineers should be allowed to do work based on their education, experience, or testing. In talking to legislators and to our lobbyist, we learned that we needed to approach any change in engineering statute with baby steps. Adopting generic licensure language (similar to other states or similar to Model Law proposed by the National Society of Professional Engineers (NSPE) was too much at one time. Because of the heavy influence by civil engineering companies in the state, we drafted bills that would continue all the advantages afforded to civil engineering. Unfortunately, our opposition would not negotiate.

In 2000, CLCPE had caught the attention of the legislature sufficient enough that they commissioned a study of the two-tier approach (Title Act/Practice Act) used in California. The study considered natural overlap between the professions, whether there was enough distinction between the engineering disciplines to support this unique regulatory structure, and whether the three Practice Act disciplines (civil, mechanical, and electrical engineering) posed more of a risk to the public (relative to the other engineering disciplines) and thus justified the existing two-tier structure. (See **www.CLCPE.org** for study by the Institute for Social Research.)

The study strongly supported changing the existing regulatory structure. The task of making this change fell to the Engineering Board. The bill it sponsored was opposed by American Consulting Engineers Council (ACEC) - California (a contractors' association) and Professional Engineers in California Government (PECG, a public employee union). Both were well-funded with strong ties to legislators — and the bill failed. [PECG is allowed to donate directly to candidates and its 13,000 members individually campaign for legislators. That has a big influence on those legislators who might support our position.]

The study marked a shift in how we approach changing legislation; CLCPE did not sponsor bills after the study. Instead, CLCPE and our individual engineering societies supported a succession of bills that were sponsored by the Engineering Board or by companies and organizations that use engineering services.

The disadvantage of this approach is that we had less influence on making amendments to the language in the bills. So while sponsoring our own bills had not been successful, this other approach allowed us to continue to pursue fixing the Professional Engineers Act.

Another challenge we had was finding companies and organizations that would commit their lobbyists. Many of these companies were very willing to prepare letters of support, but were not ready to involve their lobbyists.

End Results

This paper provides numerous lessons learned from decades of experience in trying to change California law to allow all engineers to practice in their own professions without finding a civil (typically) or other Practice Act engineer to take responsible charge of the work. Despite years of effort, we have not succeeded in changing the Professional Engineers Act. We are debating our next steps. It is our hope that the lessons we learned will be helpful to others who find the need to engage their state legislators.

(79a) Scaling Biofuels Technology

David Edwards, Sales & Marketing, Zeton Inc, Burlington, ON, Canada

The recommended approach for scaling bioenergy and biofuels technology follows a similar stage gate process to that used in traditional Chemical Process Industry (CPI) processes. But the processing of solids in bioenergy applications leads to some subtle differences, and unique challenges, that need to be considered.

(79b) Challenges in Guard Bed Design for a Biofuel Hydrotreater

Maoqi Feng, Div. of Chemistry and Chemical Engineering, Southwest Research Institute, San Antonio, TX

Bio-oils derived from fragmentation of biomass building blocks (cellulose, hemicellulose, and lignin) are unstable. Hydrotreatment of the oxygenated bio-oils to drop-in hydrocarbon fuels is necessary. Traditional hydrotreating catalysts, such as sulfided CoMo or NiMo supported on alumina have been reported in hydrodeoxygenation of biofuels and the process conditions are similar to those used in the refining of petroleum cuts. Besides the high oxygen content in bio-oils, bio-oils also differ from petroleum feedstock in sulfur, phosphorus and metal content. To prevent the hydrotreating catalyst fast deactivation and prolong the catalyst lifetime, a guard bed is necessary, and the guard bed design is very important. This presentation will address some challenging issues in the guard bed design for bio-oils hydrotreating.

(79c) Non-Geometric Scale-up to Meet Mixing Requirements

David S. Dickey, MixTech, Inc., Dayton, OH

Scale-up is a term most often used for mixing processes, because mixing is about both process results and fluid flow patterns. The interaction between process results and flow patterns can be a complicated relationship. Simple rules apply to some applications and are often combinations of factors related to both fluid motion and mixing intensity. As mixer geometry changes with different vessel sizes and impeller types, flow patterns change. As flow patterns change, so does mixing intensity and often process results.

Keeping important mixing parameters the same in different size vessels requires maintaining some similarities in flow patterns and mixing intensity to achieve comparable process results. As mixer and vessel geometry changes between the pilot scale and the production equipment, adjustments need to be made in the mixing behavior.

Geometric similarity provides a convenient way of achieving similar flow patterns, but large scale stirred tanks rarely have the same geometry as the pilot plant equipment. Finding ways of maintaining similar flow patterns and mixing intensity in not only different size equipment, but with different tank geometries and impeller types requires multiple adjustments to achieve desired results.

A systematic scale-up method combining geometric similarity and non-geometric adjustments provides a step-by-step process to observe and confirm changes in mixing intensity and flow patterns. No single method always works, but the chances of success increase when the effects of changes can be estimated and observed by basic mixing calculations. A multi-step method for non-geometric mixing scale-up will be presented.

(79d) Polycondensation Process System Design and Scale-up Challenges

Frank P. Lee, Xerox Research Centre of Canada, Mississauga, ON, Canada

Polycondensation reactions have many industrial applications and are commonly employed in syntheses of polymers such as polyesters. A typical Polycondensation process takes place at relatively high temperature and can involve exothermic and/or endothermic reactions. Small molecules are formed by the condensation reaction and are usually removed by distillation in order to drive the reaction to completion and to isolate the products. Once the small molecules are removed, the bulk of the product can become very viscous even at high temperature. The viscosity of molten raw materials and products can range from water-like to more than 10,000 Centipoises.

An institution actively engaged in development and scale up of specialty polymer can benefit from having versatile Bench Engineering and Pilot Plant process systems custom designed and built specifically for carrying out Polycondensation. A typical Polycondensation reaction system should incorporate the following components and features:

- A reactor which can handle a broad range of mixing (low to very high viscosity materials) and heat transfer (exotherm and endotherm) requirements, as well as having adequate numbers of nozzles sized for feeding bulky raw materials and discharging viscous products

- Auxiliary equipment including:

High-temperature, high-power circulation baths for the reactor and condensers
Refrigerated water or glycol supply system(s) for condensers and receivers

Atmospheric, or preferably, vacuum distillation equipment such as reflux and take-off condensers, packed columns, vacuum pump, splitter and phase separation equipment

Vessels and pumps for feeding and collecting raw materials, products and byproducts

- Process Monitoring, Control and Data Acquisition components and systems for monitoring and control of process parameters, including temperature, pressure, viscosity/torque/power, flow and agitator speeds

- Sampling and analytical instruments for end point determination

- Post reaction processing equipment such as chilled rolls or belts

In this presentation, examples of several custom Bench Engineering and Pilot Plant process system, ranging in size from less than 1 litre to over 1,000 litres, which were designed and built specifically for carrying out a broad range of Polycondensation reactions in a materials research centre are described. Challenges and lessons learned from designing, operating and scale-up of Polycondensation processes using these custom reaction systems should interest those who are involved with development and scale-up of polymer syntheses.

(79e) Pilot Plants in Transition

Mike J. Doll, HODer, Shell Global Solutions, Houston, TX

We, as chemical engineers, discuss pilot plant design, construction, operation, and management in great detail. However, we seldom, if ever, discuss the pilot plant in transition: that is --- we never discuss moving pilot plants from one location to another location.

This presentation discusses the reasons for moving a pilot plant, how to move a pilot plant safely and expeditiously, how to rebuild it and how to safely restart it. This presentation also discusses designing pilot plants with the view that they may be moved at some future date.

On the Sensitivity of "Overshoot" to Initial and Boundary Conditions and Engineering and Mathematical Perpsectives

ABSTRACT:

Output response of "temperature overshoot"" during transient non-Fourtier, heat conduction in a finite slab in response to a step change at the boundary is studied systematically. Why the overshoot appears in some systems and why it disappeares in others is explained thoroughly. A conclusion is drawn that the "overshoot pheonemena" is more a mathematical artifact rather than depiction of transient heat transfer events that can be expected in practice. The initial accumulation time condition was evaluated side by side for values of 0 initial accumulation, when initial accumulation is equal to the dimensionless group –Ve, Venrnote number respectively for a finite slab subject to the constant wall temperature boundary condition. For a material with large relaxation time the overshoot was found for the model results when zero initial accumulation time condition. For the same set of material and parameters when a physically reasonable time accumulation condition was used the overshoot disappears. However, the transient temperature was found to be subcritical damped oscillatory. Lumped analysis was used to study the average temperature in a finite slab subject to convective heating using; (i) Fourier parabolic model. The model solution was found to rise monotonically to a constant asymptotic value; (ii) Hyperbolic model with the first derivative of temperature with respect to time set to zero. Solution was obtained by the method of Laplace transforms. This model solution appears to have an overshoot; (iii) Hyperbolic model with the initial temperature at T_0 and the additional constraint that the average transient temperature should obey the energy balance equation from a engineering lumped analysis. Closed form model solutions were obtained after making the equation homogenous and illustrated in Figures. As shown in Figures 6.0-8.0, the model solution does not exhibit any overshoot when the boundary condition includes a finite wave term contribution.

A dimensionless group, *Sharma Number (heat)* is introduced. This appears to be an important parameter in determination of the average transient temperature in a finite slab during damped wave conduction and relaxation. Expression for the time taken to attain steady state was derived. The maxima in the transient temperature were found to increase with decreasting S* starting with large values such as 10. A cross-over was found after S* became less than about 2.2. Then the maxima in the average transient temperature was found to decrease with decrease in Sharma Number (heat), S*. Eq. (62) can be seen to go to zero in the asymptotic limits of infinite Sharma Number (heat), $S^* \rightarrow \infty$.

Keywords: Taitel Paradox; Overhoot Phenomena; Damped Wave Conduction and Relaxation; Wave Term in Boundary Condition; Intial and Final Time Conditions; Engineering Analysis

---

1. INTRODUCTION

There are reports in the literature [1,2] about occurrence of temperature overshoot upon use of non-Fourier heat conduction equation. The "overshoot phenomena" is studied herein a systematic manner. The Taitel paradox [3] is an anamoly seen in the model solution for transient temperature, in a finite slab, heated by isothermal-hot walls, btained by the method of separation of variables using the hyerpbolic PDE, partial differential equation. For certain values of the thermophysical properties of the finite slab the interior temperature of the slab was found to be greater than the wall temperature. This "overshoot" lead practioners to abandon the non-fourier heat conduction equation concluding that the equation violated second law of thermodynamics and hence cannot be used for engineering applications. A careful side by side study of this problem is presented below. The time condition $\frac{\partial T}{\partial t}$ was assumed to be 0 in Taitel [3].

Consider a finite slab of width 2a, at a initial temperature of T_0 heated by hot isothermal walls brought a temperature, $T_s < T_0$ for times $t > 0$, greater than 0. As can be seen from Figure 1.0 below only the *left hand limit* of the rate of change of average temperature of the slab with respect to time is zero. A lumped analysis on the heating process leads to the expression for the rate of change of average temperature of the interior of the slab as shown in the ordinate of Figure 1.0.

$$\frac{\partial \langle T \rangle}{\partial t} = \frac{\alpha}{a^2} e^{-\frac{\alpha t}{a^2}} \qquad (1)$$

The *right hand limit*, of the rate of change of average temperature in the finite slab is a maximum! With increase in time this decreases exponentially to a asymptotic zero at large times or upon attainment of steady state.

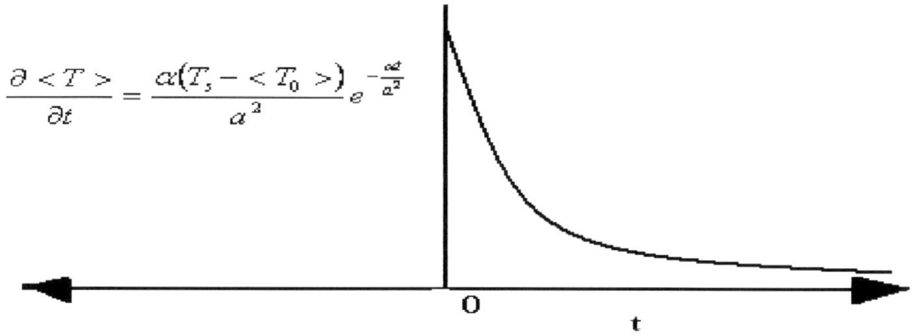

Figure 1.0 Rate of Change of Average Temperature in Interior of Finite Slab as a Function of Time, t

The right hand limit was used in the analysis. For a material with large relaxation time the overshoot was found for the model results when initial accumulation time condition is taken as 0. For the same set of material and parameters when a physically reasonable time accumulation condition was used the overshoot disappears. The transient temperature was subcritical damped oscillatory. A steady state temperature was attained after a said time. Lumped analysis was further explored.

The average temperature in a finite slab subject to convective heating was obtained using:

(i) Fourier parabolic model. The model solution rises monotonically to a constant asymptotic value.

(ii) Hyperbolic model with the first derivative of temperature with respect to time set to zero as u by the method of Laplace transforms. This model solution appears to have an overshoot;

(iii) Hyperbolic model with the initial temperature at T_0 and the additional constraint that the average transient temperature should obey the energy balance equation from a lumped analysis. The dimensionless temperature was expressed as a sum of a steady state temperature and a transient temperature. The transient temperature was expressed as a product of wave temperature and decaying exponential in time. The model solution was found to be;

$$< u >' = \left(\frac{S^*}{\omega\left(1+\beta S^*\right)} \right) e^{-\left(\frac{\tau}{2}(1+\beta S^*)\right)} \left(\sin(\omega\tau)\right)$$

(2)

Where, S* is the dimensionless Sharma Number (heat). As shown in Figure the model solution does not exhibit any overshoot. It appears that the damped wave conduction and relaxation equation can be applied to transient heat conduction problems without violation of the second law of thermodynamics. The Sharma Number (heat) appears to be an important parameter in determination of the average transient temperature in a finite slab during damped wave conduction and relaxation. The time taken to attain steady state was found to be ;

$$\tau_{ss} = \left(\frac{\pi}{\omega} \right)$$

(3)

The maxima in the transient temperature were found to increase with decreasing S* starting with large values such as 10. A cross-over was found after S* became less than about 2.2. Then the maxima in the average transient temperature was found to decrease with decrease in Sharma Number (heat), S*. The average transient temperature becomes zero in the infinite limit of S*→∞.

2. LITERATURE REVIEW

The damped wave conuction and relaxation equation was sought over Fourier's law of heat conduction for eight reasons by Sharma [4]. These reasons include:

(i) contradiction with the principle of microscopic reversibility by nobel laureate, Onsager, [5]);

(ii) neglects the time needed for the acceleration of heat flow by free electrons, Sharma [6];

(iii) occurrence of "blow-up" in the solutions obtained using Fourier parabolic model for industrially important systems such fluidized bed heat transfer [7], CPU overheating [8]), gel electrophoresis, [9], restriction mapping [10], adsorption [11], nuclear fuel rod [12] and drug delivery systems [13] ;

(iv) heat transfer in the Casimir limit [14] or nanoscale systems cannot be described using Fourier's model;

(v) basis for Fourier's law was from empirical observations at steady state;

(vi) oscillatory discharge of heat in good thermal conductors at low temperature as poinetd out by nobel laureate, Nernst [15];

(vii) ultrafast laser heating of metals [16], delayed ignition of solid propellant [17] cannot be explained using Fourier's model;

(viii) speed of heat cannot be greater than the speed of light as discussed by nobel laureate Landau and Lifshitz, [18].

The damped wave conduction and relaxation equation was originally suggested by Maxwell [19], and postulated independently by Cattaneo [20,21] and Vernotte [22]. The damped wave conduction and relaxation equation in one dimension across constant area may be written as follows;

$$q_x = -kA\frac{\partial T}{\partial x} - \tau_r A\frac{\partial q_x}{\partial t} \tag{4}$$

Where q_x is the heat transfer rate in x direction in (watts, w), A is the cross-sectional area across which the heat conduction occurs in (m^2), k is the thermal conductivity of the material in ($w.m^{-1}.K^{-1}$), τ_r is the relaxation time (s).

Reviews of the use of this equation have been presented by Joseph and Preziosi [23,24] and Ozisik and Tzou [25]. Extensive theoretical treatment of the equation have been reported by Tzou [26] and Sharma [27]. Experimental measurement of relaxation times has been reported by Mitra et. al. [28] recently for biological materials. Taitel [3] found a overshoot in his transient temperature solution for a finite slab subject to constant wall temperature boundary condition. Bai and Lavine [29] was concerned about Eq. (4) violating the second law of thermodynamics. Zanchini [30], Barletta and Zanchini [31], calculate a entropy production term and are concerned of a violation of Clausius' inequality. Al Nimir et. al. [32,33] discuss an "overshoot" and equilibrium entropy production. Haji Sheik et. al. [34] point out some anomalies in Eq. (4). Tzou [26] has found Eq. (3.68) to be admissible within the framework of second law of thermodynamics. Sharma [27,36-40] have presented closed form analytical solutions for different geometries that are within the bounds of second law of thermodynamics. He has derived the damped wave equation by accounting for acceleration of the molecules in the Stokes-Einstein formulation [41]. Antaki [42] has discussed some analytical solutions for convective boundary condition.

Why the "overshoot" manifests in such systems from the model development and implications on the physical insight to the phenomena are discussed in the following subsection. The use of physically realizable time condition such as obeyance of hyperbolic transient temperature to the energy balance equation from a lumped analysis approach can lead to the *disappearance* of the overshoot is shown . The time condition used in this analysis is *de novo*.

3. MATHEMATICAL MODEL

3.1.1 REVISIT OF TAITEL PARADOX

Taitel [3] considered a finite slab (Figure 1.0) with two boundaries of width 2a heated from both sides. Both the sides are maintained at a constant temperature Ts for times, t > 0. At initial time, t = 0, the temperature at all points in the slab is T_0. The governing equation is given by Eq. (5).

$$\frac{\partial u}{\partial \tau}+\frac{\partial^2 u}{\partial \tau^2}=\frac{\partial^2 u}{\partial X^2} \qquad (5)$$

Where u is given as $u=\left(\frac{T-T_s}{T_0-T_s}\right)$

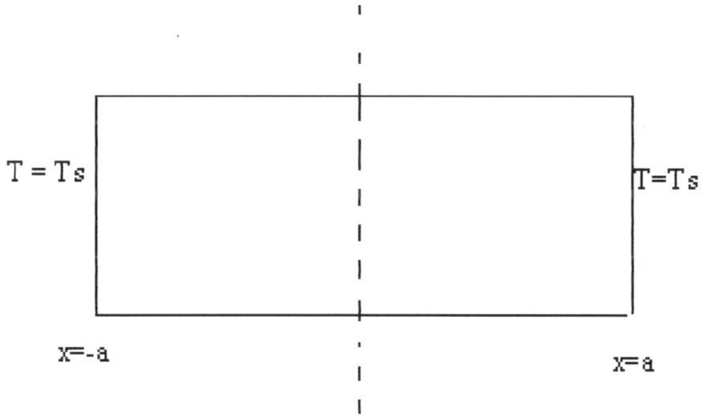

Figure 2.0 Finite Slab with Two Boundaries Maintained at Constant Temperature, Ts

The four conditions used by Taitel, two in space and two in time that are needed to completely describe a hyperbolic PDE that is second order with respect to space and with respect to time were;

$$t = 0,\ -a < x < +a,\ T = T_0, \text{ or } u = 1 \qquad (6)$$

$$t > 0,\ x = \pm a,\ T = T_s,\ u = 0 \qquad (7)$$

$$t = 0, \quad \frac{\partial u}{\partial \tau} = 0 \tag{8}$$

Taitel presented a solution for Eq. (5) for the conditions stated above obtained by analytical methods in a finite slab. The solution obtained by Taitel, for the centerline temperature of the finite slab is given below. They considered a constant wall temperature and the initial time conditions included a $\frac{\partial T}{\partial t} = 0$ term in addition to the initial temperature condition. They found that for certain values of the parameters used, i.e., relaxation time, width of the slab the transient temperature within the slab was found the exceed the wall temperature. Sharma [9] used a final condition in time and developed well bounded solutions . For materials with large relaxation times, the transient temperature exhibited subcritical damped oscillations. Multiplying throughout Eq. (2) by exp(nτ); Let $w = ue^{n\tau}$. Then,

$$\begin{aligned} \frac{\partial w}{\partial \tau} &= e^{n\tau}\frac{\partial u}{\partial \tau} + ne^{n\tau}u = nw + e^{n\tau}\frac{\partial u}{\partial \tau} \\ \frac{\partial^2 w}{\partial \tau^2} &= n\frac{\partial w}{\partial \tau} + ne^{n\tau}\frac{\partial u}{\partial \tau} + e^{n\tau}\frac{\partial^2 u}{\partial \tau^2} \end{aligned} \tag{9}$$

Combining Eq. (9) and Eq. (5),

$$\frac{\partial^2 w}{\partial X^2} = \frac{\partial w}{\partial \tau} - nw + \frac{\partial^2 w}{\partial \tau^2} - 2n\frac{\partial w}{\partial \tau} + n^2 w \tag{10}$$

For n = ½, Eq. (5) becomes;

$$\frac{\partial^2 w}{\partial X^2} = \frac{\partial^2 w}{\partial \tau^2} - \frac{w}{4}$$
(11)

w in Eq. (9) is the wave temperature. Eq. (9) can be solved by the method of separation of variables.

Let $u = V(\tau)g(X)$ (12)

Eq.(9) becomes,

$$\frac{g''(X)}{g(X))} = \frac{V''(\tau)}{V(\tau)} - \frac{1}{4} = -\lambda_n^2$$
(13)

$$g(X) = c_1 Sin(\lambda_n X) + c_2 Cos(\lambda_n X) \quad (14)$$

From the boundary conditions,

At $X = 0$, $\partial g/\partial X = 0$, So, $c_1 = 0$ (15)

$$\phi(X) = c_1 Cos(\lambda_n X) \quad (16)$$

$$0 = c_1 Cos(\lambda_n X_a) \quad (17)$$

$$\lambda_n = \left(\frac{2n-1}{2a}\right)\left(\frac{\pi\sqrt{\alpha\tau_r}}{2}\right)$$
(18)

n = 1, 2, 3.......

The time domain solution would be, for, $\lambda_n \leq 0.5$,

$$V = e^{-\frac{\tau}{2}}\left(c_3 e^{\tau\sqrt{0.25-\lambda_n^2}} + c_4 e^{-\tau\sqrt{0.25-\lambda_n^2}}\right)$$
(19)

and

$$\omega_n = \sqrt{\lambda_n^2 - 0.25}$$

The dimensionless temperature can be seen to be for materials with large relaxation times when Eq. (19) is applicable is given by an infinite series as follows;

$$u = \sum_{n=1}^{\infty} e^{-\frac{\tau}{2}} \cos(\lambda_n X)(c_n \cos(\omega_n \tau) + d_n \sin(\omega_n \tau)) \tag{20}$$

For the set of time conditions given in Taitel and by Eqs. (3,5) the cn and dn were solved for by use of principle of orthogonality and found to be;

$$c_n = \frac{4(-1)^{n+1}}{(2n-1)\pi}; d_n = \frac{2(-1)^{n-1}}{(2n-1)\pi} \tag{21}$$

The transient temperature u was computed as a function of time using MS Excel for windows 2007 spreadsheet in a desktop computer with 1.8 GHz microprocessor speed. 12 Terms in the infinite series in Eq. (20) were taken for a material with a relaxation time, $\tau r = 1.4$ s at dimensionless distance $X = 0.1$; half-width $a = 5$ mm; thermal diffusivity, $\alpha = 10^{-5}$ $m^2.s^{-1}$. The Vernote number corresponding to this simulation was;

$$Ve = \frac{\alpha \tau_r}{a^2} = 0.56 \tag{22}$$

The results of u vs. τ for the above set of parameters is shown in Figure 2.0. An overshoot can be seen.

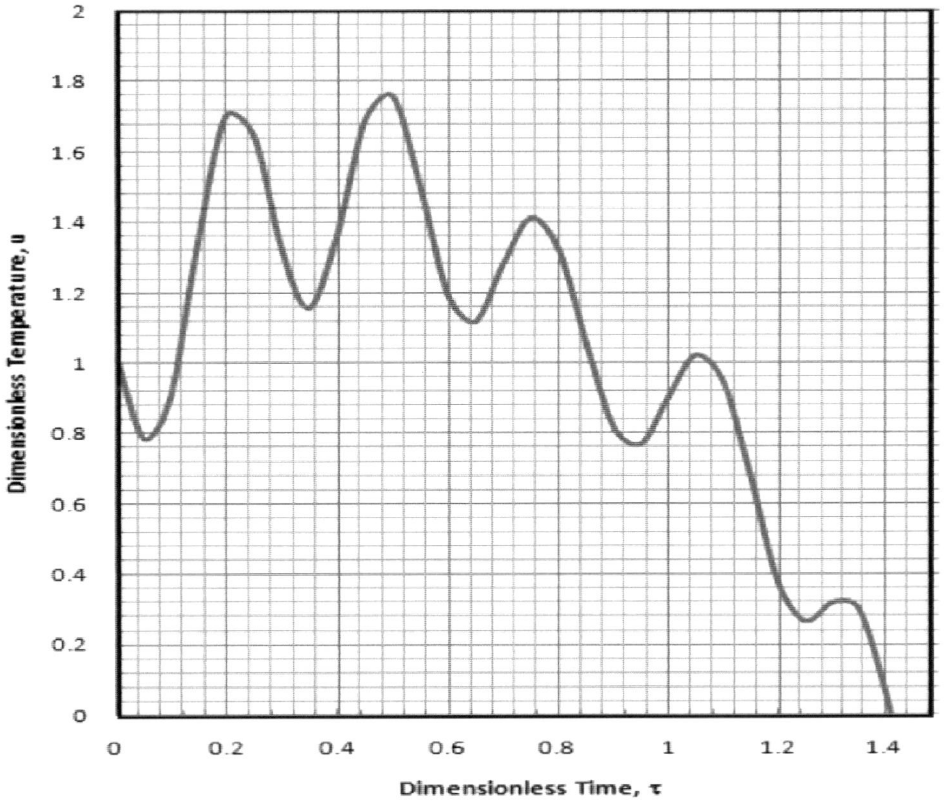

Figure 3.0 Transient Temperature inside the Finite Slab with Time Condition = 0 for Parameters: Location X = 0.1; Relaxation Time τ_r = 1.4s;Half-width of Slab, a = 5 mm; Thermal Diffusivity, $\alpha = 10^{-5}$ m2.s^{-1}

3.1.2 ACCUMULATION TIME CONDITION FROM LUMPED ANALYSIS AND ENERGY BALANCE

Examination of Figure 1.0 indicates that the value 0 for the initial accumulation of temperature in time is the "left hand limit". At t =0 the two sides of the slab abruptly brought to a higher temperature, $T_s > T_0$ where, T_0 is the initial slab temperature. The right hand limit may be more appropriate for purposes of obtaining exact closed form analytical solutions. Consider a lumped analysis as follows; Let the entire temperature of the slab other than the surfaces be lumped into an average temperature <T>. Let this be assumed to be at the center of the slab (centroid). Then in order to keep the surface temperature at T_s energy needs to be supplied from the surroundings. In turn the interior of the slab is heated by the hot surfaces. Consider the finite slab less the surface as the control volume. The energy balance can be written as follows;

$$\rho V C_p \left(\frac{\partial < T >}{\partial t} \right) = kA \left(\frac{T_s - < T >}{a} \right)$$

$$\frac{d < u >}{dt} = -Ve < u >$$

(23)

Where, Vernotte number, Ve is given by . $Ve = \left(\frac{\alpha \tau_r}{a^2} \right)$. This is assuming that Volume to Area ratio is given by the half-width a. So the "right hand limit" of the initial accumulation time condition closer to Eq. (23). For time, t =0, $< u > = 1$. Hence a more realistic time condition would be;

$$\frac{\partial u}{dt} = -Ve$$

(24)

Eq. (24) was used instead of Eq. (5) in the second time condition. The dimensionless temperature remains the same as given in Eq. (20) and is;

$$u = \sum_{n=1}^{\infty} e^{-\frac{\tau}{2}} \cos(\lambda_n X)(c_n \cos(\omega_n \tau) + d_n \sin(\omega_n \tau))$$

(25)

The constants cn and dn are recalculated for the time accumulation condition given by Eq. (23) and found to be;

$$c_n = \frac{4(-1)^{n+1}}{(2n-1)\pi}; d_n = (1 - 2Ve)\frac{2(-1)^{n-1}}{(2n-1)\pi}$$

(26)

The transient temperature u was computed as a function of time using MS Excel for windows 2007 spreadsheet in a desktop computer with 1.8 GHz microprocessor speed. 12 Terms in the infinite series in Eq. (26) were taken for the same material with a relaxation time, $\tau_r = 1.4$ s at dimensionless distance X = 0.1; half-width a = 5 mm; thermal diffusivity, α = 10-5 $m^2.s^{-1}$. The Vernote number corresponding to this simulation was;

$$Ve = \frac{\alpha \tau_r}{a^2} = 0.56$$

The results of u vs. τ for the above set of parameters is shown in Figure 3.0. An overshoot cannot be seen! The oscillations in temperature are subcritical. The transient temperature reaches a steady state value! after a said time.

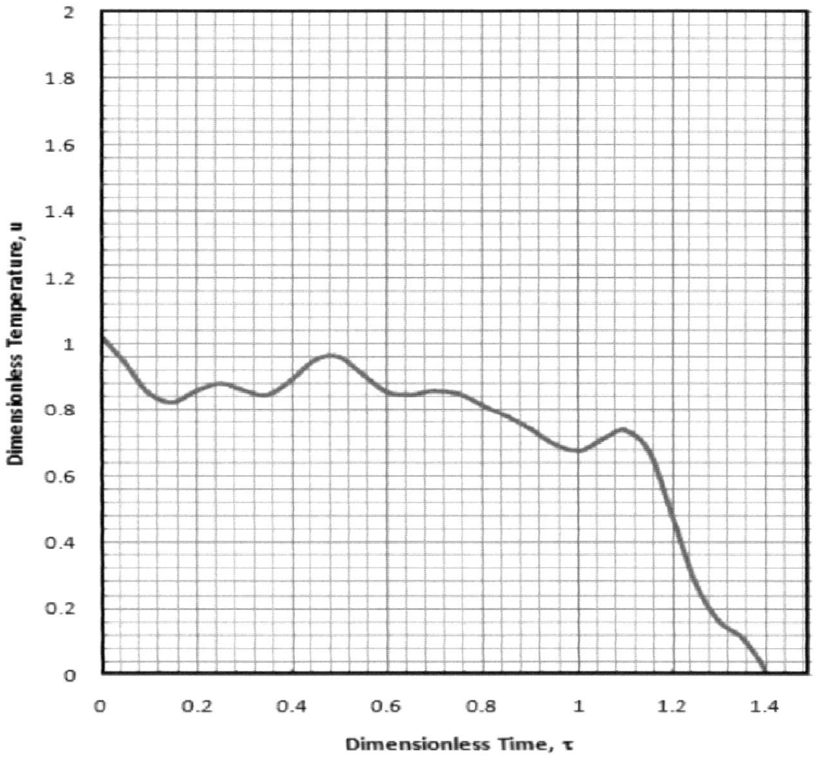

Figure 4.0 Transient Temperature inside the Finite Slab with Time Condition *Ve*for Parameters: Location X = 0.1; Relaxation Time τ_r = 1.4s;Half-width of Slab, a = 5 mm; Thermal Diffusivity, $\alpha = 10^{-5}$ $m^2.s^{-1}$

3.2 AVERAGE TEMPERATURE IN A FINITE SLAB

3.2.1 LUMPED/ENGINEERING ANALYSIS

Consider a finite slab with width 2a at a initial hot temperature T_h. This hot slab part of which is shown in Figure 5.0 is subject to a sudden cooling by convection by the fluid blown from the fan as shown in Figure 4 for times t > 0. The fluid temperature is at T_f where $T_f < T_h$. Both sides of the slab are cooled by the fluid. The transient temperature as a function of space and time need be obtained. The two time and two space conditions are given as follows;

$$t = 0,\ -a \le x \le +a,\ T = T_h \tag{27}$$

$$t > 0,\ x = 0,\ \frac{\partial T}{\partial x} = 0 \tag{28}$$

$$t > 0,\ x = \pm a,\ -k\frac{\partial T}{\partial x} = h(T - T_f) \tag{29}$$

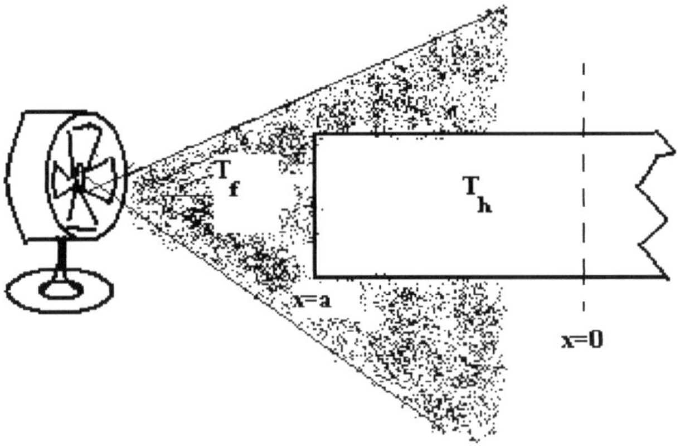

Figure 5.0 Finite Slab with Half-Width a Subject to Convective Heating

In one dimension the energy balance equation on a thin slice of thickness Δx can be written as;

$$-\frac{\partial q}{\partial x} = (\rho C_p)\frac{\partial T}{\partial t} \tag{30}$$

Combining Eq. (30) with the damped wave heat conduction and relaxation equation the governing equation can be written as;

$$\frac{\partial u}{\partial \tau} + \frac{\partial^2 u}{\partial \tau^2} = \frac{\partial^2 u}{\partial X^2} \tag{31}$$

where, $u = \frac{(T - T_h)}{(T_f - T_h)}; \tau = \frac{t}{\tau_r}; X = \frac{x}{\sqrt{\alpha \tau_r}}$ (32)

The time and space conditions are as follows;

$\tau = 0, \ \ u = 0$ (33)

$X = 0, \ \ \frac{\partial u}{\partial X} = 0$

$$X = \frac{a}{\sqrt{\alpha \tau_r}}$$
$$-h^*(u-1) = \frac{\partial u}{\partial X} \tag{34}$$

Where, $h^* = \left(\frac{h}{\sqrt{\frac{k\rho C_p}{\tau_r}}} \right)$

It can be noted that the heat transfer coefficient is defined with respect to the average temperature in the slab <T>. The energy balance on a thin spherical shell at x with thickness Δx is written. The governing equation can be obtained after eliminating q between the energy balance equation and the derivative with respect to x of the flux equation and introducing the dimensionless variables;

$$\frac{\partial u}{\partial \tau} + \frac{\partial^2 u}{\partial \tau^2} = \frac{\partial^2 u}{\partial X^2} \tag{35}$$

Eq. (35) is integrated between $-X_a$ and $+X_a$ and give;

$$\frac{\partial < u >}{\partial \tau} + \frac{\partial^2 < u >}{\partial \tau^2} = -S^*(< u > -1) \tag{36}$$

Where the Sharma Number (heat) S^* is defined as $S^* = \left(\frac{h}{Sa} \right) = \left(\frac{h\tau_r}{\rho C_p a} \right)$.

S is the storage number with units of $(w.m^{-3}.K^{-1})$ and is given by;

$$S = \left(\frac{\rho C_p}{\tau_r} \right)$$

The physical significane of storage number S is that is a material property such as thermal conductivity with units of $(w.m^{-1}.K^{-1})$ and system parameter such as convective heat transfer coefficient, h with units of $(w.m^{-2}.K^{-1})$. Storage number is a measure of energy stored in the material. The use of storage number is illustrated in another study on CPU overheating [8]. It is the ratio of the thermal mass of the material to its relaxation time. What is the material's thermal capacitance in relation to the speed of heat conduction is captured in this one number. Sharma Number (heat) is the dimensionless ratio of the convective heat transfer to the wave heat propagation heat transfer and storage in a finite region a.

3.2.2 FOURIER PARABOLIC MODEL NEAR STEADY STATE

Eq. (35) in the limit of zero relaxation time reverts to the Fourier parabolic model and the governing equation becomes;

$$\frac{\partial < u >}{\partial \tau} = -S^* < u > -1) \tag{37}$$

Obtaining the Laplace transforms of Eq. (37) ;

$$s < u > (s) - 0 + S^* < u > (s) = \frac{1}{s} \tag{38}$$

or,

$$< u > (s) = \frac{1}{s(s + S^*)} \tag{39}$$

Obtaining the inverse Laplace transform of Eq. (39);

$$< u >= \int_0^\tau e^{-\frac{hS}{a}p} dp = 1 - e^{-\frac{ht}{\rho C_p a}} \tag{40}$$

Eq. (40) is the average temperature of the finite slab in response to a step change in convective cooling at both sides of the finite slab. Eq. (40) is not a function of the relaxation time and is a function of the heat transfer coefficient, h, density, ρ, and heat capacity, C_p, and half-width of the finite slab, a.

3.2.3 **INITIAL ACCUMULATION CONDITION ZERO**

For the hyperbolic model other investigators such as Taitel [3] use the following initial condition;

$$\tau = 0, \quad \frac{\partial u}{\partial \tau} = \frac{\partial < u >}{\partial \tau} = 0 \tag{41}$$

Obtaining the Laplace transform of Eq. (37) realizing the initial condition from Eq. (41)) the Laplace domain expression for <u> (s) can be seen to be;

$$< u > (s) = \frac{1}{s\left(\frac{s^2}{S^*} + \frac{s}{S^*} + 1\right)} \tag{42}$$

Where, $S^* = \left(\frac{h}{Sa}\right)$

Eq. (42) can be compared with the prototypical second order systems with a step input [42,43] For underdamped systems, i.e., when the damping factor, $\zeta < 1$ the time response to Eq. (42) can be seen to exhibit a "overshoot". Without *any heat sources* the internal temperature greater than the boundary temperature can be a violation of second law of thermodynamics. When the damping coefficient,

$$\tau_p^2 = \frac{1}{S^*} \tag{43}$$

$$2\zeta \frac{1}{\sqrt{S^*}} = \frac{1}{S^*} \tag{44}$$

So, the damping factor is;

$$\zeta = \frac{1}{2\sqrt{S^*}}$$

For large values of the dimensionless ratio, S^*, the damping coefficient will become less than one and the output response appears to have a overshoot. This can be expected for materials with large relaxation time,

$$\tau_r > \left(\frac{\rho C_p a}{h}\right) \tag{45}$$

3.2.4 PHYSICALLY REALIZABLE STEADY STATE/FINAL TIME CONDITION

The boundary condition used during convective cooling is a constant value for the heat flux rate from the air flowing past the finite slab as given in Eq. (29). The governing equation (35) is the damped wave conduction and relaxation equation taking into account the effect of acceleration of electrons. A boundary condition that includes the wave term, i.e, $\tau_r^{air}\frac{\partial q}{\partial t}$ would be a better representation of transient events at the air solid interface. Thus. Eq. (3.141) need be modified as;

$$-\frac{\partial u}{\partial X} = h^*(<u>-1) + h^*\left(\frac{\tau_r^{air}}{\tau_{rs}}\right)\left(\frac{\partial <u>}{\partial \tau}\right) \tag{46}$$

Eq. (35) is integrated with respect to X and the governing equation written in terms of the average temperature in the slab <u> as follows;

$$\frac{\partial <u>}{\partial \tau}(1+\beta S^*) + \frac{\partial^2 <u>}{\partial \tau^2} + S^* <u> = S^* \tag{47}$$

where, $\beta = \left(\frac{\tau_r^{air}}{\tau_{rs}}\right)$ the ratio of the relaxation time in the air and the relaxation time in the solid. Let the <u> be expressed as a sum of transient temperature and steady state temperature.

$$<u> = <u>^{ss} + <u>^t \tag{48}$$

Substituting Eq. (48) into Eq. (47);

$$\frac{\partial <u>^t}{\partial \tau}(1+\beta S^*) + \frac{\partial^2 <u>^t}{\partial \tau^2} + S^* <u>^t + <u>^{ss}(1-S^*) = 0 \tag{49}$$

Eq. (49) is valid when the following two equations are valid;

$$(1+\beta S^*)\frac{\partial <u>^t}{\partial \tau} + \frac{\partial^2 <u>^t}{\partial \tau^2} + S^* <u>^t = 0 \tag{50}$$

$$+S^*(<u>^{ss} - 1) = 0 \tag{51}$$

The steady state temperature can be seen to be;

$$<u>^{ss} = 1 \tag{52}$$

Eq.(50) is a second order differential equation that has been made homogeneous. Eq. (50) is multiplied by $e^{n\tau}$ throughout. The terms group such that,

$$<W>^t = e^{+n\tau} <u>^t \tag{53}$$

and for $n = \left(\frac{1+\beta S^*}{2}\right)$ Eq. (50) can be seen to become;

$$\frac{\partial^2 <W>^t}{\partial \tau^2} = -<W>^t \left(\left|S^* - 1\right|^2\right) \tag{54}$$

The solution to Eq. (54) can be seen to be;

$$<W>^t = c_1 \cos(\omega\tau) + c_2 \sin(\omega\tau) \tag{55}$$

Where $\omega = \left|S^* - 1\right|$

Two time conditions are needed to fully describe the problem. The accumulation condition equal to zero appears to be physically unreasonable. When a slab with a initial temperature is immersed in a hot fluid with convective heat transfer coefficient the temperature in the slab would increase, rapidly in the beginning of the process and can be expected to slow down with the passage of time. At steady state the temperature would be steady and invariant with time. From the initial condition given by Eq. (32) the integration constants in Eq. (55) can be seen to be;

$$c_1 = 0 \tag{56}$$

Then Eq.(53) becomes;

$$(<u>^t = e^{-\frac{(1+\beta S^*)\tau}{2}} (c_2 \sin(\omega\tau)) \tag{57}$$

The transient average temperature $<u>^t$ also has to obey the following equation that can be obtained from a lumped analysis of the finite slab from energy balance;

$$\frac{\partial <u>^t}{\partial \tau}(1+\beta S^*) = S^*\left(1-<u>^t\right) \tag{58}$$

Eq. (58) at zero time, or τ=0, can be seen to be!

$$\frac{\partial <u>^t}{\partial \tau} = \left(\frac{S^*}{1+\beta S^*}\right) \tag{59}$$

The integration constant c_2 can be obtained by applying Eq. (3.166) to Eq. (3.167).

$$\frac{\partial <u>^t}{\partial \tau} = -\frac{c_2\left(1+\beta S^*\right)}{2} e^{-\left(\frac{\tau}{2}(1+\beta S^*)\right)} \sin(\omega\tau) + c_2 \omega e^{-\left(\frac{\tau}{2}\right)} \cos(\omega\tau) \tag{60}$$

and c_2 seen to be; $c_2 = \dfrac{S^*}{\omega\left(1+\beta S^*\right)}$ (61)

The average transient temperature in the slab is then written as;

$$<u>^t = \left(\frac{S^*}{\omega\left(1+\beta S^*\right)}\right) e^{-\left(\frac{\tau}{2}(1+\beta S*)\right)} \left(\sin(\omega\tau)\right) \tag{62}$$

Eq. (62) is plotted in Figures 6 – 8, for different values of S^* and β. The time taken for the slab to reach steady state is finite and can be calculated when ($\omega\tau_{ss}$) becomes π in Eq. (62). This happens when;

$$\tau_{ss} = \left(\frac{\pi}{\omega}\right) \tag{63}$$

For the special case when S* = 1, Eq. (54) becomes;

$$\frac{\partial^2 <W>^t}{\partial \tau^2} = 0$$

$$<W>^t = c_1\tau + c_2 \qquad (64)$$

Or,

$$<u>^t = e^{-\frac{(1+\beta)\tau}{2}}\left(c_2 + c_1\tau\right) \qquad (65)$$

From the initial condition given by Eq.(8), $c_2 = 0$. c_1 can be solved for the energy balance equation given by Eq. (3.166) at $\tau = 0$,

$$\frac{\partial <u>^t}{\partial \tau} = \left(\frac{1}{1+\beta}\right) = -\frac{c_1(1+\beta)\tau}{2} + c_1 \quad ; \qquad c_1 = \frac{1}{(1+\beta)}.$$

Thus for the special case when S*=1,

$$<u>^t = e^{-\frac{(1+\beta)\tau}{2}}\left(\frac{\tau}{(1+\beta)}\right)$$

4. CONCLUSIONS

Reports in the literature about appearance of an "overshoot" in the transient temperature for a finite slab were analyzed for the case of convective boundary condition with the wave term, . The solution from Fourier parabolic model predicts a monotonic rise in temperature with an asymptote of $\langle u \rangle^{ss} = 1$ at steady state. The solution from the hyperbolic damped wave conduction and relaxation for average temperature of the finite slab was solved for by method of Laplace transforms for the time conditions such as the first derivative of temperature in time equal to zero at time zero as used by other investigators in similar problems with constant wall temperature boundary condition. The Laplace domain expression compares from this model compares well with the expression for prototypical second order systems as discussed in a process dynamics and control textbooks. This expression when inverted has a overshoot.

The temperature in a finite slab subject to convective boundary condition was assumed to comprise of a transient component and steady state component. The steady state component temperature was obtained and given by Eq. (52). The governing equation for transient temperature is rendered homogeneous upon obtaining the solution of steady state temperature. The transient temperature was expressed as a product of decaying exponential and wave temperature. The governing equation for the wave temperature was solved for. The integration constants were obtained from the initial temperature condition of $\langle u \rangle^{t} = 0$ at zero time. The second integration constant was calculated by verifying that the model solution obeys the energy balance equation for average transient temperature. This leads to a solution for transient temperature that has a decaying exponential term and a cosinuous term with a phase lag, ϕ. Eq. (62) does not result in any overshoot as shown in Figures 3.0 – 5.0 for various values of frequency and ratio of thermal relaxation times between air and the solid. It was found that the dimensionless *Sharma Number* (heat) is given by $S^* = \left(\frac{h}{Sa}\right)$ is a significant parameter in the analysis. Expression for time taken to steady state was found and given by Eq. (63). No overshoot was found when the wave term was taken into account even for small values of β, the ratio of the relaxation time of air with that of the solid. This is an indication that the fundamental phenomena such as "acceleration of free electron" as discussed before [12] can be used to model transient heat conduction at short times and be within the second law of thermodynamics. The maxima in the transient temperature were found to *increase* with decreasing S* starting with large values such as 10. A cross-over was found after S* became less than about 2.2. Then the maxima in the average transient temperature was found to decrease with decrease in storage number, S*. Eq. (62) can be seen to go to zero in the asymptotic limits of infinite storage number, S* → ∞.

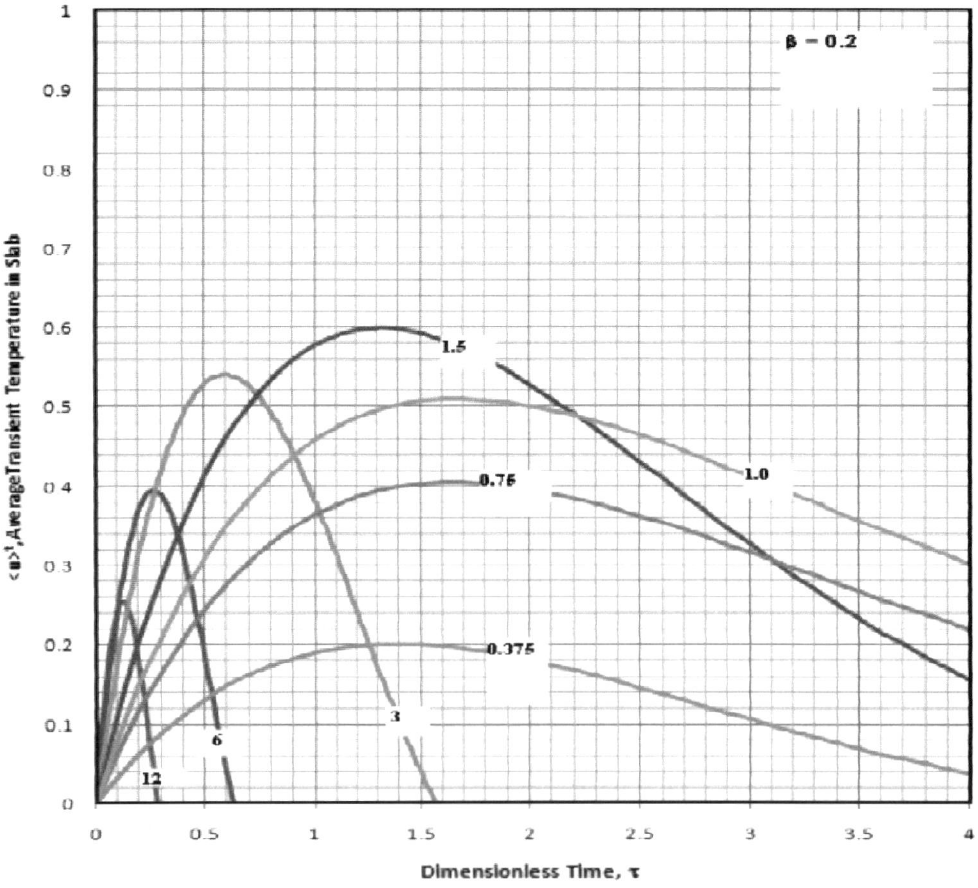

Figure 6.0 Transient Average Temperature in Finite Slab from Damped Wave Conduction and Relaxation (β =0.2)

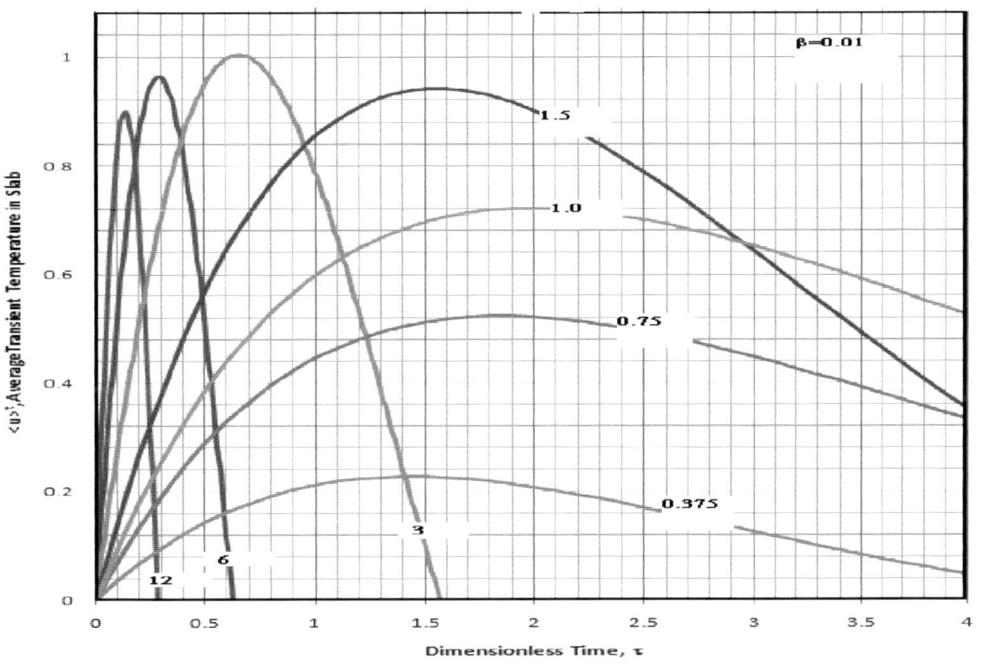

Figure 7.0 Transient Average Temperature in the Finite Slab Subject to Convective Boundary Condition with Derivative of Heat Flux with Time (β = 0.1)

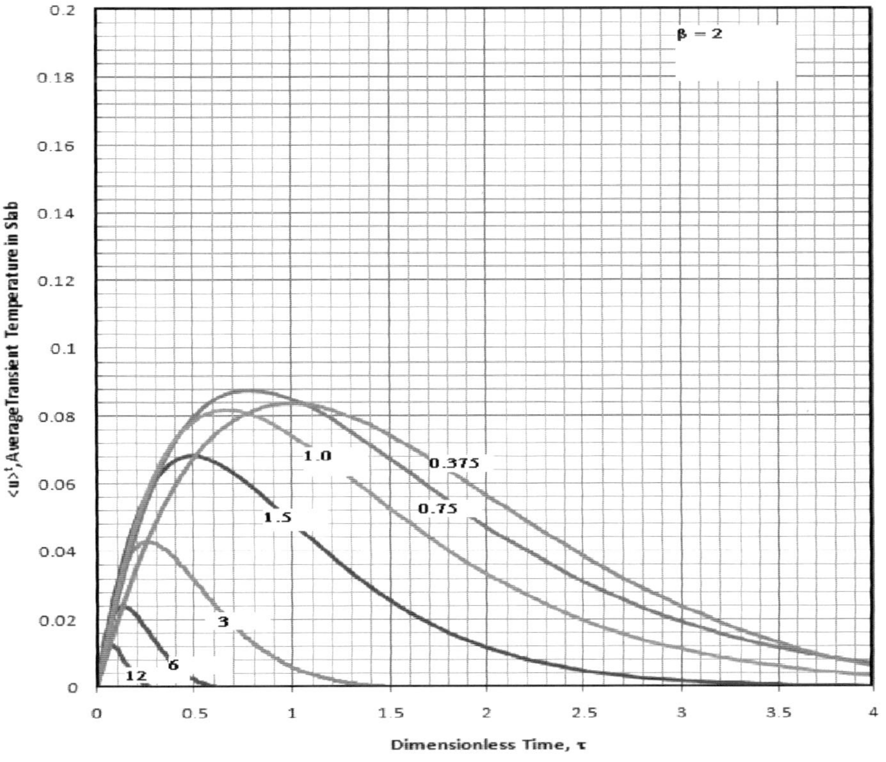

Figure 8.0 Transient Average Temperature in the Finite Slab Subject to Convective Boundary Condition with Derivative of Heat Flux with Time (β = 0.5)

5. Nomenclature

a	half-width of finite slab (m)
C_p	heat capacity ($J.Kg^{-1}.K^{-1}$)
g	function of X only
q	heat flux ($w.m^{-2}$)
T	temperarature (K)
T_s	surface temperature (K)
T_0	initial temperature (K)
V	function of time only
W	wave temperature, $u = We^{-\frac{\tau}{2}}$

Greek

α	thermal diffusivity ($m^2.s^{-1}$)
ρ	density ($kg\ m^{-3}$)
τ_r	relaxation time (s)
τ	dimensionless time (t/τ_r)

Dimensionless Groups

S* Sharma Number (heat), $S^* = \left(\dfrac{h}{Sa}\right)$

S Storage Number, $S = \left(\dfrac{\rho C_p}{\tau_r}\right)$

u Dimensionless temperature, $u = \left(\dfrac{T - T_s}{T_0 - T_s}\right)$

Ve Vernotte Number, $Ve = \left(\dfrac{\alpha \tau_r}{a^2}\right)$

X Dimensionless ordinate, $X = \left(\dfrac{x}{\sqrt{\alpha \tau_r}}\right)$

6. REFERENCES

[1] M. Al-Nimr, M. Naji, and S. Al-Wardat, "Overshooting Phenomenon in the Hyperbolic Heat Conduction Model", **Japanese Journal of Applied Physics, Part I: Regular Papers and Short Notes & Review Papers**, 42, 8, (2003) 5383-5386.

[2] M. A. Al-Nimr and M. Alkam, "Overshooting Phenomenon in the Hyperbolic Microscopic Heat Conduction Model", **Int. J. Thermophysics**, 24 (2), (2003), 577-583.

[3] Y. Taitel, On the Parabolic, Hyperbolic and Discrete Formulation of Heat Conduction Equation", **International Journal Of Heat and Mass Transfer**, Vol. 15, (2), (1972), 369-371.

[4] K. R. Sharma, On the Solution of Damped Wave Conduction and Relaxation Equation in a Semi-infinite Medium subject to Constant Wall Flux, **Int. Journal of Heat and Mass Transfer**, 51, (2008), 6024-6031.

[5] L. Onsager, Reciprocal Relations in Irreversible Processes, **Phys. Review**, 37, (1931), 405-426.

[6] K. R. Sharma, Manifestation of Acceleration during Transient Heat Conduction, **Journal of Thermophysics and Heat Transfer**, **26**, (4), (2006), 799-808.

[7] K. Renganathan, *Correlation of Heat Transfer with Pressure Fluctuations in Gas-Solid Fluidized Beds*, **Ph.D. Dissertation**, West Virginia University, Morgantown, WV, 1990.

[8] K. R. Sharma, Storage Coefficient of Substrate in a 2 GHz Microprocessor, *225th ACS National Meeting*, New Orleans, LA, March, 2003.

[9] K. R. Sharma, Third Order PDE in Restriction Fragment Length Gel Electrophoresis, *228th ACS National Meeting,* Philadelphia, PA, August 2004.

[10] K. R. Sharma, *Bioinformatics,* **McGraw Hill Professional**, (2009), New York, NY.

[11] K. R.Sharma, Adsorber Design for Removal of Acrylonitrile from Water, *AIChE Spring National Meeting,* New Orleans, LA, March /April, 2003.

[12] K. R. Sharma, Critical Radii Neither Greater than the Shape Limit Nor Less than Cycling Limit, *AIChE Spring National Meeting,* New Orleans, LA, March/April 2003. 23

[13] K. R. Sharma, Bessel Composite Function of the Third Order and First Kind: Solution to the Dissolving Pill Problem, *230th ACS National Meeting*, Washington, DC, August/September, 2005.

[14] H. B. G. Casimir, Note on the Conduction of Heat in Crystals, **Physica**, 5, **(**1938)**,** 495 -500.

[15] W. Nernst, W., *Die Theoretician Grundalgen des n Warmestatzes*, Kanppe Hall, Frankfurt, 1917.

[16] T. Q. Qiu, C. L. Tien, Heat Transfer Mechanisms during Short Pulse Laser Heating of Metals, **ASME Transactions, Heat Transfer Division**, Vol. 196, (1992), 41-49.

[17] Y. S. Xie, Y. X. Yuan, X. B. Zhang, Hyperbolic Heat Conduction Equation and Analytic Solution for Instantaneous Ignition of Solid Propellant, **Bimggmg Xuebao/Acta Aramamentarii**, Vol. 27, Issue Supplement, March., (2006), 24-28.

[18] L. Landau, and E. M. Liftshitz, *Fluid Mechanics*, **Pergamon**, UK, 1987.

[19] J. C. Maxwell, On the Dynamical Theory of Gases, **Phil. Trans. Roy. Soc.**, 157, (1867), 49.

[20] C. Cattaneo, C., A Form of Heat Conduction which Eliminates the Paradox of Instantaneous Propagation, **Comptes Rendus**, **247**, (1958) 431- 433.

[21] C. Cattaneo, Sulla Coduzione del Calone, Atti. Sem. Fis. Univ. **Moderna,** 3 (1948), 83.

[22] P. Vernotte, Les Paradoxes de la Theorie Continue de l'equation de la Chaleur", **C. R.. Hebd. Seanc. Acad. Sci. Paris**, 246, (22), (1958), 3154 - 3155.

[23] D. D. Joseph, L. Preziosi, Heat Waves, **Reviews of Modern Physics**, 61, (1989), 41-73.

[24] D. D. Joseph, L. Preziosi, L., Addendum to Heat Waves, **Reviews of Modern Physics**, 62, (1990), 375-391.

[25] M. N. Ozisik, D. Y. Tzou, On the Wave Theory of Heat Conduction**, ASME J of Heat Transfer**, 116, (1994), 526- 535.

[26] D. Y. Tzou, *Macro to Microscale Heat Transfer: The Lagging Behavior*, **CRC Press**, New York, 1996.

[27] K. R. Sharma, *Damped Wave Transport and Relaxation*, **Elsevier**, Amsterdam, 2005.

[28] K. Mitra, S. Kumar, A. Vedavarz, M. K. Moallemi, Experimental Evidence of Hyperbolic Heat Conduction in Processed Meat, **Journal of Heat Transfer**, Vol. 117, (1995), 568-573.

[29] C. Bai, A. S. Lavine, On Hyperbolic Heat Conduction and Second Law of Thermodynamics, **Journal of Heat Transfer**, Vol. 117, (2), (1995), 256-263.

[30] E. Zanchini, Hyperbolic Heat Conduction Theories and non-Decreasing Entropy**, Phy. Review B-Condensed Matter and Material Physics**, Vol. 60, (2), (1999), 991-997.

[31] A. Barletta, E. Zanchini, Thermal-Wave Heat Conduction in a Solid Cylinder which undergoes a change of Boundary Temperature, **Heat and Mass Transfer/ Warema-und Stoffuebertragung**, 32, 4, (1997), 285-291. 24.

[32] M. A. Al-Nimr and O. M. Haddad, "The Dual-Phase-Lag Heat Conduction Model in Thin Slab under Fluctuating Thermal Disturbance", **Heat Transfer Engineering,** 24, 5, (2003), 47-54.

[33] M. A. Al-Nimr and M. Naji, "The Hyperbolic Heat Conduction Equation in an Anisotropic Material", **Int. J of Thermophysics**, 21, (2000), 281-287.

[34] A. M. Haji-Sheik, W. J. Minkowycz, and E. M. Sparrow, **J. Heat Transfer**, 124, 2, 307-319 (2002).

[35] P. J. Antaki, Solution for non-Fourier Dual Phase Lag Heat Conduction in a Semi-infinite Slab with Surface Heat Flux, **International Journal of Heat and Mass Transfer**, Vol. 41, (14), (1998), 2253-2258.

[36] K. R. Sharma, Temperature Solution in Semi-Infinite Medium under CWT using Cattaneo and Vernotte for non-Fourier Heat Conduction", *225th ACS National Meeting*, New Orleans, LA, March 23rd - March 28th, 2003.

[37] K. R. Sharma, Finite Speed Heat Conduction - III Relativistic Transformation from Cylindrical Coordinates in Semi- Infinite Medium, *Chemistry Preprint Archive*, Volume 2003, Issue 2, February 2003, 360-377.

[38] K. R. Sharma, Finite Speed Heat Conduction - IV Relativistic Transformation from Spherical Coordinates, *Chemistry Preprint Archive*, Volume 2003, Issue 2, February 2003, Pages 378-395.

[39] K. R. Sharma, A Fourth Mode of Heat Transfer called Damped Wave Conduction, *42nd Annual Convention of Chemists Meeting*, Santiniketan, India, February , 2006.

[40] K. R. Sharma, R. Turton, Mesoscopic Approach to Correlate Surface Heat Transfer Coefficients with Pressure Fluctuations in Gas-Solid Fluidized Beds, **Powder Technology**, Vol. 99, (2), (1998), 109-118.

[41] K. R. Sharma, On the Derivation of an Expression for Relaxation Time from Stokes-Einstein Relation, 233rd ACS National Meeting, Chicago, IL, March 2007, American Chemical Society, Washington, DC.

[42] B. W. Bequette, *Process Control: Modeling, Design and Simulation,* **Prentice Hall** (2003), Upper Saddle River, NJ.

[43] F. Golnaraghi, B. C. Kuo, *Automatic Control Systmes,* **John Wiley & Sons,** (2010), New York, NY.

(84a) Chemical Engineering Careers At Independent Research Institutes

Chee Kai Tan, Div. of Chemistry and Chemical Engineering, Southwest Research Institute, San Antonio, TX

Independent research institute is established to serve industries, government, and the public in areas that need assistant in problems identification, improvement, and implementation. Chemical Engineers are always faced with challenging science and technology questions related to chemical reactions, process control and development, and scaling up laboratory bench scale to pilot plant studies. In order for chemical reactions to work properly for producing desired products, the professionals will understand the basic theories and involve in controlling the chemicals in proper quantity ratio as feedstock. In addition, other associated factors such as catalyst selection and catalyst development, controlling the processes reaction temperature and pressure, and ensuring the processes are probably designed, evaluated and safe to operate.

Chemical engineers at independent research institutes mostly serve the manufacturing industries varied from petroleum refining, natural gas processing, chemical intermediates, alternative fuels, renewable fuels, and pharmaceutical industries. The Chemical Engineer professionals will offer services in process development and implementation, troubleshooting, pilot plant operation, and small-scale laboratory tests. The Institutes must have extensive, real world expertise in chemistry, chemical engineering, as well as other specialties such as corrosion engineering and safety engineering.

To support these areas of research and development, the institutes must have extensive laboratory facilities in hydrocarbon, organic, and inorganic chemistry for analysis, synthesis and development. In addition, the professionals must be equipped with knowledge in lab tests, pilot plant and demonstration scale operations, chemical and refining process modeling, process flow diagram, safety review, and debottlenecking, etc. Communication skill and team work are also necessary for a successful Chemical Engineer career.

Managing a Process Development Career in Industy

William Hollar
Chief Scientist - MBB
SABIC
1 Lexan Lane
Mount Vernon, IN 47620

Abstract

A career in process development is an interesting and rewarding way to use your chemical engineering training. Proper management of one's process development career can make the difference between a lifelong journey of technical successes and growth or unhappy career experiences. While there are no recipes or ideal paths to a successful process development career, there are factors which can enhance your probability of success.

Important Aspects in Career Management

1. What Do You Want to Do in Your Career?

Managing your career properly requires that you understand what you want to accomplish in your career and which aspects of process development that you truly enjoy. Setting the direction and focus of your career requires understanding your answers to several questions.

What aspects of process development do you prefer?

Process development activities can cover an enormous range of scope from the bench scale all the way to the manufacturing plant scale. Understanding where your interests lie (from both a career and technical standpoint) provides insight into the aspects of career development for focus and development. For example if your interests lie in process modeling and simulation, this choice defines both skills you will need to build/develop and also impacts where in the process development cycle you would contribute. Alternatively if you enjoy working in a pilot plant or plant environment, a different training and development focus would be more appropriate. In case 1, you may focus on training in process simulators, CFD codes and economic evaluation software. Case 2 would lead to greater focus in the details of unit operations, process measurement and control, scale-up and environmental/safety considerations.

Based on your preferences you should look for assignments which build and expand your chosen skillset through experience. This may involve taking on specific assignments or incorporating desired skills into existing assignments. As an example, some process development programs may not require modeling, but the assignment results may be improved by you personally including modeling efforts as part of the scope. Keep in mind this choice may require extra effort on your part.

Do I want to manage or remain technical?

This aspect of career management is not limited to a process development career. It is an important decision for managing the steps in achieving your desired goals. Many chemical companies have both technical and management ladders for career development. While not mutually exclusive, increasing managerial responsibilities will lead to less focus on technical aspects of the job and more on the people management, program planning and communication skills. Ultimately your choice will determine how much focus you want to maintain on the technical aspects of process development.

Some of the other papers in this session go into more detail about this decision and the training and development needs to be successful.

One unique situation can arise when you reach advanced levels on the technical ladder. Many engineers in these roles will lead teams of people on projects that do not report to them. This type of project requires strong leadership skills and an ability to influence people to achieve your goals. Success in this type of role can be both challenging and very satisfying. In this case, learning interaction skills to lead and influence can help insure your success in such a role.

Technical Breadth versus Depth

For the purposes of this discussion, technical breadth refers to competencies/experience in a wide range of chemical engineering topics such as unit operations, heat and mass transfer, separation processes, etc. Depth refers to very deep knowledge in a more limited topic, such as distillation, two phase reactor design or similar more narrowly focused fields.

My experience has been that some combination of both breadth and depth can enhance a process development career. Many process development projects require knowledge of multiple unit operations, process and cost modeling, and an ability to propose alternatives and narrow down to optimum choices. This requires breadth in the field of chemical engineering. In many cases depth can be developed by focus on process technologies. In my case I've done multiple projects on processes involving solids unit operations like drying, so I have developed depth in this area that I leverage and translate where possible.

2. Tools to Build your career

Project management

Any career in process development requires strong project management skills. Process development programs are typically large and complex projects with tight time lines, cross-functional interactions and resource constraints. Effective project management will improve the chances of success. Project management training is widely available and should be considered early in your career.

Problem solving skills and systems

Engineers are by definition problem solvers. A strong problem solving approach and toolset can improve the quality and speed of resolving process development problems.

Many chemical companies use problem solving systems such as Lean Six Sigma to help provide a common language and set of tools for solving business and technical problems. Most of these tools relate back to common best practices for solving a problem efficiently – define the problem, identify solutions, test solutions and validate improvements. The name or type of the approach is less critical than it being efficient in properly defining the problem to be solved and quickly identifying and demonstrating viable solutions that are sustainable in the long term.

Mentors

A technical mentor early in your career can be a huge benefit in launching a process development career. Some companies may assign a mentor at the start of your career. Typically the mentor has significant technical experience within the company and can support you in two ways:

1. Acting as a consultant or sounding board for your early technical assignments and helping you avoid time-consuming pitfalls
2. Assisting in accomplishing goals within the company's operating systems (i.e. other functions such as Sourcing, Legal, EHS)

Getting the right mentor is important. Look for one who enjoys the mentoring process and does not treat it as another task on an already full schedule.

Keep in mind that while the goals of having a mentor change as you get more experience, there continues to be value in having either mentors or other colleagues who you can use to discuss technical or career questions. This remains true throughout your career.

Continuing Growth

Conferences

Attending conferences is an excellent way to develop new contacts, stay current in your field and in general get a fresh viewpoint from the internal work processes and viewpoints of your company. You can also take the opportunity at these meetings to attend sessions in unfamiliar areas. Many times new concepts through analogies or pieces of the unrelated work can bring innovative ideas to your assignments. I typically target 1 – 2 sessions outside my field of interest/expertise at each conference I attend.

Participation in a local or national professional society can often bring the benefit of widening your professional network and exposing you to new areas of expertise.

Technical Journals

E-Journals and e-books allow you to stay current on the latest developments in process development topics very easily. Early in my career I would devote 2 – 4 hours each Friday to scanning a wide range of technical journals. Staying current has become even easier as a corporate or University library is no longer required to access the latest books and journals. The ability to set automatic alerts in some technical data base systems makes the effort even easier by letting keyword queries do the work for you.

Conclusions

Proper management of a career in process development can be aided by two key elements:

1. Choosing your desired path for your career
2. Taking advantage of opportunities to continuously learn and grow throughout your career

The suggestions presented in this paper can help you formulate your strategy for succeeding in the path you choose for your career.

Innovation as a Pathway for Industry Career Enhancement

Joseph B. Powell, Ph.D.
Shell Chief Scientist – Chemical Engineering, Houston, TX

Abstract

Most organizations actively solicit new ideas, and most career professionals have knowledge, competencies and interests that extend beyond their current job assignment. Actively engaging in "innovation" is a means to extend one's role and visibility in the organization, and to extend one's network. It also provides a mechanism which drives lifelong learning and expansion of knowledge base, by providing a justification and outlet for professional society involvement (AIChE) and conference attendance, as well as reading of technical literature.

At Shell, innovation has been formalized into a Gamechanger™ process, which provides support for innovation and investigation of technology and opportunities beyond one's current job assignment. Informal "Hunters" networks also exist in a number of locations, created to facilitate matching of ideas with identified challenges and new opportunities. These networks provide initial screening and enhancement for inputs to the formal idea review process. Network collaboration provides the energy and support often needed to enable the pursuit of new ideas and opportunity, despite barriers and inertia that can act to suppress organizational growth.

This seminar will present an overview of corporate innovation methodologies and the use of networks and online communities to encourage development and pursuit of new ideas. It will also examine how innovation can be used to grow one's personal network, and broaden one's skill base and value to the organization. Finally, it will conclude with a review of grand challenges for chemical engineering in future energy and chemicals, which provides the impetus for new idea generation and resultant opportunities.

Career Enhancement Challenges and Opportunities

An academic career path may reward in-depth knowledge of a specific discipline area. While this may be true also of industry, quite often the successful industry career requires broadening to consider a variety of experiences or job assignments. Management and especially senior management training will typically entail a set of job assignments designed to broaden experience in research and technology, operations, and business functions.

If you are identified early on as a fast track candidate for senior management, broadening may happen with a great deal of help from the organization, in a structured way. In other situations, career broadening may occur by happenstance, via termination of projects or assignments. AIChE CheNected is currently seeking examples of career transitions and impacts on career [1].

One of the reasons a broadened career background is considered valuable both in industry and in academy, beyond its use in managing a technology space or project, is the fact that much innovation occurs at boundaries or overlaps between disciplines. In last decade universities have realized this in creation of multidisciplinary centers of excellence, which resemble those deployed in industry work teams and departments.

Broadening can be self-directed, to a degree. Involvement in AIChE through volunteer opportunities, or attending national conferences or local section meetings to gain exposure to new developments in areas related or adjacent to one's main discipline or job assignment, is one good way to self direct one's career broadening. Large companies have both internal and external speaker seminar series, which provide opportunities to both learn, and network. Setting time aside to read the relevant publications in ones broader discipline is a surprisingly simple, yet often difficult to accomplish method for career and expertise broadening.

Most companies have an explicit interest in innovation, either directly in new technology, or in more efficient systems to execute existing work processes. "Quality" [2] or "Six Sigma" [3] training involves assembly of work team subgroups to identify streamlined and more efficient ways of working. Volunteering for one of these groups or initiatives provides an excellent way to network, and learn about other functions within the company.

Disruptive innovation often requires a less structured approach, and may find resistance from existing organizational structures. In Shell, the Gamechanger^{TM}[4] program was created to drive innovation beyond the interest of existing business units. An on-line idea submission site is maintained to encourage idea development. Ideas are reviewed, and those showing sufficient promise are subjected to a panel review to provide feedback, and suggestions for next steps. Many universities are forming incubation centers to encourage cross discipline innovation, and provide business and legal advice to burgeoning idea generators and entrepreneurs. In the absence of formal innovation processes or groups, there is the self-empowered option of a business lunch with friends and colleagues, to exchange ideas and interests. Several prominent ideas are said to have been first sketched on a lunch table napkin. At Shell, informal Hunters networks exist for idea sharing, networking, as well as brainstorming specific problems or targets, typically via a lunchtime setting. These events provide not ideas for the Gamechanger program or business sponsors, but also provide networking and broadening for the participants themselves. Online communities and bulletin boards or blogs are also used to create global virtual collaboration on innovation themes.

A central theme for innovation is the connection with business value. "Innovation" means developing and progressing a new ideas into a commercial reality, often requiring a set of innovations. Participation in innovation activities can thus help instill business acumen, which can be applied either in development of the idea or concepts, or to one's current assignment. A second theme is collaboration and broader knowledge of the organization with which one is associated, which quite often can be translated into career advancement via awareness of additional opportunities, or opportunities to practice ones expertise in solving problems beyond ones current assignment.

Current innovation opportunities in chemical engineering are extensive, given new advantaged sources of feedstocks for energy and chemicals resulting from shale gas and tight or unconventional oil, development of renewable energy solutions to address energy security, climate change and the environment, advances in sensors and computing leading to new opportunities in big data management and optimization, advances in materials lead by approved ability to manipulate and model matter at the nanoscale.

Summary

The concept of the single inventor working meticulously in the lab or with theory to discover the "next big thing" may be partially true in some instances. Innovation is more likely to occur via cross fertilization and collaboration, however, and a broader group will almost inevitably be required to complete the additional steps and deliver the complementary innovations needed to achieve commercial reality. Much innovation occurs at the interface between disciplines, and may entail both technical and social factors. Participation in cross-functional innovation is a great way to succeed at innovation, while also broadening your own knowledge base, acumen, organizational visibility, and value.

References

1. Loraine Kasprzak, *ChemE's and Career Transitions, AIChE CheNected,* *http://chenected.aiche.org*
2. *Deming, W. Edwards (1986). Out of the Crisis. MIT Press*
3. *M. Barney, T. McCarty, The New Six Sigma: A Leader's Guide to Achieving Rapid Business Improvement and Sustainable Results, Prentice Hall (2002).*
4. *Shell Gamechanger™: http://www.shell.com/global/future-energy/innovation/game-changer.html*

(84d) Bridging Between Engineering Communities: A Hitchhiker's Guide

Kenneth R. Cox, Chemical and Biomolecular Engineering Department, MS-362, Rice University, Houston, TX

Research and application opportunities are continually lost by ineffective communication among engineering communities. Many of the great innovations are cross-over technologies where great ideas from one discipline finds new application with another. Communications between academia and industry are often hindered by a different sense of mission and very different reward systems. Yet, the ability to bring new technology into practice is greatly slowed by this well-documented industry-academia divide. Lack of interactions also limit the effectiveness of engineering societies which far too often function as weakly-joined collection of smaller societies.

Learning to span these divides is an important career skill every student should strive to develop. Professor Cox will share his first-hand experience working with different bodies of engineers and offer his perspective about the challenges, opportunities, trends, and strategies for bridging between the various communities.

(84e) Industry or Academia? Ph.D. or MBA? - Factors to Consider in Your Decision Making

Santiago Faucher, Technologies, Hatch, Mississauga, ON, Canada

We define our careers through our passions, interests, actions, efforts, work, and drive; but most importantly we define it by the decisions we make. For a young engineer who is early on in their career, the first fork in the road forcing a decision is often related to the pursuit or not of further education, the second fork relates to the type of academic education (MBA or Ph.D. or other) and the third (for those who have chosen a Ph.D. path) addresses the pursuit of an academic or industrial career. In this talk, the author will review the factors and benefits associated with the above potential choices and use his own career as an example to provide the audience with better insight into career decisions and outcomes.

(86a) Explosion and Fire At Caribbean Petroleum Tank Terminal in Bayamòn, Puerto Rico

Vidisha Parasram and Maria Mazzocchi, Incident Screening, US Chemical Safety Board, Washington, DC

On October 23, 2009, normal tank filling operation became catastrophic when a 5,000,000 gallon atmospheric tank overflowed for approximately 30 minutes at the Caribbean Petroleum terminal in Bayamòn, Puerto Rico. The developing vapor cloud enveloped part of the tank farm and reached the waste water treatment area where it ignited. A flash fire and explosion ensued and the contents of the tank burned for 2.5 days. Neighboring communities were evacuated and oil contaminated the nearby wetlands. This incident revealed a number of safety system deficiencies at the facility and gaps in the regulatory requirements and industry standards addressing overfill prevention and emergency response. This incident also raises questions regarding the adequacy of safety management systems in place at tank terminals and the need for automated overfill prevention alarms including high, high high level alarm and independent alarms to detect and shut down flow into a tank during filling operations. In addition, due to the close proximity of the community to the Caribbean Petroleum tank terminal, this incident further demonstrates the need for a risk assessment process that addresses high hazard facilities. This presentation will provide an overview of the US Chemical Safety Board's Caribbean Petroleum investigation and discuss recommendations made by the CSB to prevent a similar incident.

(86b) Use Dynamic Simulation to Reduce Startup Flare Emissions for Ethylene Plants

Jian Zhang, Meiqian Wang, Jie Fu, Chuanyu Zhao, Ha Dinh, Shujing Zhang, Qiang Xu and Kuyen Li, Dan F. Smith Department of Chemical Engineering, Lamar University, Beaumont, TX

Ethylene plants play an important role in petrochemical industry. The products of an ethylene plant, such as ethylene, propylene, butadiene, benzene, toluene and other co-products, are all basic feedstocks in petrochemical industry. Ethylene production process is one of the biggest and most complicated processes in petrochemical industry. Startup of an ethylene plant is also very critical. A typical startup of an ethylene plant could flare 400,000 - 1,000,000 lbs of materials, and generate 200,000 lbs of CO and VOCs (volatile organic compounds). Reduction flare emissions will help reducing environmental impact and saving materials.

In this paper, rigorous steady-state and dynamic models of an ethylene plant was built. The models include CGC (charge gas compressor), chilling train, demethanizer, deethanizer, depropanizer, C2 splitter and C3 splitter sections. Both steady-state and dynamic models have many details so that they are more close to startup procedure. For example, performance curves were used in CGC section, so that the section can simulate scenarios on different feed and bypass flowrate. In chilling train section, information of equipment size and heat capacity is considered for simulating precooling process. For distillation columns, rigorous distillation model and tray sizing calculation are employed. Both steady-state and dynamic models are tuned and validated to make them close to real plant operations.

The validated dynamic model was used to simulate startup scenarios. A startup procedure has dozen of steps, and some steps have several flare reduction options. Through dynamic simulation, all major steps and options can be simulated and evaluated, and the flare emissions can be calculated. For example, CGC startup is the most critical step and has largest amount of flare vent, because charge gas has to be continuously produced through furnaces and sent to flare to keep inlet pressure of compressors. We simulated several scenario cases with charge gas of different flowrate, methane and nitrogen, and found the best scenario, which could largely reduce flare emissions of CGC section. Another case study is regarding the precooling of chilling train. A chilling train should not be chilled very fast, otherwise it could be damaged by thermal expansion. On the other hand, chilling train cannot work properly before chilled to certain temperature. Through dynamic simulation, several chilling scenarios were simulated. For each scenario, control strategies are employed to control the chilling rate within a certain range. Simulation results showed some scenarios could even cool down chilling train without any flare emissions.

Flare calculation showed that the total flare emissions could be greatly reduced through dynamic simulation. The final scenarios have been acknowledged by ethylene plants.

(86c) Development of Novel Reducing Agents for NOx Reduction by Sncr Method

Maoqi Feng, Div. of Chemistry and Chemical Engineering, Southwest Research Institute, San Antonio, TX and Reggie Zhan, Emissions and Vehicle Research Division, Southwest Research Institute, San Antonio, TX

Nitric oxides (NO_x), mainly including nitric oxide (NO) and nitrogen dioxide (NO_2), are of the most toxic pollutants formed during combustion and refining processes. Post-combustion treatment methods include selective catalytic reduction (SCR), selective noncatalytic reduction (SNCR), reburning, and the combination of these aftertreatment technologies. Among the post-combustion technologies, the SNCR technology is considered to be an effective approach to reduce NO_x, with reasonable capital investment and operation cost. To reduce NO_x in a lean environment, a reductant is needed. For SNCR, the most commonly used reductant is ammonia or urea water solution as an alternative to NH_3. One of the disadvantages of ammonia is the cost of the injection system. Different reducing agents for NOx removal by SNCR method will be evaluated in this paper, including carbon monoxide, hydrogen, char, hydrocarbons, and oxygenates, etc.

Rotating Zigzag Bed Industrial Application Research on Ammonia Stripping

Zhi-chao Xu, Xiao-hua Li, Guang-quan Wang, Jian-bing Ji
(College of Chemical Engineering and Materials Science, Zhejiang University of Technology, Zhejiang Province Key Laboratory of Biofuel, Hangzhou, Zhejiang, China 310032)

Abstract: About 400 units of rotating zigzag bed (RZB) have been used in industrial distillation process. These successful applications fully demonstrated its advantages such as high gas-liquid mass transfer rate and small size. There will also be broad prospects of RZB applied in waste water treatment process. In this paper a preliminary attempt was made to study ammonia nitrogen removal from industrial waste water by RZB before emission. Firstly, stripping experiments were carried out to remove ammonia nitrogen from the waste water containing ammonia and ammonium sulfate in a small single-rotor RZB, obtaining ammonia solution with a proper concentration. Secondly, a unit of industrial-scale three-rotor RZB with outer diameter of 750mm was designed based on experimental data and used to treat ammonium sulfate solution with 2% ammonia nitrogen. The industrial results showed that the feed capacity was greater than 1500 kg/hr, ammonia concentration in the bottom was less than 30 mg/L, and aqua ammonia concentration in the top was greater than 25%.

With the improvement of human life, there is growing public concern about environment problems. Wastewater emission is one of the important factors contributing to environment deterioration. In the wastewater emissions, ammonia nitrogen concentration is a very important target which should be strictly controlled because excessive ammonia in the river will cause water eutrophication. The ammonia wastewater has a wide variety of sources, such as refining, fertilizer, inorganic chemicals, pesticide, ferroalloys, glass and food industries. Therefore, studies on removal ammonia from the wastewater and how to recover ammonia to reuse is of great significance. Rotating zigzag bed (abbreviated to RZB) has been successfully applied in about 400 cases of industrial distillation[1-5]. Due to its higher mass transfer rate it was expected to have a very wide application in wastewater treatment. Some traditional enterprises have demand for retrofit and optimization of the production process without occupied space increase because of limited land area. RZB can exactly meet this demand owing to its very small volume. In this paper lab-scale experiment of stripping ammonia from wastewater with RZB was conducted and ammonia solution with a certain concentration was obtained. Then, the experimental results were scaled up to design a unit of RZB successfully used in industrial production.

1.Lab RZB device and flowsheet

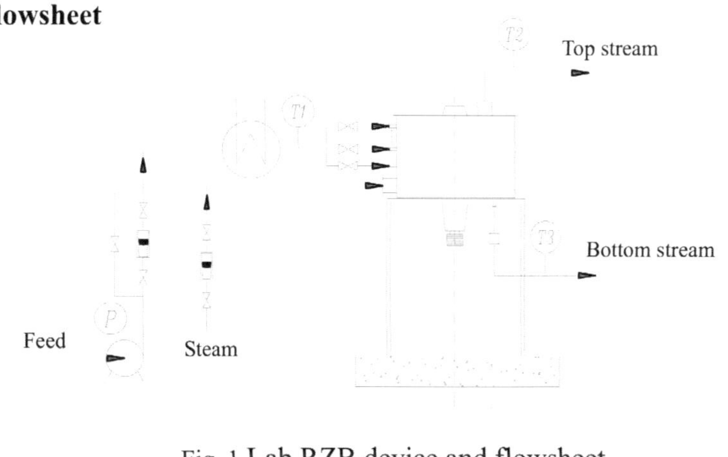

Fig. 1 Lab RZB device and flowsheet

Lab RZB device was set with multiple liquid feed inlets, which were used in variety of processes with different number of theoretical stages. For feeds with different concentration of ammonia, their optimal operation and equipment parameters can be determined by experiments.

In experimental flowsheet shown as Fig.1, the liquid feed was measured with a rotormeter and pumped into RZB device with multiple inlets after heating by a heat exchange. The steam with higher temperature enters into vapor inlet after flow measurement. The treated water was drawn from liquid outlet on the bottom of RZB. Vapor phase containing ammonia withdrawing from the top of RZB was condensed and absorbed in the next steps. T1-T3 were thermometers to measure and record the temperatures of several key positions.

2. Experiments and results analysis

In the experiments, the concentration of the prepared feed was close to that of industrial stuff, which was ammonium sulfate solution with about two percent ammonium. The feed position can be determined through the preliminary experiments. For this study, the flow ratio of feed to steam was varied and analyzed the ammonium compositions of bottom and top outlet stream as well as the power consumption of the motor. These data provided the basis for scale-up design of RZB.

Feed preparation: Ammonium sulfate solution with concentration of 15.5%, which pH was adjusted to about 13 by dripping NaOH solution, was prepared by adding ammonium sulfate powder into water. Ammonium sulfate solution has ammonium concentration of 3.3% and was chosen as the feed.

In the experiments, originally the top outlet vapor stream entered into an absorber but not a condenser. The system was maintained under vacuum condition by adjusting the relief valve at the end of absorber. The pressure of the system was 72kPa and the temperature of the top outlet vapor phase was about 80℃. The typical process and operation parameters in the experiments were shown in Table 1:

Table 1 Typical process and operation parameters

feed				steam		Bottom stream	Top stream
$(NH_4)_2SO_4$ %	Ammonium concentration %	pH	flowrate L/hr	flowrate kg/hr	pressure MPa	Ammonium concentration mg/L	Ammonium concentration(vapor) %
15.5	3.3	13	1000	120	0.35	<100	>48

If the top outlet vapor stream was connected to a condenser, solution with ammonium concentration higher than 20% can be obtained.

3. Industrial application of RZB in ammonium stripping

Fig. 2 Industrial equipment

After the results of industrial field tests were accepted by the customer, scale-up design was made for RZB. The section between the feed position and the top, corresponding to the rectifying section of distillation column, has a separation ability of 4.5 theoretical plates. This section can effectively prevent entrainment. On the other hand, if the actual ammonium concentration was lower than designed, partial reflux can be introduced to increase the ammonium concentration in the top outlet stream. The type of RZB is BZ750-3P, which means the largest one in the three rotors has a diameter of 750mm. Its appearance was shown in Figure 2.

The RZB was used in Taiwan, with industrial process flowsheet shown in Figure 3. The feed and alkali solution for adjusting the pH value, which flowrate were measured by flowmeter respectively, were mixed in a fixed proportion. The mixed stream was heated by the bottom outlet liquid and entered into the RZB. The steam was introduced by vapor inlet. The outlet vapor entered into absorption system which performs cooling and absorption steps. The absorption systems consist of four sections: pure water (absorbent), low concentrated ammonium recycle, high concentrated ammonium recycle and finished ammonium. The flow order is pure water section → low concentrated ammonia section → high concentrated ammonia section → finished ammonia section. The finished ammonia left from the bottom of the absorption system.

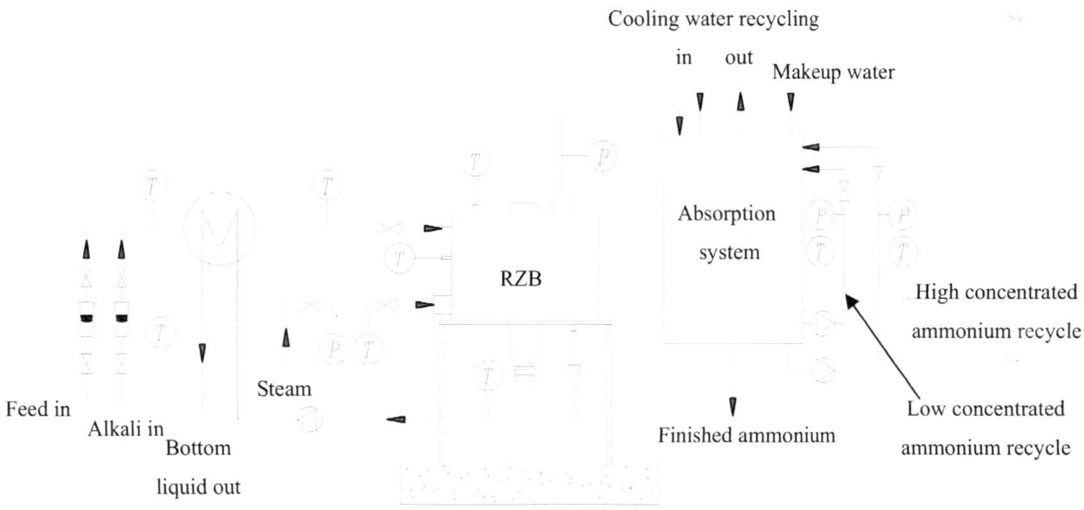

Fig. 3 flowsheet of industrial application of RZB
(pump symbol)-Pump (P)-Manometer (T)-Thermometer

Actual operating parameters are as follows: the feed rate 1500kg/hr, ammonium concentration 18g/L, pH value 11.5-12, outlet vapor temperature about 80℃, no reflux, finished ammonia concentration > 25%, ammonia in outlet liquid<30mg/L, unit consumption of steam 90kg/ton feed, bottom temperature about 98℃. the temperature approach between inlet steam and heated feed was about 10℃. The customer was very satisfied with the results.

4. Conclusion

About 400 units of RZB used in a variety of industrial distillation processes demonstrated its higher separation ability. Its successful application in stripping ammonium from the wastewater was another good example.

RZB has a small volume with a height of lower than 1m. For traditional enterprises with small unused space in the factory, this kind of device can find its place here.

In this system, feed was heated by bottom outlet stream which sensible heat was recovered. So the system shows energy-saving effect.

5. Prospects

Modification of existing RZB device can realize liquid cut off between its rotors, which can extend its applications in recovery of ammonium from the wastewater. For RZB with more than three rotors, its top and bottom section perform absorption and stripping respectively, thus these two processes can be integrated in single equipment.

Reference

1 Ramshaw, C. (1983)., "Higee distillation—An example of process intensification, " Chem. Eng., February, pp. 13–14

2 Rao, D.P., Bhowal, A. and Goswami, P.S. (2004), "Process intensification inrotating packed beds(HIGEE): An appraisal," Ind. Eng. Chem. Res., pp. 1150-1162.

3 Li, X.P., Liu, Y.Z. and Li, Z.Q. (2008), "Continues distillation experiment with rotating packed bed," Chin. J. Chem. Eng., pp. 656-662.

4 Wang, G.Q., Xu, Z.C., Yu, Y.L.and Ji, J.B. (2008), "Performance of a rotating zigzag bed—A new HIGEE," Chem Eng Process, November, pp. 2131–2139.

5 Ji, J.B. , etc. (2004), "Equipment of zigzag high-gravity rotating beds," CN Pat., 01134321.4.

(104a) Multi-Stage-Counter Current Rotating Packed Bed For Distillation

Andrzej Gorak, Dortmund University, Chair of Fluid Separation Processes, 44221 Dortmund, Germany, Philip Lutze, Department of Biochemical and Chemical Engineering, Laboratory of Fluid Separations, TU Dortmund University, D-44227 Dortmund, Germany and Daniel Sudhoff, Dortmund

The development of new apparatuses and processes for the chemical and bio-based production is necessary to meet the global future challenges. The realization of highly efficient, flexible and innovative processes contributes to achieving these goals. Apparatuses for separation that potentially fulfill these requirements are phase contactors applying centrifugal force instead of gravitational force, so called HIGEE-apparatuses ("high gravity"). A promising example are rotating packed beds (RPBs), in which a torus shaped packing is rotating while the liquid (radially outward direction) and the vapor phase (radially inward direction) are contacted counter currently. The phase contact occurs in a centrifugal field with a multiple of the gravitational force. The high acceleration enables intense mixing, large interfacial areas and quick mass transfer between the phases. This gives the opportunity for high capacities at short residence times of the chemicals inside the heat exposed areas. Additionally, a flexible layout and orientation of the apparatus independently from the gravity as well as niche-applications, for example for highly viscous media, are enabled which cannot be distilled in conventional columns.

The aim of this work is to proove the feasibility and to quantify the separation efficiency of distillation processes in RPBs. For this purpose a Multi-Stage Counter-Current Rotating Packed Bed (MSCC-RPB) has been developed. This apparatus consists of three consecutive rotors mounted onto a single shaft. In between the rotors a liquid feed can be added to the apparatus to allow variation in the feed position. In the first step the hydrodynamics of the apparatus regarding pressure drop and flooding behavior of this novel type of RPBs have been characterized. Especially the flooding behavior has to be redefined compared to conventional columns since parameters such as superficial area, centrifugal acceleration and liquid and vapor loadings are strongly dependent on the radial position inside the rotor. In the second step distillation experiments with finite reflux ratio using an ethanol water mixture have been carried out to characterize the separation efficiency of the MSCC-RPB. The influence of the rotational speed, liquid and vapor loadings and feed position have been characterized.

(104b) Effects of Gas Jet in the Downcomer On Liquid Circulation in Internal Loop Airlift Reactors

Luhaibo Zhao[1], Bo Liao Sr.[1], Xiaohao Liu[1], Xiaoquan Wang[2] and Yuhan Sun[3], (1)Low Carbon Conversion Center, Shanghai Advanced Research Institute, Shanghai, China, (2)Low Carbon Conversion Center, Shanghai Advanced Research Institute,Chinese Academy of Sciences, Shanghai, China, (3)State Key Laboratory of Coal Conversion, Institute of Coal Chemistry, Chinese Academy of Sciences, Taiyuan, China

In recent years, internal loop airlift reactors (ILAR) is widely used in petrochemical, biochemical and energy processes due to its simple mechanical design, effective interphase contact and efficient slurry circulation. However, the liquid velocity in ILAR is limited because the liquid circulation is driven by the density differences of flow in riser and downcomer, which in turn imposes a limitation on the gas velocity and the reactor yield, thus hinders the broader application of ILAR. To solve this problem and increase the yield of ILAR, the present paper provides a promising method by designing a gas jet in the downcomer of ILAR to promote the liquid circulation, which allows the reactor to operate with higher superficial gas velocity, and thus increases the yield of ILAR. In addition, computational Fluid Dynamics (CFD) three-dimensional modeling is employed to investigate the influences of key parameters of jet (such as size, angle, location and flowrate etc.) on the hydrodynamics in ILAR involving liquid circulation, and to determine the optimal configure and design of gas jet. The results are validated by the cold model experiments conducted in an industrial-scale ILAR. It is shown that with optimal design, the liquid velocity is improved by 20% by using only 2% of total inlet gas to inject into the downcomer, and the yield is increased by 5% at a high superficial gas velocity (ug=0.4m/s).

(104c) Comparison of the Thermal Performance of Aluminum and Stainless Steel Compact Heat Exchange Reactors

Steven J. Vallee, Hani Gadalla and Zhijun Jia, Chart Energy & Chemicals, La Crosse, WI

High oil prices have led to increased interest in the conversion of syngas (a mixture of CO and H_2) to liquid fuels. Conventional fixed bed reactor technology for this application suffers from heat transfer limitations. A Compact Heat Exchange Reactor (CHER) combines a reactor and compact heat exchanger into a single device, and provides a solution to the heat transfer limitations commonly observed for conventional fixed bed reactors.

The thermal performance of aluminum and stainless steel CHER were compared using a combination of experimental results and modeling. The aluminum CHERs are well suited for the Fischer-Tropsch reaction with cobalt catalyst, and provide superior thermal performance and ease of manufacturing scalability. The greater structural strength of stainless steel CHERs are suitable for other highly exothermic reactions at higher temperatures and pressures.

(104d) Electrodeposition of Aluminum From AlCl3-BMIC Ionic Liquids in High Gravity Fields

Lianbin Xu[1,2], Nannan Shan[1,2], Zhiyun Pan[1,2] and Jian-Feng Chen[1,2], (1)Research Center of the Ministry of Education for High Gravity Engineering and Technology, Beijing University of Chemical Technology, Beijing, China, (2)State Key Laboratory of Organic-Inorganic Composites, Beijing University of Chemical Technology, Beijing, China

In high gravity fields,the electrodeposition of aluminum (Al) on stainless steel electrodes from $AlCl_3$-BMIC ionic liquids with 2:1 molar ratio were studied at 75 . Analyses of the cyclic voltammetry indicate that the deposition of Al is a diffusion-controlled quasi reversible process either in normal or in high gravity fields and the high gravity coefficient (***G***) has no influence on the diffusion coefficient of Al ions. The electrodeposition process is strengthened with the intensification of convection of electrolyte in high gravity fields. The results of the chronoamperograms suggest that the deposition process of Al is instantaneous nucleation in both normal and high gravity fields. The density of crystal nucleus increases from $8.02*10^8$ cm^{-2} in normal gravity field to $20.34*10^8$ cm^{-2} with ***G***=200, which means high gravity field could accelerate nucleation to enhance the electrochemical reaction rate. The influence of high gravity coefficient (***G***) on surface morphology and crystal structure of the Al coatings was also investigated. SEM results show that high gravity field can refine the grain and avoid dendrite, resulting in good adhesive coatings. XRD analysis reveals that all of the electrodeposits exhibit a preferred (111) crystallographic orientation, enhanced with the increase of ***G***.

(104e) MASS Transfer Enhancement Due to the Presence of Rotating Single Twisted Tape Promoter in an Electrolytic Cell

Nalluri chitti Babu and Paladugu Venkateswarlu, Chemical Engineering, Andhra University, Visakhapatnam, India

The presence of turbulence promoters in an electrolytic cell can increase the intensity of local turbulence and hence the mass transfer coefficient at the transfer surface. Limiting current data are obtained using diffusion- controlled cathodic reduction of ferricyanide ion at the microelectrodes fixed on the electrode support in an electrolytic cell in the presence of a rotating single twisted tape promoter. The mass transfer coefficient is increased with increases in rotational speed, tape width, tape pitch of the promoter and flow rate of the electrolyte. The correlation developed is

$$Sh_e/Sc^{1/3} = 49.24\ Re_r^{0.187}\ Re_f^{0.134}$$

ENHANCEMENT OF MASS TRANSFER WITH SQUARE-GROOVED SERRATED DISC AS TURBULENCE PROMOTER IN CIRCULAR TUBE

D.S.Seetha Rama Raju[a]*, *M.Gangadhar*[b], *V Nageswara Rao*[a], *N.Chitti Babu*[a]
[a]*Department of Chemical Engineering, College of Engineering(A), Andhra University, Visakhapatnam-530 003, Andhra Pradesh, INDIA*
[b]*Department of Chemical Engineering, G.M.R. Institute of Technology, GMR Nagar, Rajam, Srikakulam-532127, Andhra Pradesh, INDIA*
**E-mail:* sai_dandu@yahoo.co.in

ABSTRACT

Mass transfer data were obtained in circular conduit using an electrochemical technique with a potassium ferri-ferro cyanide couple. In circular conduit coaxially placed entry region square-grooved serrated disc was used as turbulence promoter. The study comprised of evaluation of mass transfer rates at the outer wall of the electrochemical cell. Mass transfer coefficients were evaluated from the measured limiting currents. The study covered a wide range of geometric parameters such as diameter of the disc (d_d), thickness of the disc (d_t) and distance of the disc from the entrance of the test section (h). The results revealed that the mass transfer coefficient increased with increase in velocity, diameter of the disc (d_d), thickness of the disc (d_t) and decreased with increase in distance of the disc from the entrance of the test section (h). Within the range of variables covered, the augmentation achieved in mass transfer coefficients were up to 2.1 fold over the tube flow in absence of promoter. The entire mass transfer data were correlated with Colburn J_D factor and roughness Reynolds number. The following correlation was reported out of the study.

$J_D = 128.40\ (Re^{+})^{-0.9753}\ (\phi_1)^{0.9224}(\phi_2)^{0.1785}\ (\phi_3)^{0.4109}$

Where $\phi_1=d_d/d$, $\phi_2=d_t/d$, $\phi_3=h/d$ are dimensionless groups. d is diameter of test section

KEYWORDS: Mass transfer, square- grooved serrated disc, turbulence promoter

1. INTRODUCTION:

Researchers will always aim at the maximum output with reduced equipment size so as to minimize the unit product cost. This is particularly true in the design of electrolytic cells where the mass transfer limiting conditions exist. Use of turbulence promoter is one such technique that increases the transfer coefficients by several folds over the smooth flow. Disc placed across the flow inside the tube generates wakes and eddies within the tube. This disturbance continues to its effect on either side of disc and continues to a longer distance which in turn influences the wall mass transfer coefficient over the length of the tube. The study was conducted to know the effectiveness of disc in augmenting mass transfer coefficient. Burns and Jachuck[1] determined liquid–solid mass transfer performance on a 30cm diameter spinning disc reactor by use of the limiting current technique. Ravi et al [2] did experiments and obtain mass transfer data on the walls of the fludized beds using an electrochemical technique with a potassium ferro-ferri cyanide couple. The disc promoter assembly was used as internal in fludized beds. Sherwood and Stone [3] determined the flow of a viscous fluid around disc in a pipe and the added mass of the accelerating disc and flow profiles were also presented. Podesta et al [4] obtained a dimensionless correlation for ionic mass transfer at horizontal disc electrodes under longitudinal vibration. Escudie et al [5] carried out experiments in water fluidized binary mixtures of Teflon spheres, discs and rods. Venkateswarlu et al [6] studied the effect of string of discs placed co-axially in a circular conduit on mass and momentum transfer. Data on mass transfer in circular conduit with coaxially placed square-grooved serrated disc as turbulence promoter has been taken up since the data on the present system has not been reported in literature. Based on the data a semi-theoretical model has been developed. Parameters covered in the present study are shown below in Table 1

Table 1

Variable	Minimum	Maximum	Max/Min
Diameter of the disc, (d_d), m	0.025	0.045	1.8
Disc thickness, (d_t), m	0.001	0.005	5
Distance of the disc from the entrance of the test section, (h),m	0.10	0.26	2.6
Velocity, V, m/s	0.03289	0.3289	10
Reynolds number, Re	1933	19337	10
Schmidt number	982.2	1040	1.0588

2. EXPERIMENTAL PROCEDURE:

The schematic diagram of experimental setup is shown in figure 1. It is similar in layout to that was used in earlier studies [6,8,12]. It essentially consists of a storage tank (T_S), centrifugal pump (P), rotameter (R), entrance calming section (E_1), test section (T) and exit-calming section (E_2). The storage tank is cylindrical copper vessel of 100 liter capacity with a drain pipe and a gate valve (V_1) to facilitate periodical cleaning. A copper coil (H) with perforations is provided to bubble nitrogen through the electrolyte. The tank is connected to the pump of 2 HP motor with a 0.025m diameter copper pipe on the suction line of the centrifugal pump. The suction line is also provided with a gate valve (V_2). The discharge line from the pump splits into two. One served as a bypass line and controlled by valve (V_3). The other connects the pump to the entrance calming section (E_1) through rotameter. The rotameter is connected to a valve (V_4) for adjusting the flow at the desired value. The rotameter has a range of $0\text{-}166\times10^{-5}\ m^3/s$.

The entrance calming section consisted of 0.05 m ID circular copper pipe with a flange and is closed at the bottom with a gland nut (G). The up-stream side of the entrance calming section is filled with capillary tubes to damp the flow fluctuations and to facilitate steady flow of the electrolyte through the test section.

Test section is made of a graduated perspex tube of 0.36m length with point electrodes fixed flush with the inner surface of the tube. The point electrodes are made out of a copper rod and machined to the size. They are fixed flush with the inner surface of the test section at equal spacing of 0.01m. Exit calming section is also of the same diameter copper tube of 0.5 m long, and it is provided with a flange on the upstream side for assembling the test section. It has gland nuts (G) at the top and bottom ends to hold the central tube. Two thermo wells (t_1, t_2) are provided, one at upstream side of the entrance calming section and the other at the downstream side of exit calming section for measurement of temperature of the electrolyte.

Square- grooved serrated disc serving as turbulence promoter is made of copper of various sizes with a provision to fix it rigidly within the test section. The serrated disc is placed concentrically in the test section.

The limiting current measuring equipment consisted of multimeter of 'Motwane' make that has 0.01mA accuracy and vacuum tube voltmeter is used for potential measurements. The other equipment used in circuit is rheostat, key, commutator, selector switch, and a lead acid battery as the power source. The commutator facilitated the measurement of limiting currents for oxidation and reduction process under identical operating conditions by the change in polarity while the selector switch facilitated the measurements of limiting currents at any desired electrode along the test section. The circuit diagram used for the measurement of limiting currents is shown in the figure 3.7.

Data on limiting currents for the case of reduction of ferricyanide ion is obtained for fluid flow in circular conduits in the presence of square- grooved serrated disc as insert promoter. The following electrode reaction is involved.

Cathodic reduction of ferricyanide ion:

$$[Fe(CN)_6]^{-3} + e^- \longrightarrow [Fe(CN)_6]^{-4} \quad ...... (1)$$

Eighty liters of equimolal solution of 0.01M Potassium ferricyanide and Potassium ferrocyanide with 0.5N NaOH as indifferent electrolyte were prepared. The electrolyte is analyzed for ferrocyanide ion concentration by volumetric titration method using standard potassium permanganate solutions (1) and for ferricyanide ion using idometric method (2). The viscosity of the solution at different temperatures was measured with Ostwald Viscometer and densities were measured using specific gravity bottle.

The point electrodes are polished using four zero emery to get a smooth surface without protuberances. The size of the electrode is measured with a two-tank micrometer and also with a traveling microscope.

Initially blank runs are conducted with indifferent electrolyte (sodium hydroxide solution) alone to ensure that the limiting currents obtained in the subsequent runs are due to diffusion of reacting ions (Ferri cyanide ion) only.

The electrolyte is pumped at a desired flow rate (through the test section) by operating the control and by-pass valves. After steady state is attained, potentials are applied across the test electrode and wall electrode in small increments of potentials (100mV) and the corresponding currents are measured for each increment. In view of the large area of the counter electrode in relation to the test electrode nearly constant potential is maintained at the test electrode. Since the potential values are not of criteria in the present study, the limiting currents are determined from the measurements of applied potentials and currents as has been done in several earlier works [6,8,12].

The attainment of limiting current is indicated by the constancy of current with a large increase in the potential. Mass transfer coefficients are computed from the measured limiting currents by the following equation:

$$k_L = \frac{i_L}{n.F.A.C_0} \quad ......(2)$$

Pressure drop measurements are also taken simultaneously using a U-tube manometer with Carbon tetrachloride as manometric liquid. The turbulence promoters are shown in figure 2.

3. RESULTS

Mass transfer coefficients obtained were arithmetically averaged and average mass transfer coefficients are computed. Mass transfer coefficients were influenced by dynamic and geometric parameters. The geometric parameters covered in the present study are square- grooved serrated disc diameter (d_d), serrated disc thickness (d_t) and distance of the serrated disc from entrance of the test section (h).

Effect of disc diameter:

Disc diameter has strong influence on mass transfer coefficient K_L. K_L versus velocity (V) was drawn for different disc diameters and is shown in figure 3.Various disc diameters used in the present study are d_d=0.025m, 0.030m, 0.035m, 0.040m and 0.045m.Mass transfer coefficient increases with increase in disc diameter. The augmentation in mass transfer coefficient is 1.25 times over the smooth tube [7] for disc diameter 0.025m at a velocity of 0.3289m/s while the augmentation is 2.08 times over the smooth tube [7] for the disc diameter 0.045m at the same velocity 0.3289m/s.

Effect of disc thickness (d_t):

In figure 4, mass transfer coefficient was plotted against velocity for the different disc thickness values of d_t=0.001m, 0.002m, 0.003m, 0.004m and 0.005m by keeping all the other geometric parameters constant at d_d=0.045m,h=0.18m.From the figure it was revealed that the mass transfer increases from 1.87 times 2.63 times over Lin et al[7] for the smooth tube as the thickness of the disc (d_t) increases from 0.001m to 0.005m at the velocity of 0.3289m/s.

Effect of distance of the serrated disc from the entrance of the test section (h):

As the distance from the entrance of the test section (h) varied, the extent of turbulence also varied because of the change in circulatory pattern. Mass transfer coefficient versus velocity data for different values of 'h' were plotted and shown in figure 5.Mass transfer coefficients are decreased from 1.48 times to 2.28 times over smooth tube of Lin et al[7] while increasing 'h' from 0.10m to 0.26m at the velocity of 0.3289m/s.

Table 2 indicates the exponent on velocity of the present study together with the other works. The exponent on velocity is comparable to that in the studies on mass transfer with different turbulence generating systems.

Table 2

Comparison with other works:

Author	Promoter	System	Exponent on velocity	Range of Re
Klaczack [10]	Spiral coil	Mass transfer	0.520	1700-20000
Sujatha [12]	Tapes mounted on a rod	Mass transfer	0.490	1348-30605
Venkateswarlu [6]	String of discs	Mass transfer	0.498	3300-18650
Sitaraman [11]	String of spheres	Mass transfer	0.556	100-34000
Sarveswara Rao [8]	String of cones	Mass transfer	0.431	690-20200
Changal Raju [9]	Wires wound on a rod	Mass transfer	0.490	1500-20000
Nageswara Rao [13]	Single disc	Mass transfer	0.405	1300-12000
Present work	Square- Grooved Serrated disc	Mass transfer	0.431	1933-19337

4. CORRELATION DEVELOPED:

The data on mass transfer with single serrated disc as turbulence promoter could well be calculated in the lines done in earlier studies [8,12].Correlation of data using Colburn J_D factor with roughness Reynolds number have yielded the following equation

$$J_D = 91.846(Re^{+})^{-0.9723} \qquad \text{.....................(3)}$$

Average deviation = 20.93, Standard deviation = 24.15

By incorporating dimensionless geometrical groups, the following correlations are yielded

$J_D = 128.40\,(Re^+)^{-0.9753}\,(\phi_1)^{0.9224}(\phi_2)^{0.1785}\,(\phi_3)^{0.4109}$ …………...(4)

Average deviation = 1.69,

Standard deviation = 2.14

Where $\phi_1=d_d/d$, $\phi_2=d_t/d$, $\phi_3=h/d$, which are dimensionless groups and S_c is Schmidt number.

Correlation graph for equation 4 is given in the figure 6.

Comparison of correlations:

For a selected set of geometric parameters correlation factor in mass transfer (Y_3) is plotted against Re+. For comparison of data with the other studies, Nageswara Rao data [13] having comparable geometric parameters, is taken and computed with present method and shown in figure 7. The data falls close to the present study indicating correlation presented in the present work is comparable and better correlating.

Figure 8 shows efficiency of promoter against Reynolds number for the same set of data points which augment best of present study and compared with data of Nageswara Rao [13]. The figure shows efficiency of promoter is increasing with increase in Reynolds number of present study. More over efficiency of promoter of present study is found to be higher when compared to the studies of Nageswara Rao [13]

5. CONCLUSIONS:

Mass transfer coefficient increases with increase in velocity, serrated disc diameter (d_d), thickness of the serrated disc (d_t) and decreases as distance of the disc (h) from the entrance of the test section increases. In the present study, it is found that disc diameter (d_d) of 0.045m, disc thickness=0.005m, distance of the disc from the entrance of the test section (h) =0.10m, offered maximum augmentation. Maximum augmentation of 2.1 folds is observed over smooth tube flow. This is valid for the range of variables covered in this study, but probably capable of predicting data at higher ranges. The data generated and correlation developed is helpful in the design of efficient electrolytic reactors for organic synthesis and other applications.

Dimensionless Groups:

J_D	=	Mass Transfer Factor $(k_L/V).Sc^{2/3}$
Re	=	Reynolds number = $dV\rho/\mu$
Re^+	=	Roughness Reynolds number = $(d_d / d).Re.\sqrt{(f/2)}$
St	=	Stanton number = k_L/V
St_o	=	Stanton number for conduit without internals
Sc	=	Schmidt number $\mu/\rho D_L$
Sh	=	Sherwood number $k_L d/D_L$
u^+	=	dimensionless velocity, u/u^*
y^+	=	dimensionless radial distance from the wall, $y\,u^*/\nu$

Nomenclature:

d	=	Diameter of test section, m
D_L	=	Diffusivity of reacting ion, m^2/sec
d_d	=	Disc diameter, m
f	=	Friction factor, $\Delta p\, d\, g_c /2LV^2 \rho$
ΔP	=	Pressure difference, N/m^2
F	=	Faraday's constant = 96,500 coulombs/g-mol
g	=	Acceleration due to gravity, m/sec^2

g_c = Gravitational constant.
i_L = Limiting current, amp
k_L = Mass Transfer coefficient, m/s
k_o = Mass transfer coefficient of the empty conduit, m/s
L = Length of test section, m
n = Number of electrons transferred
Q = Volumetric flow rate, m^3/s
d_d = diameter of the disc, m
d_t = thickness of the disc, m
h = Location of the disc from the entrance of the test section, m
u = Local velocity, m/s
u^* = Friction velocity = $\sqrt{(\tau_w g_c / \rho)}$, m/s
V = Average velocity, m/s
y = Radial distance from the wall, m
Y_3 = $J_D / (d_d/d)^{0.9224} (d_t/d)^{0.1785} (h/d)^{0.4109}$

Greek letters:

$\in$ = Eddy viscosity, m^2/s
$\in_D$ = Eddy diffusivity, m2/s
μ = Viscosity of fluid, Kg/m. sec
ν = Kinematic viscosity, m2/s
ρ_c = Density of manometer fluid, Kg/m^3
ρ = Density of fluid, Kg/m^3
τ_w = Shear stress , N/m^2

References:

[1] J.R.Burns and R.J.J.Jchuck,
Determination of liquid –solid mass transfer coefficient for a spinning disc reactor using limiting current technique, Int.J.Heat and Mass Transfer,48,2005,2540-2547

[2] T.Ravi,B.Srinivasa Rao,P.Gopala Krishnaand P.Venkateswarlu
Ionic mass transfer studies in fluidized beds with co-axially placed discs on a rod as internal Chem.Engg.Process, 35, 1996, 187-193

[3] J.D.Sherwood ,H.A.Stone
Added mass of a disc accelerating within a pipe
Phys.Fluids., 9,1997,3141-3148

[4] J.J.Padesta,G.S.Paus,A.J.Arvia
Ionic mass transfer at horizontal disc electrode under longitudinal vibration
Electrochem.Acta, 19, 1974, 583-589

[5] R.Escudie,N.Epstein,.R.Grace,and H.T.Bi
Effect of particle shape on liquid fluidized beds of binary solids mixtures,
Chem.Engg.Science., 61,2006,1528-1539

[6] P.Venkateswarlu,T.Gopichand,G.J.V.J.Raju
Increased mass transfer in a circular column in the presence of disc promoter
J.Energy Heat Mass Transfer, 2000, 22,195-203

[7] Lin C.S.,E.B.Denton,H.S.Gaskill and G.L.Putnam
Diffusion controlled reaction, Ind.Engg.Chemistry, 1951, 43, 2136-2143

[8] S.Sarveswara Rao, G.J.V.J.Raju
Studies on Ionic Mass Transfer with co-axially placed cones on a rod in homogeneous fluid and fluidized beds, Paper presented at World Congress III of Chemical Engineers, 1986,Tokyo,p481
[9] D.Changal Raju,C.Bhaskara Sharma,G.J.V.Raju,
Ionic mass transfer at the outer wall of concentric annuli in the presence of fluidizing solids- Effect of wires wound on central rod, J.Energy Heat Masss Transfer, 1992, 14, 151-156
[10] Klaczac,A
Trans. A.S.M.E., J. Heat Transfer, 1973, 95, 557
[11] Sitaraman, T.S
Augmentation of mass transfer with coaxial string of spheres as internals in tubes and fluidized beds, 1977, Ph.D. Thesis, University of Madras, India
[12] V.Sujatha,C.Bhaskara Sarma,G.J.V.Jagannadha Raju
Studies on ionic mass transfer with coaxially placed helical tapes on a rod in forced convection flow, Chem.Engg.Process.,1997,36, 67-73
[13] Nageswara Rao, V.and Chitti Babu, N.
Studies on ionic mass transfer with coaxially placed disc as turbulence promoter in circular Conduits, Caledonian journal of Engineering, Vol.4 (2009)

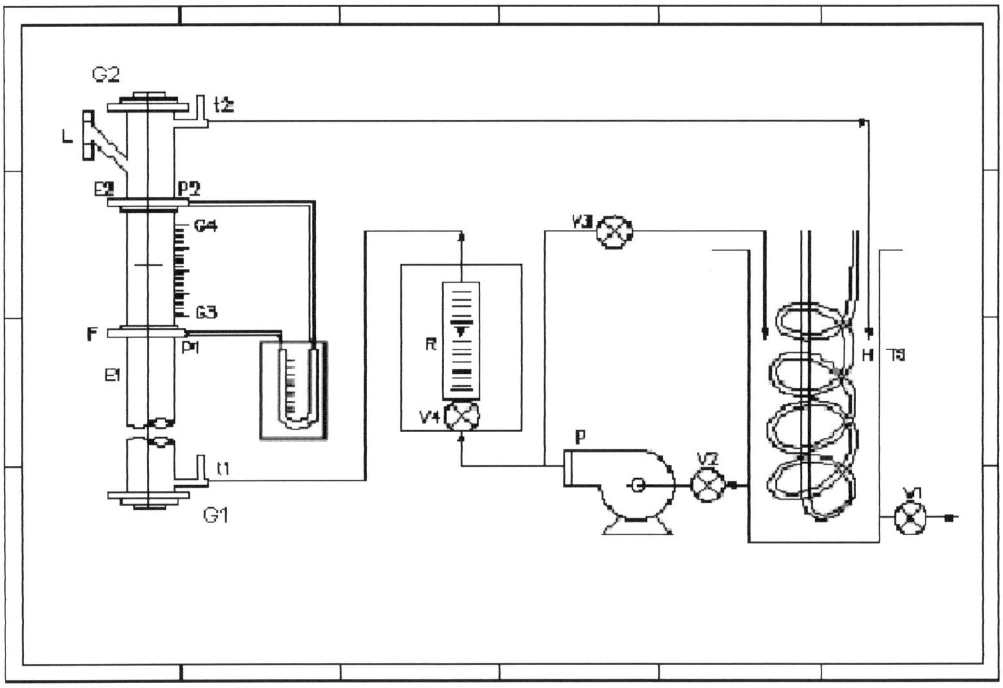

Figure 1 Schematic Diagram of Experimental Setup

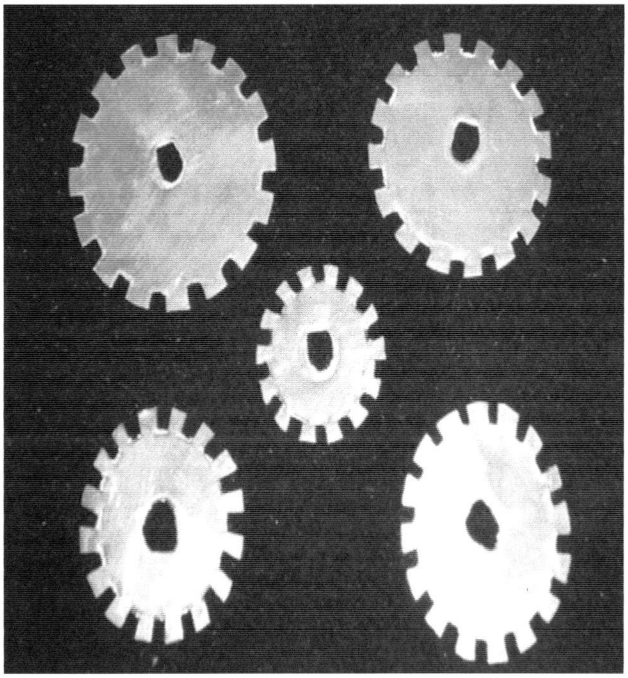

Figure 2 Turbulence Promoters

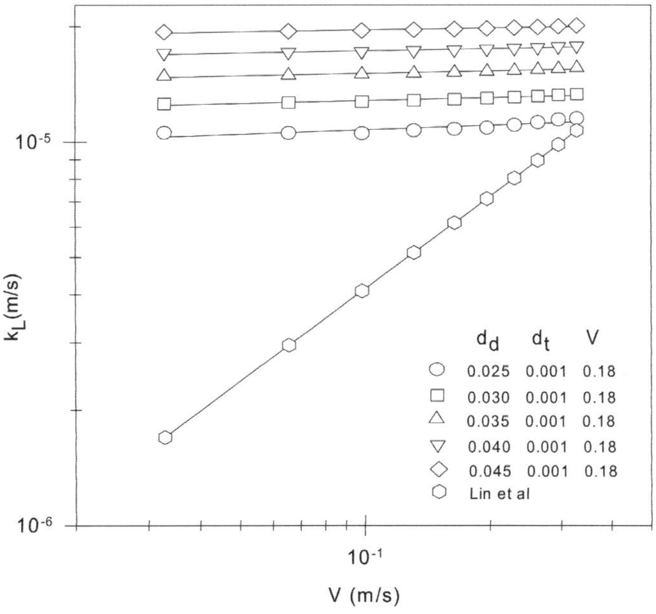

Variation of Mass Transfer Coefficient with Velocity
-Effect of Disc Diameter

Figure 3

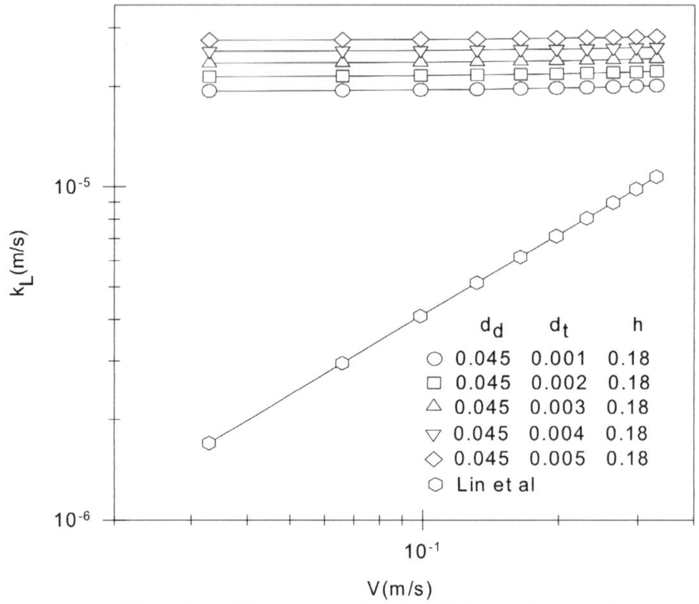

Variation of Mass Transfer Coefficient with Velocity

-Effect of Disc Thickness

Figure 4

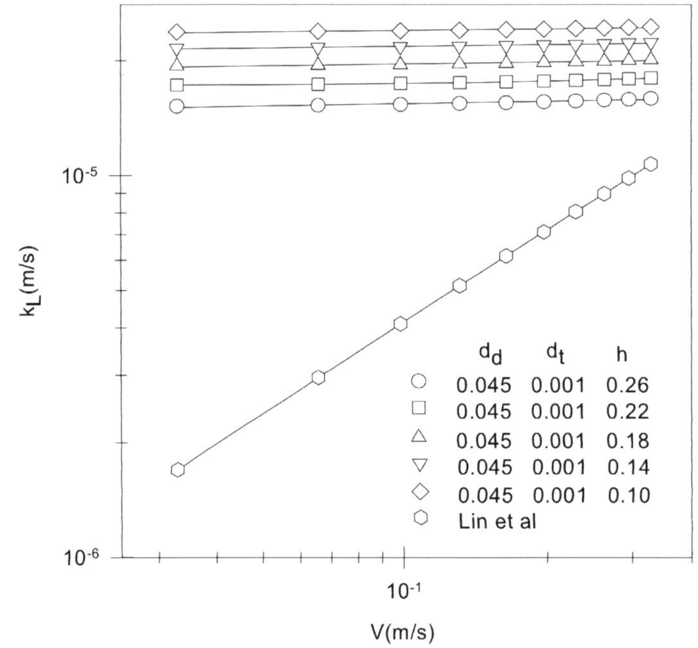

Variation of Mass Transfer Coefficient with Velocity

-Effect of Distance of Disc from Entrance of the Test Section

Figure 5

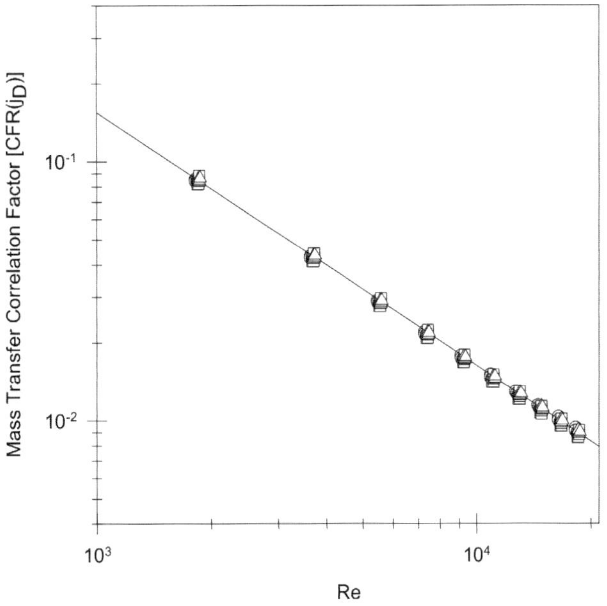

Correlation plot for equation (4)

Figure 6

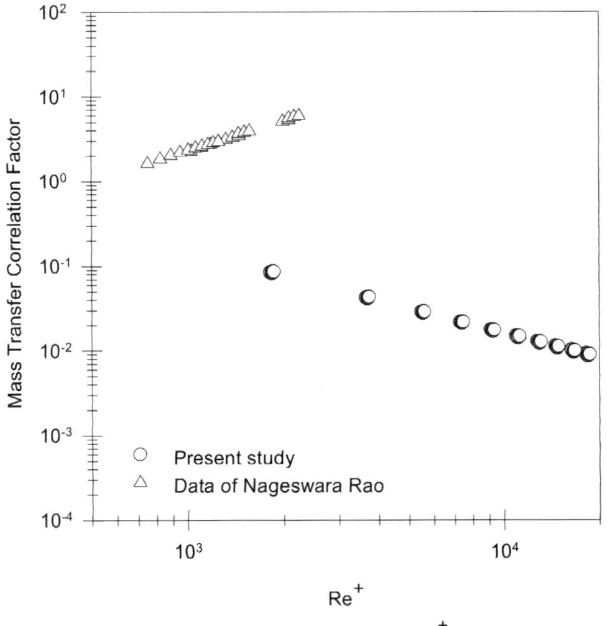

Variation of CFR with Re^+

- Comparision plot

Figure 7

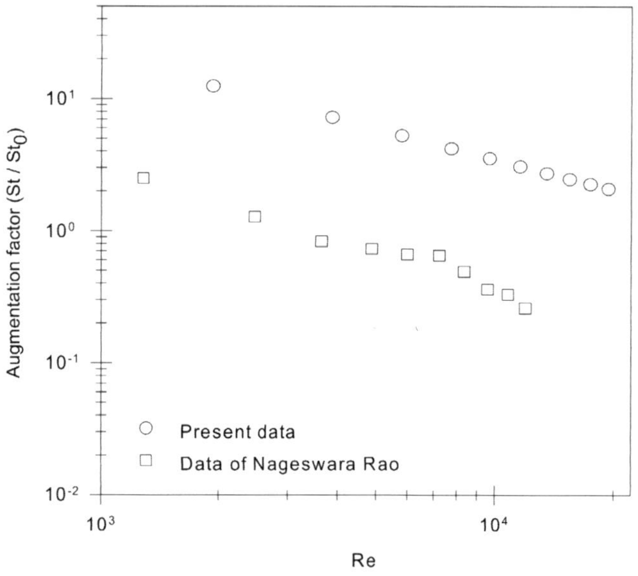

Variation of St/St_0 with Reynolds Number

-Comparision plot

Figure 8

Use of Mechanical Agglomeration for Rubber Particle Size Control in Emulsion ABS Production

Jianhua Xu[1], Vern Lowry[2], Satish K. Gaggar[1] and Dane M. Ferraris[1],
(1) Innovative Plastics, SABIC, Washington, WV,
(2) Innovative Plastics, SABIC, Ottawa, IL

Abstract

Poly(acrylonitrile-butadiene-styrene) or ABS is an engineering thermoplastic used in a wide range of applications like automotive parts, business machines, household appliances, toys, etc. Rubber particle size (PS) and particle size distribution (PSD) in emulsion ABS play a key role in determining the final performance of the ABS material. Typically there are 3 ways to control the PS and PSD: 1 direct growth, 2 chemical agglomeration, and 3 mechanical agglomeration. This article briefly compares the benefits and disadvantages of each method showing that a mechanical agglomeration technology is superior in operational and capital costs by reducing the reactor cycle time and thus increasing overall line productivity. Direct growth method has more flexibility in controlling particle size distribution.

Introduction

Poly(acrylonitrile-butadiene-styrene) or ABS is an engineering thermoplastic used in a wide range of applications like automotive parts, business machines, household appliances, toys, etc. General properties of ABS and major production technologies have been extensively reviewed in open literature (1) (2) (3).

Both the average particle size (PS) and the particle size distribution (PSD) of the rubber particles have a strong effect on the final properties of the ABS thermoplastic (2). As an example, obtaining high gloss can be achieved with an average rubber particle size of approximately 100 nm, but will also yield low Izod impact performance and poor tensile elongation (4). On the other hand, a similar balance of properties is possible by varying other ABS structural factors independent of latex technology utilized (3) (5).

Rubber used in ABS is typically polybutadiene (PBD) for its low glass transition temperature. However other rubber has also been widely used such as copolymers of styrene-butadiene and acrylonitrile-butadiene, but tends to result in poorer low temperature impact performance. For simplicity of discussion PBD is used for rubber phase in this article.

Most commercial ABS made from emulsion technology has average particle size of 300 nm to 400 nm. The effects of average particle size on notched Izod impact as shown in Figure 1 indicates that impact strength maximizes at average PSD of 300 – 400 nm.

Typically a certain particle size distribution consisting of both large and small particles can achieve better balance of impact performance, flow, and gloss, etc., than a narrowly distributed monomodal particle size (6). It is known that small particles are more efficient in inducing shear deformation and large particles are known to cause craze deformation in the matrix phase (7) (8). It has been indicated that having both deformation modes in the material results in enhanced toughness. Some technologies make rubber latex with small and large particles separately and then blend in latex (9) (10) (11) or in a compounding extruder (6). More specific PSD's like tri-modal rubber particle size blends have been described in some patents but requires an investment of two manufacturing processes which include ABS made from both emulsion and bulk processes (6).

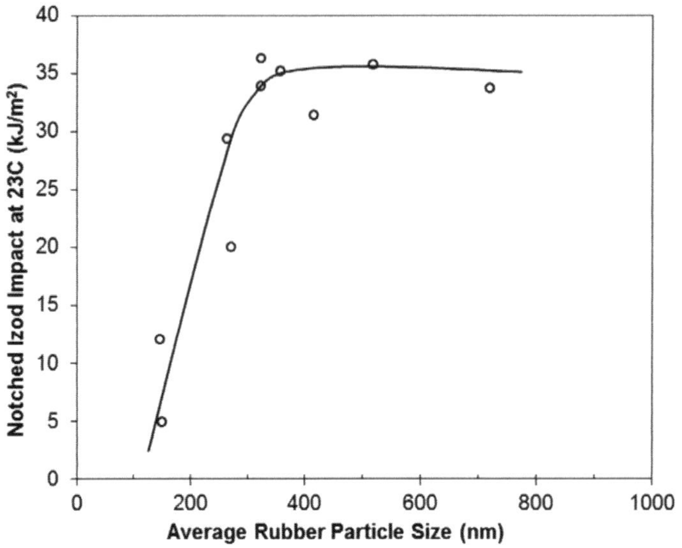

Figure 1. Effects of average rubber particle size on notched Izod impact.

Technologies to Increase Particle Size

The most common technologies for achieving both PS average and PSD are 1) direct growth, 2) chemical agglomeration utilizing acid, anhydrides, and milder hydrophilic & hydrophobic monomer(s), and 3) mechanical agglomeration including "Freeze Thaw", "Agitation", and "Pressure Homogenization".

Direct growth

One way to create bigger particle size is by direct growth method when less emulsifier is used at extended reaction time. According to Smith-Ewart theory, reducing emulsifier loading and extending reaction time will increase particle size. It was reported from pilot scale experiments that particle size can grow to about 100 nm at 12 hours, and 300 nm at 72 hours (12). In order to reduce reaction time, latex having small particle size can be used as a seed. When more monomers are added to the seed latex, the particles continue to grow into much larger size (13). Recently it was reported that 300 nm could be achieved within 30 hours of reaction at industrial scale by adding batches of emulsifier during the reaction and optimizing usage of electrolyte (14) (15).

Typically the particle size distribution of the latex produced from direct growth is narrow. Some companies chose to blend latexes with different particle size in order to achieve a bi-modal particle size distribution in the final product (11). A schematic process diagram of a typical direct growth technology is shown in Figure 2.

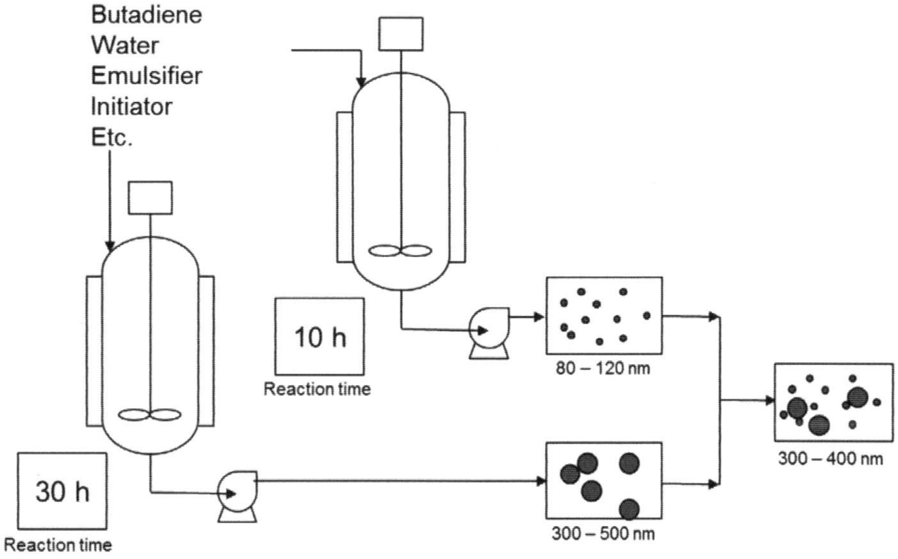

Figure 2. Schematic flow diagram of direct growth technology.

Chemical agglomeration

Acid and anhydride

There is a long history in the use of acid in agglomerating particle size in synthetic latexes. In 1948 Standard Oil Company patented a technology of using a combination of salt and acid to agglomerate the particles in latex but not causing much coagulation (16). In 1966 Shell Oil Company patented a process where a combination of salt and acetic acid were used (17).

In addition to acid, anhydride could be used to agglomerate the particles. In 1970 Monsanto patented a process where acid anhydride was used in a batch process (18). The water soluble acid anhydride is mixed with the latex, and hydrolyzed anhydride reduces the stability of latex causing particles to agglomerate. After the agglomeration is completed, emulsifier is added to stabilize the agglomerated latex. In 1977 Monsanto patented a continuous process utilizing acid anhydride (19). However since large amount of coagulum was formed using acid or anhydride method, GE Plastics patented a technology that uses a mixture of acid and anhydride that reduced the amount of coagulum to essentially zero (20).

Latex

Acid agglomeration has some disadvantages including large amount of coagulum and difficult process control. So a milder agglomerating agent was desired. One of the well-studied agent consists of a copolymer with both hydrophilic and hydrophobic properties (4) (21) (22) (23) (24) (25) (26). Basic hydrophilic characteristics are achieved using functional monomers such as acrylic acid, methacrylic acid, maleic anhydride, itaconic acid, cortonic acid, methacrylamide, ethacrylamide, and n-butyl acrylamide. The hydrophobic characteristics are achieved with monomers such as methylacrylate, methyl methacrylate, ethyl acrylate, butyl acrylate, 2-ethylhexyl acrylate, octyl acrylate, styrene, vinyl acetate, and butadiene. The polymer can either be made into a latex using conventional emulsifier (21) (22) or grafted onto a rubber substrate forming a core-shell structure (4). Some studies suggested that the selection of surfactant of the agglomerating agent is also important (27). The usage of the agglomerating agent is generally 2-5% of the total polymer (22) (23). Timing of addition can be during the PBD reaction (23), after PBD reaction is completed but still in reactor (4) (25), or after reaction and in a separate agglomerating vessel (24) (26). The duration of agglomeration is typically 1 to 4 hours (24) (26) or even up to 12 hours (25). The resulting particle distribution can be monomodal (26) or bimodal depending on quantity of agglomerating agent being used (4).

Due to its complicated chemical structure, the agglomerating agent is most often synthesized in-house in a separate reactor (4) (24) (25) using a free radical emulsion polymerization process. The emulsion and initiation system for the agglomerating agent preparation may or may not be the same as those used for the PBD latex. However one article pointed out that commercial grade poly acrylic acid appeared to work as well at pilot scale (28). A schematic process diagram of a typical latex agglomeration is shown in Figure 3.

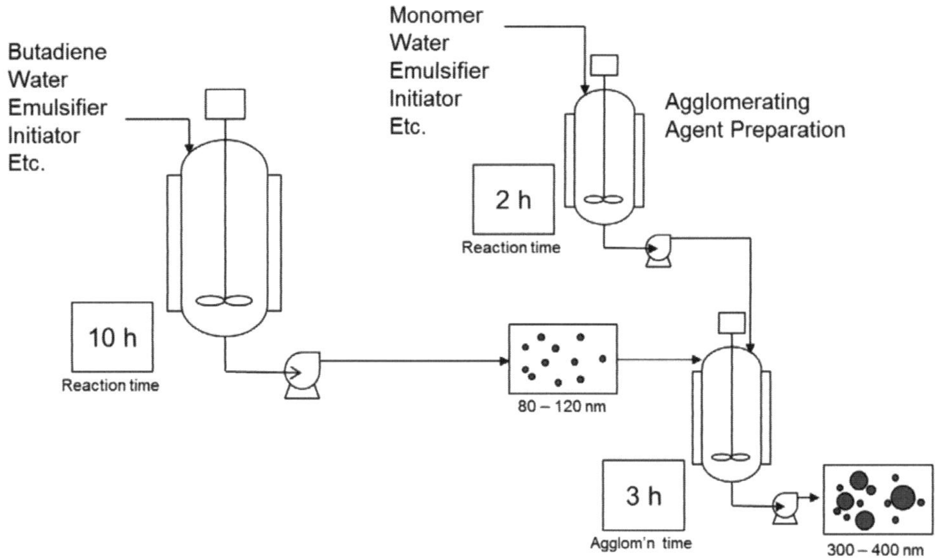

Figure 3. Schematic flow diagram of latex agglomeration.

Other

Other chemicals have also been used for agglomerating rubber latex to form large particles. In one patent polyethylene oxide is oxidized by air with the aid of small amount acetic acid. The product is then added to a synthetic rubber latex. Under agitation the rubber particle size can be increased from 50 nm to 1000 nm (29).

Mechanical agglomeration

Freeze-thaw

The freeze-thaw technology was development mainly for producing synthetic latex containing about 60% solid. Latex at about 35-45% solid content was frozen at about -20 to -40°C and then thawed. The rubber particles agglomerate and form larger particles (30) (31). According to the experience from a commercial ABS company, the average particle size could be increased from 50-60 nm to 160 nm using a freeze-thaw technology. Due to the limitation of the size in the agglomerated particles as well as the long process and high energy consumption, the company abandoned the technology (15).

Agitation

Vigorous agitation can also create large rubber particles. In order to achieve this, the latex polymerization is designed to have multiple monomer additions. The rubber particles are swollen with monomers added in later additions, and under vigorous agitation, the particles collide and form agglomerate. The particle size distribution is broad and the technology can generate extremely large rubber polymer particles with size up to approximately 15,000 nm (32).

Pressure gradient (homogenizer)

Pressure gradient has been used to overcome the repulsing force between particles created by the emulsifiers. The rubber particles collide and combine, forming larger particles. This technology is also referred as pressure agglomeration or homogenization. The latex is pushed through a small orifice or gap under high pressure such that a large pressure gradient is created (33). The pressure can range from 0.5 MPa (34) (35) to 70 MPa (33) (36). The particle agglomeration takes place in small chamber within the homogenizer. The gap between the valve seat and the valve is typically 10-500 μm and overall capacity of the working volume is around a few cubic centimeters (37) (38). Figure 4 shows a typical homogenizer valve assembly.

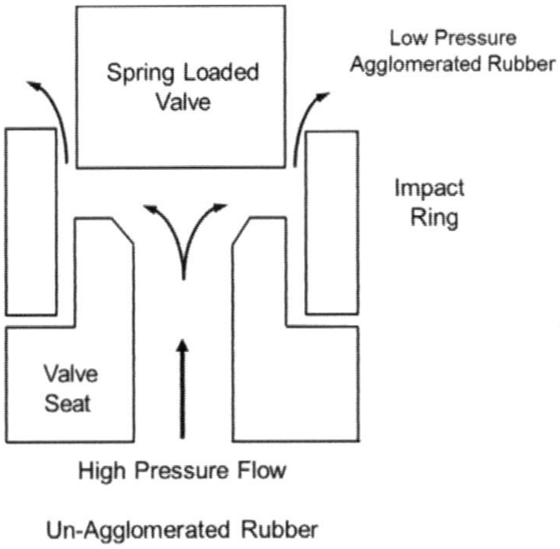

Figure 4. Schematic drawing of a homogenizer valve, seat and impact ring.

Typically pressure-homogenized latex has bimodal or tri-modal PSD with average PS of 200 nm to 500 nm. Several methods can be used to control the average PS and PSD. Higher agglomerating pressure, latex solids, lower temperature, and lower pH usually lead to larger particles (2) (36). Figure 5 shows PSD generated at various process conditions illustrating the flexibility of PSD control.

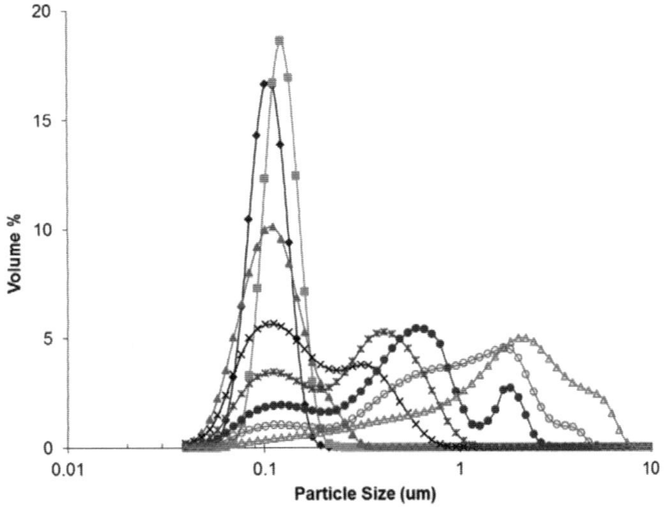

Figure 5. Typical PSD made by pressure homogenization technology.

This method has many advantages. It's a continuous process so the inherent variation associated with batch process is reduced. The operating temperature to achieve the desired particle size average can have a wide operating window typically above room temperature. Also there is no additional chemical used, so the cost is lower and the product contains less impurity. The main disadvantage is that the particle size distribution in the final latex is broader, but can be controlled to meet critical performance requirements. Regardless how the process is controlled, there are always some initial particles in the final latex. A schematic flow diagram of pressure homogenization is shown in Figure 6.

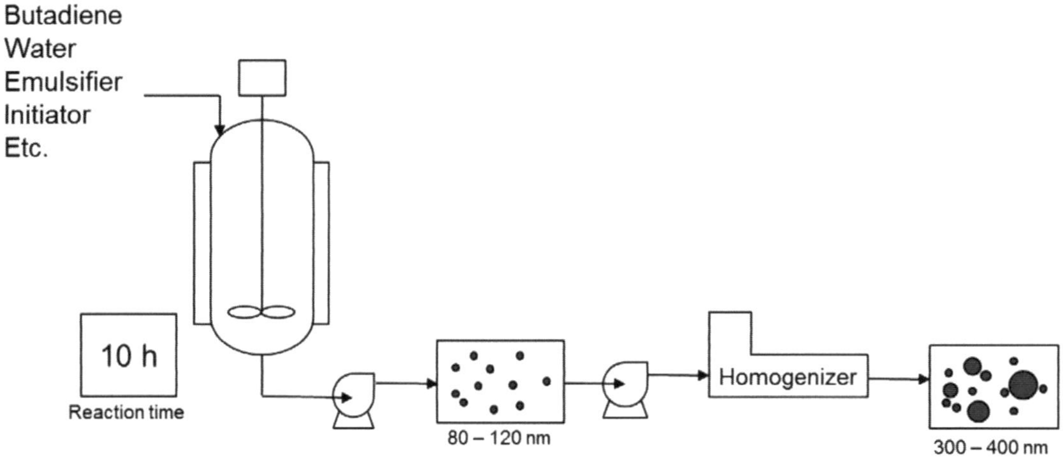

Figure 6. Schematic flow diagram of pressure homogenization technology.

Graft Process

After rubber latex is prepared, it is typically sent into a grafting reactor where acrylonitrile and styrene are added and grafted to the rubber particles (39). The grafted poly-acrylonitrile-styrene (SAN) serves as a compatibilizer such that the rubber particles can be well-bonded to the SAN matrix. This grafting is a necessary step for making a high performance ABS material.

Discussions

Among all particle size increasing technologies described above, 3 main technologies are chosen and a preliminary economic analysis was done to compare the equipment and operating cost. Since the information disclosed in the open literature did not contain enough technical and production information for a full economic evaluation, the following assumptions were made here so that some major aspects of the particle size increasing technologies can be compared. Other assumptions, such as cost of reactors, are listed in the table.

1. Latex stability is not a concern for all technologies during and after the agglomeration.
2. Similar amount of coagulum is generated, if any, in all technologies.
3. Agglomerated particles can be further processed to produce ABS with equivalent quality.

Equipment Cost

Assuming PBD reactors are 50 m^3 each and reaction cycle time is 10 hours for small particles. For direct growth method reactor cycle time is 30 hours for large particles. For chemical agglomeration method the agglomeration occurs in reactors at the end of reaction and it takes 3 hours to complete. For pressure homogenization method 2 homogenizers are needed. The number and cost of key equipment for making PBD latex for a 200 thousand ton per year ABS plant is listed in Table 1. The equipment cost for each reactor system is estimated to be approximately $0.75 million, and each homogenizer is about $0.4 million. Total installed cost (TIC) is estimated to be 6 times of equipment cost (40).

Table 1. Cost of key equipment used by various technologies.

	Direct Growth	Chemical Agglomeration	Pressure Homogenization
Number of reactor systems	7	4	3
Batch Cycle Time (h)	10 & 30	13	10
Additional Equipment	None	1 small reactor systems for making agglomerating agent	2 homogenizers
TIC (million $)	30	20	18

Energy Consumption

For direct growth and chemical agglomeration technologies, major electricity consumption for the reactor systems are reactor agitators, discharge pumps, and jacket cooling water circulation pump. The homogenizers use relatively large feed pumps to provide high pressure gradient. Total amount of electricity usage for direct growth technology is estimated to be about 40% higher than electricity consumption in chemical agglomeration of pressure homogenization. A comparison of estimated utility consumption (40) between the technologies is shown in Table 2.

Table 2. Energy consumption comparison between various technologies.

	Direct Growth	Chemical Agglomeration	Pressure Homogenization
Reactor Agitator Motors (kW)	$7 \times 40 = 280$	$5 \times 40 = 200$	$3 \times 40 = 120$
Reactor Discharge Pump Motors (kW)	$7 \times 20 = 140$	$5 \times 20 = 100$	$3 \times 20 = 60$
Reactor Jacket Circulating Pump (kW)	$7 \times 40 = 280$	$5 \times 40 = 200$	$3 \times 40 = 120$
Homogenizer Feed Pump Motors (kW)	0	0	$2 \times 100 = 200$
Total power (kW)	700	500	500

Chemical Consumption

No new chemicals are used in direct growth and pressure homogenization technologies. However chemical agglomeration typically uses a small amount (typically 2-5%) of new chemicals. Depending on the chemistry of the agglomerating agent and the synthesizing process, the cost of the agent typically increases the overall cost of the ABS material. Also, the new chemicals will lead to additional costs associated with inventory, material handling, and possible environmental and safety concerns. Table 3 shows the usage of new chemicals by various technologies.

Table 3. New chemicals used in various technologies.

	Direct Growth	Chemical Agglomeration	Pressure Homogenization
New Chemicals	None	2-5%. Typically acrylate type polymers or acetic acid.	None.

ABS Properties

ABS properties depend on many material structural features like grafting efficiency, rubber content and type, SAN matrix molecular weight and distributions, rubber particle size and distributions and additives. For commercial ABS materials available, various factors are optimized to obtain a competitive property balance for all rubber latex types. Some properties like chemical resistance of ABS polymers are mostly dominated by composition of matrix and graft SAN and the molecular weight with very little influence of rubber particle size and preparation techniques. Similarly, heat distortion temperature (HDT) of ABS is mainly affected by matrix SAN and the rubber content. When properties of various ABS materials are compared at equivalent flow and rubber content, one finds only slight differences in other properties like toughness and mechanical strength properties.

Figure 7 and Figure 8 compare Izod impact vs modulus and Izod impact vs melt flow data estimated from published data sheet properties of various commercial emulsion ABS materials. Since the modulus values for emulsion ABS are expected to be related to rubber content in the formulation, it is clear from Figure 7 that Izod impact values vary with rubber content and not with latex (rubber particles) preparation techniques. A similar conclusion can be drawn from Figure 8 indicating no significant effect of latex preparation techniques on impact/ flow balance. High gloss and gloss stability may be superior for mechanical agglomeration methods whereas colorability may be slightly superior for chemical and direct growth latex processes.

In general, based on commercial data sheet properties and applications of various ABS products it can be stated that overall balance of properties is very similar for all latex technologies commercially employed.

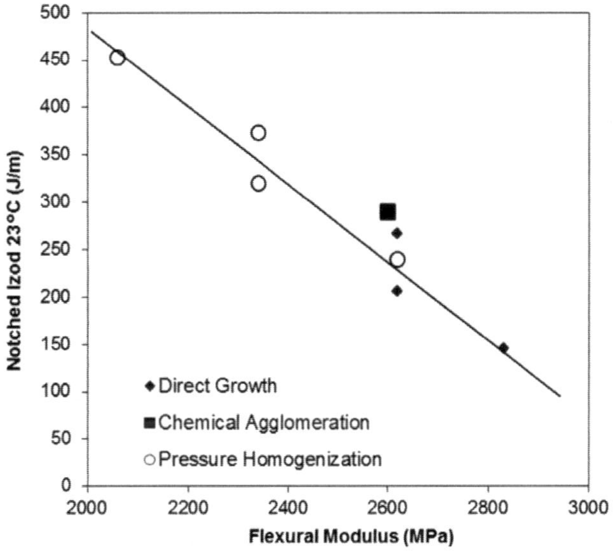

Figure 7. Impact vs flexural modulus.

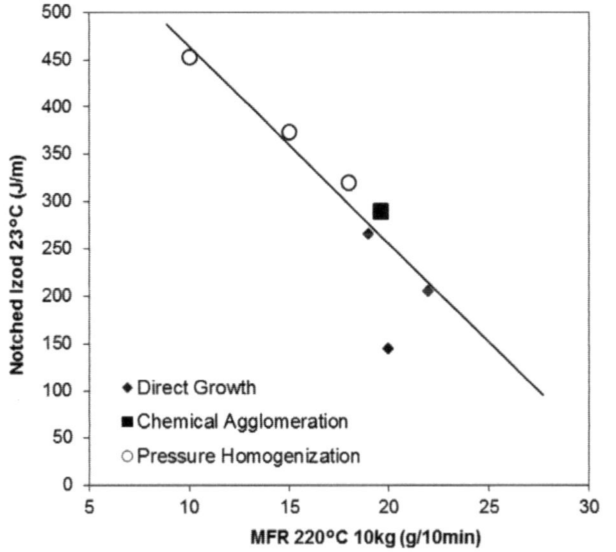

Figure 8. Impact vs melt flow.

Transmission Electron Microscopy (TEM)

TEMs of ABS made from these 3 technologies are shown in Figure 9. The PSD of rubber particles made from direct growth technology is narrow and close to a monomodal distribution. The chemical agglomeration clearly shows a bimodal distribution showing a mixture of many small but uniform-sized particles and some large particles. The PSD of pressure homogenization is broad, showing un-agglomerated as well as larger agglomerated particles up to 0.5 micron.

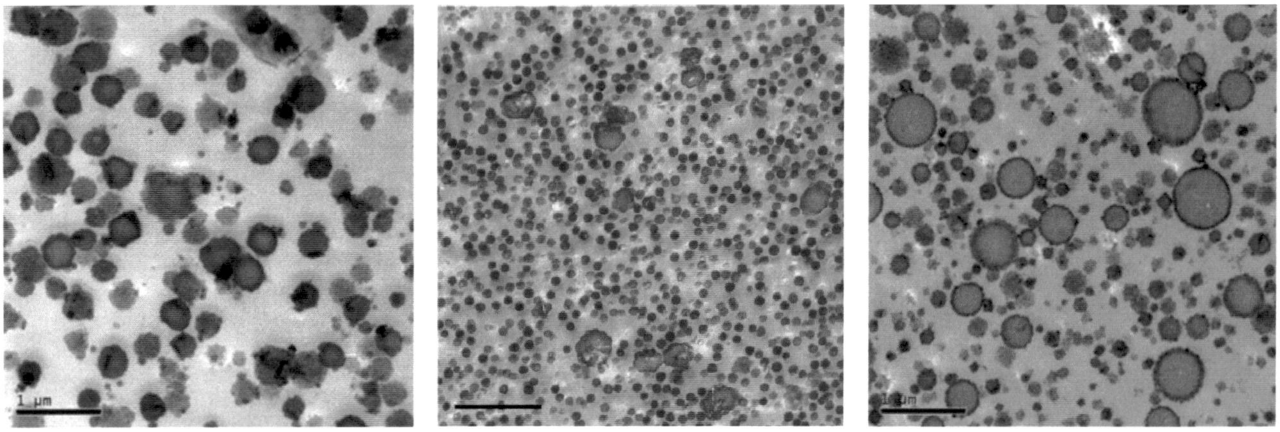

Direct Growth | Chemical Agglomeration | Pressure Homogenization

Figure 9. TEM of ABS made from various technologies (scale bar: 1 µm).

Conclusions

In this article various technologies for increasing PBD particle size are reviewed. Among them three most common technologies were compared side-by-side with respect to equipment and operating cost. It is clear that pressure homogenization technology has the lowest equipment and operating cost, followed closely by direct growth, then chemical agglomeration method. From ease of operation point of view, direct growth has more advantages as the particle size can be precisely controlled over a wide range of sizes. Chemical agglomeration can be advantageous if reducing initial equipment cost is more important comparing to direct growth technology while still maintaining good particle size control.

References

1. **Kulich, Donald M, et al., et al.** Acrylonitrile–Butadiene–Styrene Polymers. *Encyclopedia of Polymer Science and Technology.* s.l. : John Wiley & Sons, Inc., 2002.
2. **Berry, VK, et al., et al.** Acrylonitrile-Butadiene-Styrene. [ed.] Joseph C Salamone. *Polymeric Materials Encyclopedia.* s.l. : CRC Press, 1996, Vol. 1, pp. 82-101.
3. **Kulich, Donald M., Kelley, Paul D and Pace, John E.** Acrylonitrile-Butadiene-Styrene Polymers. *Encyclopedia of Polymer Science and Engineering.* New York, NY : John Wiley & Sons, 1985, Vol. 1, pp. 388-426.
4. **Henton, David E and O'Brien, Teresa M.** *Particle Agglomeration in Rubber Latices. 4,419,496* United States of America, December 6, 1983. Dow Chemical Company.
5. **Dion, Robert P, Mitchell, David W and Leng, Patricia B.** *Process for the Preparation of Multimodal ABS Polymers. 5,955,540* United States of America, September 21, 1999. Dow Chemical Company.

6. **Henton, David E.** *ABS Compositions Having Trimodal Rubber Particle Distributions. 4,713,420* United States of America, December 15, 1987. Dow Chemical Company.
7. **Donald, AM and Kramer, EJ.** *Journal of Material Science.* 1982, Vol. 17, p. 1765.
8. **Bucknall, CB and Clayton, D.** *Journal of Material Science.* 1972, Vol. 7, p. 202.
9. **Ott, Karl-Heinz, et al., et al.** *ABS Moulding Compositions. 4,009,227* United States of America, February 22, 1977. Bayer Aktiengesellschaft.
10. **Kawashima, Kouzo, et al., et al.** *Thermoplastic Resin Composition. 5,008,331* United States of America, April 16, 1991. Japan Synthetic Rubber Co., Ltd..
11. Samsung ABS Technology Licensing Introduction. *Samsung Cheil Industries.* [Online] Samsung. http://www.starex.co.kr/jsp/eng/tech_center/sms_abs.jsp.
12. **Guan, Guomin.** Effect of electrolyte on particle size of polybutadiene latex. *China Elastomers.* February 25, 2010, Vol. 20, 1, pp. 77-79.
13. **Kasai, Kiyoshi, et al., et al.** *Process for Preparing large-sized Polymer Particles. 4,694,035* United States of America, September 15, 1987. Japan Synthetic Rubber Co., Ltd..
14. **Sun, Yanjun and Wang, Jun.** Influence factors on polymerization and product quality of emulsion butadiene rubber. *China Synthetic Rubber Industry.* September 15, 2001, Vol. 24, 5, pp. 265-267.
15. **Zhang, Chuanxian.** Research and Development History of Petrochina Lanzhou's ABS Business. *Petrochemical Technology & Application.* January 2003, Vol. 21, 1, pp. 29-36.
16. **Arundale, Erving.** *Creaming of Synthetic Rubber. 2,444,801* United States of America, July 6, 1948. Standard Oil Development Company.
17. **Moss, Fred Dudley.** *Process for the Preparation of Synthetic Rubber Latex. 3,281,386* United States of America, 10 25, 1966. Shell Oil Company.
18. **Dalton, William O.** *Method For Agglmerating Latices With An Anhydride. 3,358,541* United States of America, January 26, 1971. Monsanto Company.
19. **Paster, Mark D.** *Continuous method of agglomerating aqueous latices. 4,043,955* United States of America, August 23, 1977. Monsanto Company.
20. **Kulich, Donald M.** *Agglomeration of latices by organic acid--anhydride mixtures. 5,633,304* United States of America, May 27, 1997. General Electric Company.
21. **Ford, Emory A.** *Method of Graft Polymerization and Polymers Produced Thereby. 3,825,621* United States of America, July 23, 1974. Monsanto Company.
22. **Ide, Fumio and al, et.** *Process for Preparing an Impact Resistant Thermoplastic Graft Copolymer Composition. 3,944,630* United States of America, March 16, 1976. Mitsubishi Rayon Co., Ltd..
23. **Kidder, Kevin R.** *Method for Preparing Emulsion Polymerized Polybutadiene of Increased Particle Size. 5,294,659* United States of America, March 15, 1994. General Electric Company.
24. **Lu, Shulai.** *A Method for Preparing Super Large Particle Size Polybutadiene Latex. 200910237011* China, May 11, 2011. PetroChina Company Ltd..
25. **Lee, Byeong-do and Jin, Dong-un.** *Method for Preparing Polybutadiene Latex. 6,365,669* United States of America, April 2, 2002. Cheil Industries Inc..
26. **Hu, Wenxi, Wu, Guangfeng and Zhang, Huixuan.** Synthesis and application of agglomeration agent to latex. *China Elastomers.* June 25, 2012, Vol. 22, 3, pp. 61-64.
27. **Pippa, Roberto, et al., et al.** *Process for the Agglomeration of Rubber Latexes. 4,357,270* United States of America, November 2, 1982. Montedison S.p.A.
28. **Zhang, Dongmei, et al., et al.** Research of Special Polybutadiene Rubber Latex for Sheet ABS Resin. *Engineering Plastics Applications.* 2009, Vol. 37, 9, pp. 26-28.
29. **Shluter, Herburt.** *Latex Agglomeration Using Oxidized Polyalkylene Oxide. 3,330,795* United States of America, July 11, 1967. Chemische Werke Huls Aktiengesellschaft.

30. **Sutherland, John D.** *Agglomeration of Synthetic Rubber Latex. 3,296,178* United States of America, January 3, 1967. Copolymer Rubber & Chemical Corporation.
31. **Carpenter, Paul G.** *Process for Enlarging the Average Particle Size and Concentrating Synthetic Rubber Latex and Products Obtained Thereby. 2,993,020* United States of America, July 18, 1961. Copolymer Rubber & Chemical Corporation.
32. **Gauslaa, Kjetil.** *Process for Preparing High Solids Concentrated Latex. 3,318,831* United States of America, May 9, 1967. Copolymer Rubbere & Chemical Corporation.
33. **Bennett, DA.** *Synthetic Rubber Latices. 976,212* Great Britain, November 25, 1964.
34. **Maeda, Tetsuro and Okamoto, Akihiro.** *Thermoplastic Resin Composition. 4,965,315* United States of America, October 23, 1990. Denki Kagaku Kogyo Kabushiki Kaisha.
35. **Mimori, Mitsumuki.** *A method for increasing particle size in synthetic latices. 48-4458* Japan, Feburary 8, 1973. Tokyo Denki Kagaku.
36. **Kim, Young Min and al., et.** *A process for preparing thermoplastic resins. EP 0473400 B1* Europe, January 31, 1996.
37. **Kleinig, Andrew R and Middleberg, Anton PJ.** The Correlation of Cell Disruption With Homogenizer Valve Pressure Gradient Determined by Computational Fluid Dynamics. *Chemical Engineering Science.* 1996, Vol. 51, 23, pp. 5103-5110.
38. **Casoli, Paolo, Vacca, Andrea and Berta, Gian L.** A Numerical Procedure for Predicting the Performance of High Pressure Homogenizing Valves. *Simulation Modelling Practice and Theory.* 2010, 18, pp. 125-138.
39. **Calvert, William C.** *Polymerization of Acrylonitrile-Alphamethyl Styrene Mixtures onto Polymerbutadiene and Products Products Therby. 2,908,661* October 13, 1959. Borg-Warner Corporation.
40. **Bell, Susan L.** *Process Economic Program Review No. 2010-07 ABS Resin Production.* s.l. : SRI Consulting, 2010.

Ensuring Lower Life-Cycle Costs by Integrating Maintenance Execution and Reliability Management Work Processes

David A. Rosenthal, PE, CMRP
Reliability Delivery and Asset Management Manager
Jacobs Engineering Group
Houston, TX

Summary

The paper will detail how the Maintenance execution and Reliability work processes interact with each other in a manufacturing setting to improve asset condition and reduce their life-cycle costs. These life-cycle costs include the purchase price, maintenance, parts, utilities, etc. This paper will detail where both processes interface, what information is exchanged, the tasks involved, and the consequences of not leveraging this integration.

Background

Manufacturing competitiveness relies on first meeting and then exceeding productivity goals. Supply chain systems deliver customer demands based on assumptions of reliability for the manufacturing assets at each site. Business models that predict production, distribution and delivery capacity contain base maintenance assumptions for each facility including times for repair, preventative maintenance downtime, overhaul frequency, and turnaround duration.

Maintaining the condition of these assets to meet the above objectives is a strong function of the continuous improvement of equipment strategies aimed at reducing Mean Time between Failures (MTBF). These strategies contain maintenance tasks needed to maintain their intended function. Maintenance care for critical assets is generally based upon vendor recommendations, historical experience and addressing known equipment failure modes. When MTBF rates fall outside of excepted industry benchmarks, manufacturers must analyze failure history and the effectiveness of their equipment strategies.

Maintenance is in charge of only restoring function while Reliability designs care to ensure lower life-cycle costs. The continuous improvement of asset care depends on the exchange of information between these two work processes. One can draw the analogy of the application of Plan-Do-Check-Act cycle. Key "touch points" exist between processes to ensure correct asset care is provided to eliminate failure modes seen during operation. This paper identifies each of these interfaces and discusses the importance of managing the information that passes between these work processes that are managed by Maintenance and Reliability personnel. Poor execution of the above will lead to shorter MTBF times providing increased risks to supply chain assumptions which ultimately could affect the customer.

Maintenance Work Process

The Maintenance Execution work process (MEWP) has not changed much over the past several decades. A typical process is shown below. The mission of the MEWP is to safely restore function of assets that are about to or have failed. Failure in the present day context is when the asset no longer performs the function it was designed to meet. For instance, a pump designed for 100 GPM has failed when it only produces 50 GPM. In another example, a conveyer belt designed for 100 FPM has failed that is now only moving at 75 FPM.

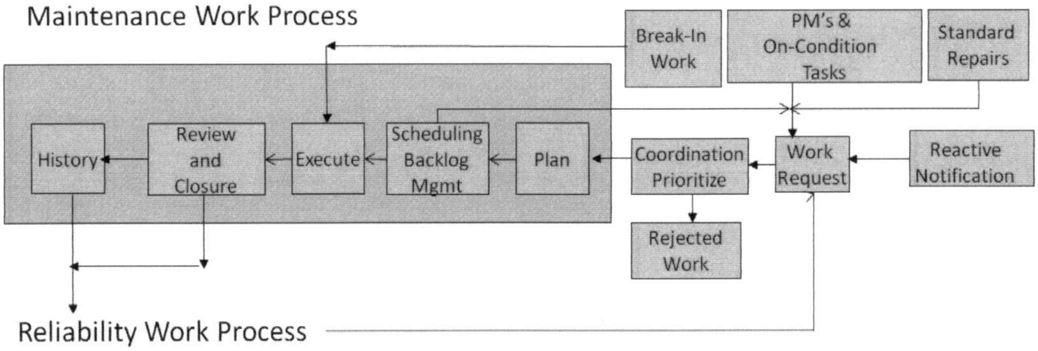

The MEWP receives many inputs from facility stakeholders. The obvious input is a work notification (work order) for a failed piece of equipment. Another could be a scheduled preventative maintenance task or another could be urgent work that must be addressed immediately. This process must then assess the priority of each type of work, plan what needs to be done (parts, resources, tools, procedures, permits) and then schedule resources to execute the work. However, that is not the end of its functionality.

After the work was performed, an analysis and history must be provided as an output for the record noted against the assigned component or asset. This history is vital for later analysis. It contains information of why it failed, what was done, and recommendations from the field to prevent future failures. This information is a key input for the Reliability work process which will be discussed. This information informs those who design and manage the care of the asset of the effectiveness of the current maintenance strategy.

Before we move on, there is a second interface as an input to the MEWP from the Reliability work process. Planned work such as preventative, predictive, time-based, and standard work prescribed by the Reliability work process drives another key interface. This tactical work is developed from the maintenance strategies for each critical piece of equipment. Failure to execute this work as prescribed increases the risk for failure.

The management of these two critical interfaces depends upon the asset care culture of the facility. Reactive organizations that are responding to excessive breakdowns do not allot time to their management. Proactive organizations drive recognition of the early signs of failure before function is lost. These organizations devote focus on continual improvement in asset care. Unfortunately reactive organizations are more plentiful than those that are proactive in their approach.

Reliability Work Process

The Reliability work process is shown below. It is shown tied to the Maintenance Execution work process. The Reliability work process is responsible for establishing asset maintenance, engineering, and operating strategies and continually improving them. The maintenance strategies are comprised of preventative, predictive, repair and life extension (overhaul) tasks. The mission of this combination of tasks is to enable the plant assets to perform at their desired function and thus meet the goals of the manufacturing organization.

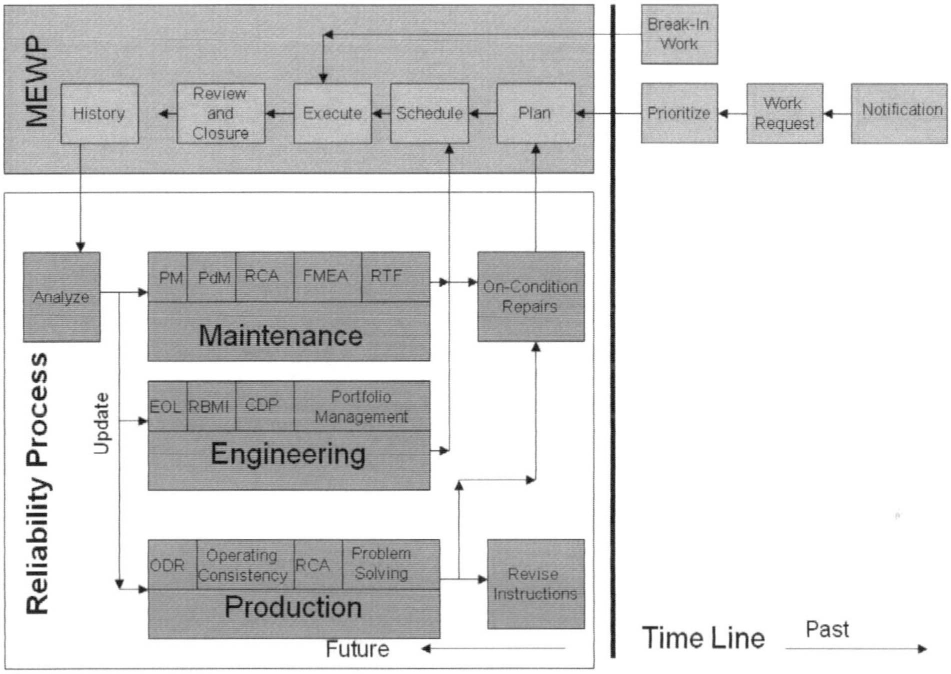

Note: PM – Preventative, PdM – Predictive, RCA – Root Cause Analysis, FMEA – Failure Modes and Effects Analysis, RTF – Run to Failure, EOL – End of life, RBMI – Risk Based Mechanical Integrity, CDP – Capital Deployment, ODR – Operator Driven Reliability

This process also includes Engineering and Production practices since the "triad" is responsible for the life-cycle practices associated with any asset. Although, the focus of this paper is the Maintenance interfaces, Reliability has equally important interfaces with both Engineering and Production. Especially, Production, since industry experience shows that operating practices have the greatest impact on maintaining the function of any asset. Operators have the most "exposure" to the assets and act as the "front line" to detect the earliest signs of failure. Their actions to adjust, calibrate, lubricate, monitor, and replace are essential to maintaining asset function. While, the Engineering practices have the next level of impact since it is estimated that 80% of the reliability of any asset is "built in" with the design.

Key Interfaces

There are two interfaces that will be described. They are the maintenance strategy – planning and scheduling interface and the work history - failure mode analysis interface. These two interfaces connect the Maintenance and Reliability work processes. These processes work in tandem to ensure the assets perform as per design. Their operation leads to the higher equipment availability, less downtime and a predictable supply chain performance for the business. Shown below are the interfaces and the typical components of the configuration "engines" for Production, Maintenance, and Engineering.

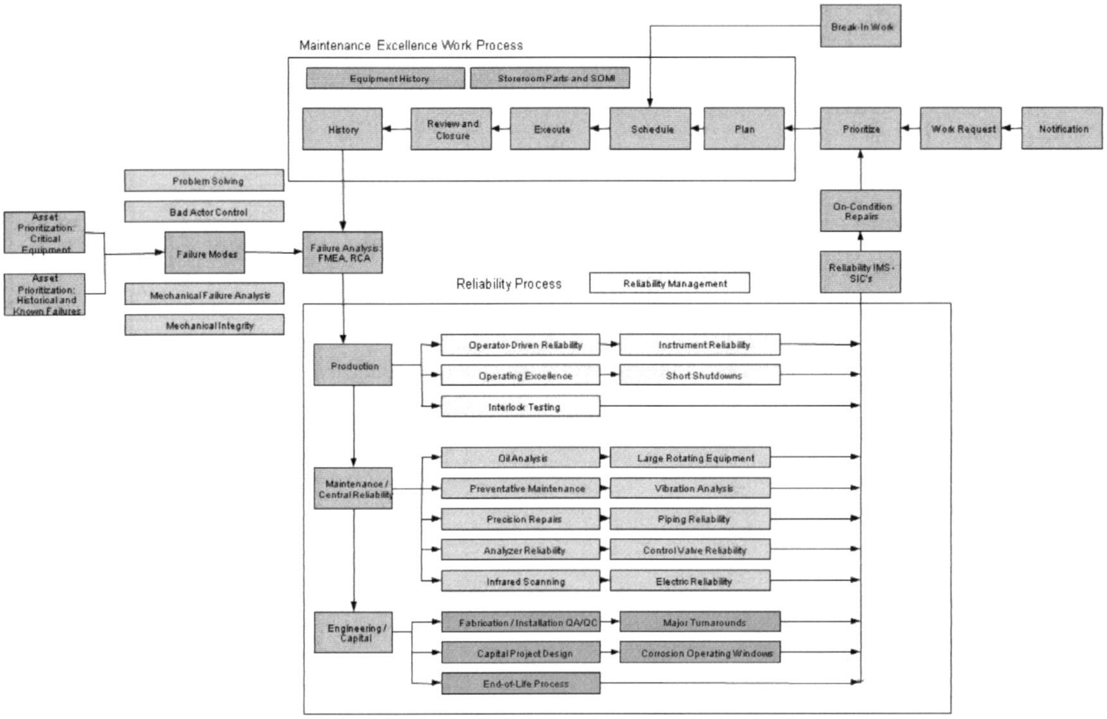

Maintenance strategy – Planning and scheduling interface

This interface between these processes ensures that prescribed care for first the critical assets and then the run-to-failure assets is executed. The planning and scheduling function receives this information through work orders, third-party work (Vibration, oil analysis), exceptions found during operator rounds, standardized work, etc. It is communicated electronically through the site's CMMS. Once it appears as a work order the planning function arranges the work for the maintenance schedule as planned work. Planned work is defined through several criteria including the presence of procedures, parts (both critical and non-critical), permits, tools, lifting and rigging plans (if needed).

If this interface is not working effectively, backlogged work grows, unplanned maintenance expands, maintenance costs rise, and the supply chain predictability is threatened. One major impediment seen at many sites is the prevalence of unplanned work on the schedule that "crowds out" the planned work that is needed to ensure high equipment availability. Moving towards performing more planned work is the subject of analysis, use of subject matter experts, and reliance on methods from previous case studies seen in the process manufacturing industry.

Work history - Failure mode analysis interface

The second interface allows for the continuous improvement in asset care. At the conclusion of most work orders, mechanics will write down what they found to repair the failure. Work history is vital for later analysis. It too though is a victim of a lack of understanding of how it is used and the information thought to be controlled only by mechanics. This work history must be recorded in the CMMS against not only the asset but against the specific component that has failed. The importance of having a complete asset hierarchy in the CMMS now comes into play. If work history is not recorded against the correct asset component or misplaced at a higher level in the hierarchy, an opportunity for later improving care is lost.

Work history includes not only what is recorded on work orders but what is recorded as "root cause" of the failure in the CMMS. Most CMMS have standard codes for failure type, cause, and remedy. These codes are often forgotten by those who "close out" work orders after their completion. This information truly aids those who are performing this later analysis.

For critical assets, work history should be analyzed to determine the existence of failure modes not previously used to design the current care plan. Generally, maintenance managers, maintenance engineers, or reliability engineers perform this task. Maintenance strategies for equipment (predictive, preventative) are designed from the both vendor supplied failure modes and those historically seen. Industry databases do exist which can provide a comprehensive failure mode view. The analysis involves reviewing equipment failure data, performing a failure modes and effects analysis (or based on experience) to determine changes in the maintenance care. For instance, a failure of a bearing on a rotating piece of equipment with no history might indicate the need for a revised bearing design, addition to the vibration route, or a note on an operator route to observe any abnormalities.

These changes to the asset care plan which are recorded in the site's CMMS will then be executed in the coming days, weeks, or months depending on frequencies. These changes represent a continual improvement in the care given to that asset.

Now that the two interfaces have been described, the implementation as per best industry practices will be detailed.

Implementation Scheme

The two interfaces described above are executed with a series of industry accepted practices. These practices must include assigned accountabilities and responsibilities for carrying them out. These are usually detailed in each site's maintenance execution and reliability work flows. These practices will now be described as observed over a multitude of process manufacturing industries.

Maintenance strategy – Planning and scheduling interface

Management of this interface generally is directed by the Maintenance Execution workflow and enabled through the site's Computerized Maintenance Management System (CMMS). Preventative maintenance work prompts are generated by the CMMS for the attention of the planner and scheduler. The majority of these tasks have standard work, parts, and procedures already attached to them. These components comprise what is termed a planned job. It is the job of the planner to order parts or reserve them if they are already in stock. Once this standard job is ready for scheduling, the scheduler determines where in the week's schedule this standard job will "fit" given current work and available resources.

Predictive work is generated both from the CMMS and from other information systems. Operator rounds are generally prompted through operations management systems. Fixed equipment inspections may be generated through the site's mechanical integrity management system. Third-party inspections for such tasks as vibration, oil analysis, thermography rounds are setup independent of the schedule unless they interfere with the operation of the facility. In that case, the inspections are placed in the maintenance schedule for the week. Some mechanic rounds are prompted through the CMMS for the planning scheduling function to manage.

All manufacturing sites struggle with scheduling all the work required to care for its assets. High levels of reactive (breakdown) work places the completion of these tasks in jeopardy since they cannot be accommodated in the schedule. The industry benchmark of greater than 80% weekly preventative maintenance work completion is a typical goal.

Work History - Failure mode analysis interface

This interface is managed both in the field and completed in the CMMS. It is the responsibility of first-line craft leaders to ensure that craft include the work history after work order completion. Especially for high priority breakdown work, this information is needed for analysis. Some sites attach a sheet to each work order with failure codes and reasons to assist field compliance. Utilizing troubleshooting tools like "5 Why's" can be helpful for field determination of the root causes of failures.

The next step involves the task of a planner, first-line craft leader, mechanic, or clerk to close out the work order. Accepted industry practice directs the mechanic to close-out work. However, this practice is dependent upon the staffing level of the organization and the labor culture. This involves ensuring all charges for parts and labor are assigned to the specific component that failed. This includes completing the failure codes and reasons that most CMMS offer in their functionality. Often these charges are not recorded against the right asset or at the right component in the asset hierarchy. Many times, information from the field is missing which leads to a missed opportunity to improve asset care to prevent the failure from re-occurring.

In most cases, a failure of an asset is from a known or forecasted failure mode for which the care tasks are designed. The reliability or maintenance engineering function of the site is responsible for determining if this is not the case and a new failure mode was observed. This detection can be done by reviewing regularly all high priority work orders completed. Some sites have a trigger system in place which is activated by the impact of the failure to either productivity or maintenance cost. Once the trigger is surpassed, the appropriate tool (Root cause analysis, failure modes and effects analysis) is applied. The purpose of these tools is to determine if the failure mode is new or previously observed.

Another technique is the use of a Pareto chart. This analysis used to identify "problematic assets" is generated from maintenance cost data and production loss time data. Assets are arranged from highest impact to the least against those two criteria. Generally, the work history from the top 3-5 assets is reviewed to determine the cause or causes of this higher impact.

Once the appropriate analysis tool is used, the appropriate change in the asset care plan is proposed. These changes can be change in the parts stocking, increased frequency of inspection, addition of preventative or predictive care tasks, or a design change. It is generally believed that two-thirds of these changes should be a movement toward more condition-based care moving away from preventative or time based care tasks. Only a third should be considered for design changes.

In closing, both the Maintenance Execution and Reliability work processes must be effectively linked to ensure manufacturing can reliability provide product to its customers. The failure to manage the two interfaces described above will put the supply chain at risk.

(108b) The Application of Lean Six Sigma within the Specialty Chemicals Industry

Darren Clark, Momentive Performance Materials, Waterford, NY

Lean Six Sigma (LSS) is a powerful and flexible methodology that has been utilized by a range of companies for delivering significant cost and growth productivity via simultaneous reduction of waste and defects from internal processes. LSS principles have been successfully applied to a full range of both batch and continuous processes in the chemical industry.

This presentation will provide an overview of the application of both the Lean and Six Sigma Methodologies within Momentive, a global manufacturer of specialty chemicals and materials. Examples will be shared on how this world class framework can promote safety, quality and operational effectiveness within the specialty chemicals industry.

(108c) Environmental Compliance Approaches in Manufacturing, and the Future of EPA Enforcement

Mary Ellen Ternes, McAfee& Taft, Oklahoma City, OK

The manufacturing industry sector is incredibly diverse, yet basic compliance approaches to achieve and maintain compliance are shared among industries. This presentation will review compliance assurance approaches, as well as methods of resolving issues of noncompliance and addressing potential enforcement actions arising from noncompliance. This presentation will also review the latest developments in EPA enforcement approaches and the future of EPA enforcement as conveyed by EPA Management.

ON THE CHALLENGE OF PRESENTING MEANINGFUL ENERGY DATA

Rein de Jonge, SABIC, Bergen op Zoom, The Netherlands
Deborah Downey, SABIC, Burkville, Alabama

Introduction

How do you present your site's energy data? Is it meaningful? Does it drive actions and progress? Finance, operations, business leadership may all have slightly different perspectives or methods to represent a plant's energy usage and performance. All of these perspectives and reporting methods may have value to their audiences. However, as this paper discusses, presenting meaningful energy data to drive improvement for a plant requires rigor, detail and analysis far beyond monthly utility bills.

Background

Energy has always been an important topic for the chemical industry – the chemical and petrochemical industries account for 30% of global industrial energy use, between 5-6 Exajoules/yr.[1] Publicly stated energy reduction targets are common today and have increased with the prevalence of Sustainability efforts and the advent of Corporate Social Responsibility (CSR) reporting. The world's top 10 Chemical companies[2] all published a CSR in 2011. CSR reporting for Fortune 500 companies grew from a meager 19% in 2010 to 53% in 2011[3]. This represents a staggering one year increase of nearly three-fold. Public interest in sustainability has traditionally been spurred on by significant and visible events, such as the first Earth Day in 1969, UN Earth summit Day in 1992, publication of the Kyoto Protocol and most recently, supplier sustainability demands from companies like Sony and Walmart[4].

Similarly, interest in energy efficiency is also on the rise; however, its importance can be tied to the more mundane topic of spikes in energy prices (as shown by the historical trend of Henry Hub Gulf Coast Natural Gas Spot Price[5] (Figure 1)). Energy intensive industries are both blessed and cursed with energy costs being a significant portion of controllable, operating costs. The "curse" is having to absorb fluctuating energy costs like NG spikes in the US to over $13/MMBTU just a few years ago. The "blessing" lies in the sheer magnitude / size of our utility bills. The gross dollar savings opportunity is simply too significant for management to ignore and affords leadership the opportunity to dedicate resources, time and efforts to a regimented approach to energy efficiency.

---

[1] Tracking Industrial Energy Efficiency and CO2 Emissions (2007), Industrial Energy Agency
[2] http://www.chemistryviews.org/details/ezine/2517771/Top_Ten_Chemical_Companies_in_2011.html
[3] Georgia Institute.com www.gainstitute.com
[4] Davies, John and the Editors of GreenBiz.com, "State of the Profession", January 2013
[5] US Energy Information Agency

Figure 1, Henry Hub Gulf Coast Natural Gas Spot Price

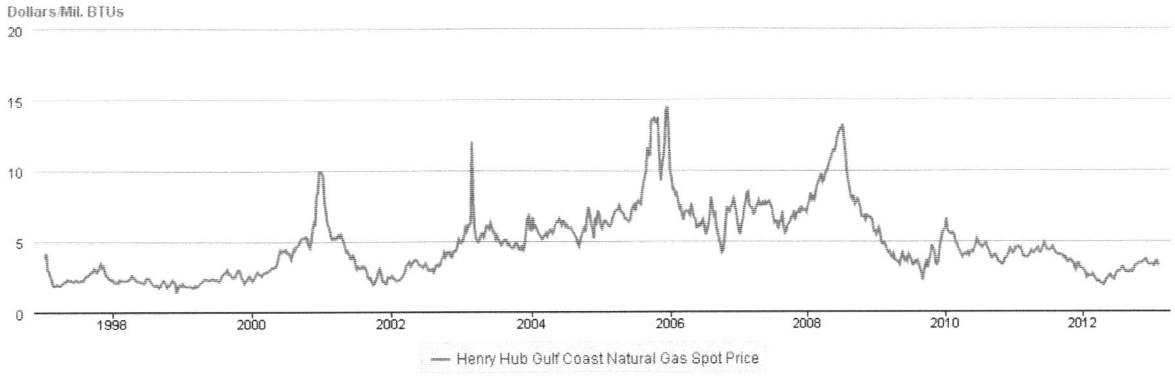

Fundamentally, chemical industry is interested in energy reduction and energy costs to become more profitable and increase competitiveness. There is a desire to improve energy usages, reduce utility consumption and conserve energy across our facilities. In order to make improvements, regimented techniques are not only helpful, but necessary to make lasting shifts in performance.

Process Improvement – Plan-Do-Check-Act

The Plan Do Check Act cycle, also called PDCA, Deming cycle or Shewhart Cycle can be found in nearly all process improvement designs and is clearly evident in today's Energy Management Improvement processes. As examples, Six Sigma DMAIC (Define, Measure, Analyze, Improve and Control), EPA Energy Star's Guidelines for Energy Management[6] and ISO50001 Energy Management improvement processes are all fundamentally based on variants of the PDCA cycle (Figure 2).

Figure 2, Shewhart Cycle, EPA Energy Star Guidelines for Energy Management

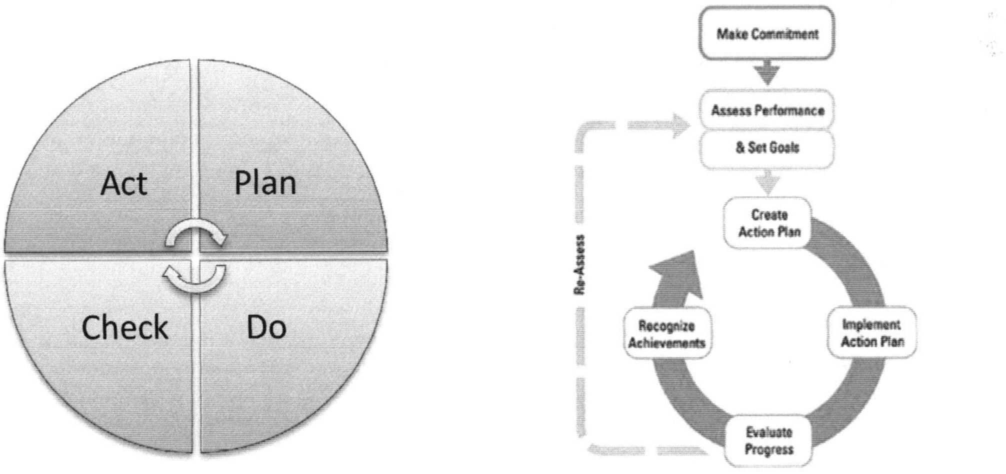

[6] **EPA Energy Star, energystar.gov**

In operations, the PDCA cycle fundamentally relies on having a valid metrics by which to measure progress. The below three elements of energy data are keys to maximizing the impact of energy efficiency improvement efforts. More accurately stated, these are the ways to keep delivering on your targets/goals:

1. Understanding the different levels of an organization and how data and metrics must be customized for each level
2. Driving a comprehensive approach -- many functions contribute to energy, relevant metrics must exist to provide feedback for each area's contribution
3. Driving a continuous process -- metrics must help maintain control, identify and eliminating leakage

The topic and title of this paper is critical for all three of these elements - Presenting meaningful Energy data to all levels of a corporation. The PDCA cycle may be revisited at different time intervals as appropriate for the different functions and/ or levels of business.

Different levels of an organization require different levels of data.

In driving a corporate initiative, goals and targets are often set for the business at a high level and aimed at external stakeholders (shareholders, customers and the general public). These targets are cascaded down to the lower level organization. Most companies are composed of many manufacturing sites, diverse in size, complexity, geography, etc. Making the targets meaningful at every level of the organization is critical to success.

In the case of Energy and Sustainability, the corporate, business unit and site energy KPI targets for reducing Energy Intensity are in units of GJ/ton. The use of GJ as an overall metric for energy makes sense as there is a need to aggregate all of the energy used in all forms (electricity, NG, gasoline, steam, etc) into a single metric. Even at the site level, the use of GJ/ton as a metric is of value because a site is often comprised of multiple units, making different products. However, at some point, the use of GJ is of less value – operators simply do not have levers that can directly affect "GJ". At lower levels in the organization, we transition from the use of generic GJ to more relevant, recognizable energy units of kwh, tons steam, etc).

Driving a Comprehensive Energy approach

This article is on the control of progress towards energy sustainability goals. We have learned that having a good Plan, Check and Act in place is crucial in our journey. As transparency between goals and execution rises, opportunities for improvement will reveal themselves. This however is insufficient to reach our goals. To reach our Sustainability goals, a structured approach to Energy Management is required. All functions (Figure 3) need to be engaged along with valid metrics to provide necessary feedback as to where the opportunity may lie.

Figure 3, SABIC Energy Value Chain

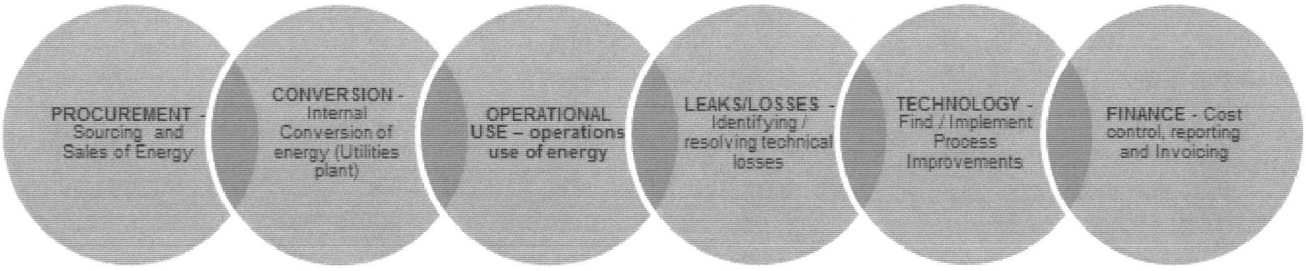

Continuous process, eliminating losses

Metrics must help maintain control, identify and eliminate the leakage of energy. So often, we focus on identifying and implementing new technology or improvement (CapEx) projects while taking an eye off the ball and allowing energy performance to slowly worsen over time. A basic, yet classic example is steam trap maintenance - steam traps have a life of approximately 10 yrs. Not having an ongoing steam trap management program will lead to deterioration in steam usage over time, often followed by a major steam trap repair capital expenditure. Other common examples are piping insulation and compressed air leaks. A continuous process will help mitigate these losses.

How to present meaningful data meaningfully?

Every company will, even if only for invoicing and accounting reasons, collect energy data. More complex sites will measure energy distribution to the various units within its boundaries. Typically this data is collected and disseminated to the plant managers and production engineers, compared to month over month, year over year, or quarter over quarter results. Within the Innovative Plastics Business Unit of SABIC (referenced as SABIC), we are no exception and until recently we would report absolute energy usage.

Figure 4, SABIC Site X, % of 2010 Energy Use

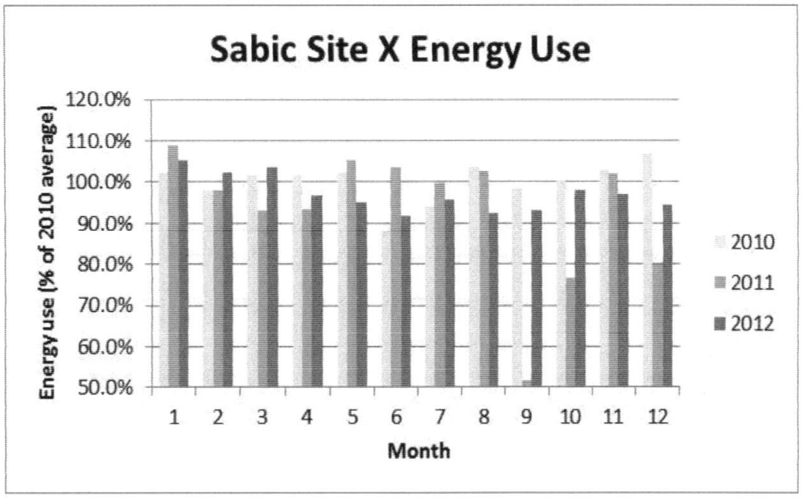

Try to assess the performance of the Site X asset from Figure 4. What conclusions would you make? Statistical tools are unable to distinguish between 2010 and 2012. We will demonstrate in this paper, how to turn this data into meaningful data.

The next evolution of measuring energy usage was to switch from absolute energy terms to energy per unit of product, intuitive since, from a business point of view, we normally look at conversion cost per unit of product. For Site X we have presented the Energy usage data in terms of Energy intensity in Figure 5. The keen observer will notice several anomalies.

1) A seasonal trend, during the mild temperatures of spring energy intensity is normally low.
2) 2011 was a year where a major turnaround was executed, with low volumes in Sept and Oct.
3) 2012 values appear to be consistently below those of 2010. Great! Statistical analysis confirms that there is indeed a significant drop in Energy Intensity.

Figure 5, Energy Intensity SABIC Site X, % of 2010 Average

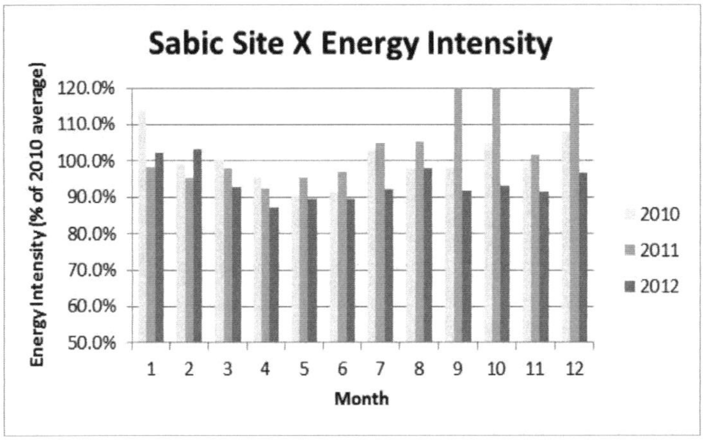

Volume Leverage

Our site energy leaders all intuitively understand that Energy Intensity has a volume component. The base energy use of a production unit is simply distributed over more units of product. Understanding the impact of volume on the energy use is a critical factor in advancing the game of the PDCA cycle in Energy Performance Management. Energy usage data requires normalization for volume and other factors not in control of the plant operators

Figure 6, Energy and Energy Intensity as a function of Volume

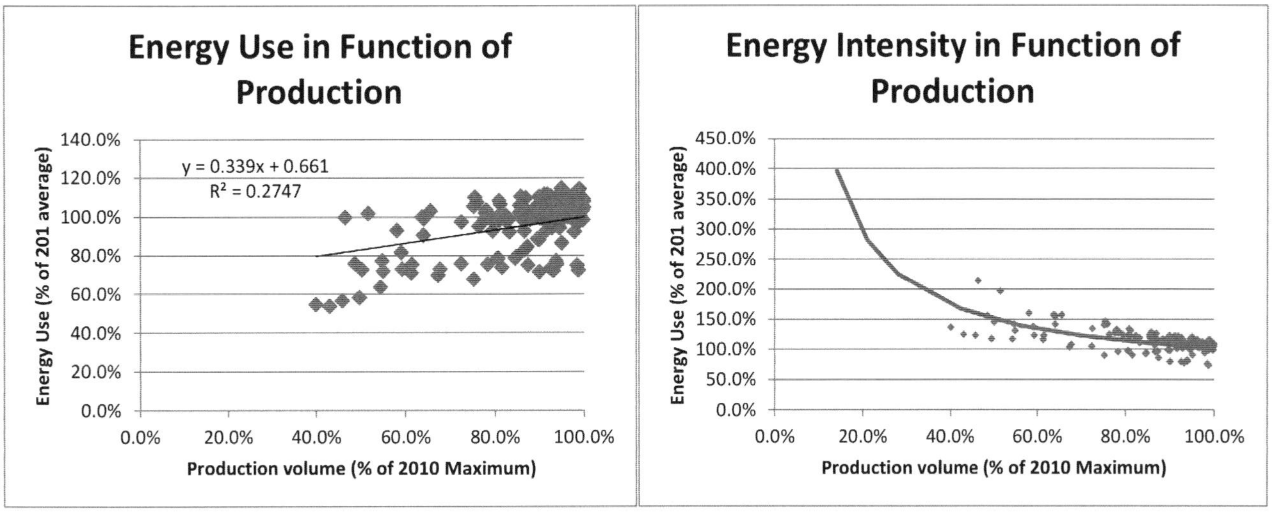

Energy Usage typically has a linear relationship with production volume. (Energy = A * Volume + B), Energy Intensity models should then have the form of (Energy Intensity = A + B / volume). Figure 6 shows the site energy use plotted against the sum of all plant's outputs. The resulting correlation is, with an R2 of 0.27, not impressive in quality. This is not surprising as we are trying to develop a "total Energy" model for a "complex" site, ten independent production facilities, and a utilities operation using CHP's to generate steam and electricity. This level of complexity is typical for a chemical site.

We have endeavored to enhance the Energy Usage model, by trying to build a linear model to the individual plants production volumes as well as external factors like temperature, imports and exports from the site. Finally, we've constructed a model that has the following form:

$$\text{Total Energy} = Base\ load + \sum_{i=0}^{n} Energy\ Factor_i Production_i - Temp_{Factor} * Ambient\ Temp$$

The result is of high quality, R2 = 0.87 and the 95% confidence interval is 1.2% of Total Energy for groups of predictions. We now have a power tool in our hands; we can accurately predict the energy consumption of the site. Understanding the usual now enables us to look for the unusual. We have observed variations in Site Energy Intensity that vary by 10% due to volume and temperature changes. These variations are in the same order of magnitude as our medium term improvement objectives (12.5% by 2015). Having a model describing the site energy intensity enables us to quantify these effects.

Looking for the unusual – Distance from Excellence

In the search for the unusual, we now look for trends in the difference between the models prediction and the actual Energy usage. For this we use the concept of 'Distance from Excellence' or as where excellence is defined as the model and the distance is the sum of residuals of the period under consideration, the sum of "actual – model" of all predictions in the period. The unit of measurement is in absolute Energy loss (GJ) and for convenience of our managers translated into currency by multiplying the Energy Loss with an average energy price. The distance to excellence principle can be adapted easily to other industries and a more generalized example can be found in the Energy Efficiency OEE methodology of McKinsey[i]

Figure 7, Distance from Excellence

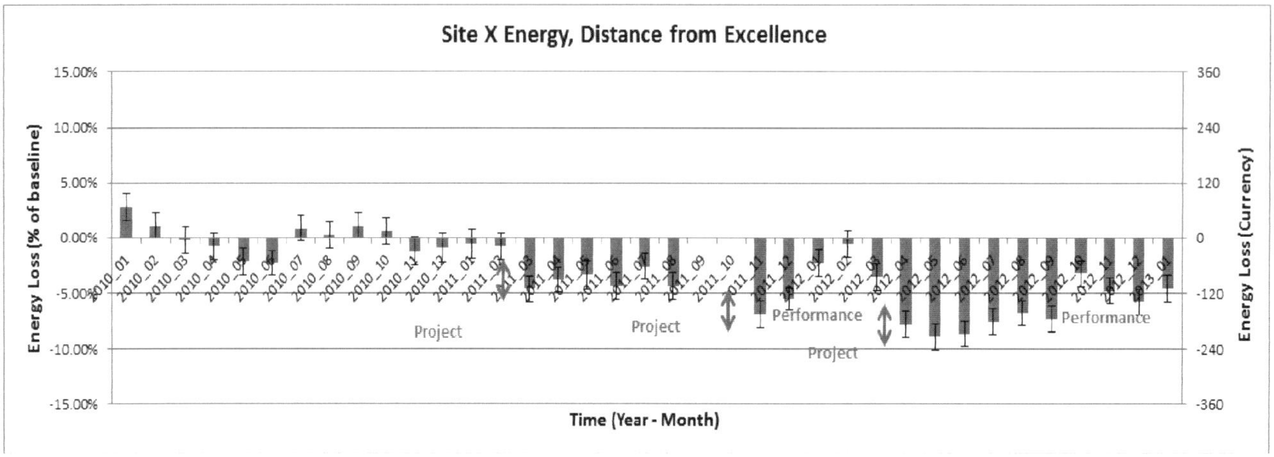

With this information in hand, we can have a robust conversation with our management. Figure 7 shows Distance from Excellence and, as we progress to ever more negative values, it means that we are improving versus the baseline. Since the graph is corrected for volume, temperature and other distorting factors, the trends shown in the graph clearly show improvements (downwards trends) and performance issues (Upwards trends). It is a solid foundation for robust conversations. Several key capex projects that were implemented are clearly visible as well as a number of significant performance and reliability issues.

Next step: moving from site to plant level

The method above is very useful for measuring progress towards the company's sustainability goals on various sites as it will give the corporate leaders insight in how each site is progressing, and where headwind towards the target is developing. Thus is obvious to try the same approach for each plant within a site. In our chemical operations, heat is the dominant force in energy usage. We therefore decided to move from "energy" to heat, the benefit of this is that we get better regressions if the energy sources are separated. For each plant we have again established models based on linear relationships.

With the exception of one plant, the modeling of the plant levels was straightforward and gave reasonable to accurate results with a basic linear model. The exception is a plant that runs two serial production lines, heavily integrated on energy, but not in production planning, using two different, exchangeable energy sources. Even though the developed model was limited in accuracy, we successfully deployed the methodology described in the previous pages and had encouraging results.

Combining "distance from excellence" graph of the site level view, with the plant levels gives an excellent understanding of which plant had a positive or negative contribution to the site's result. Site performance issues can be traced back to the originating plant. This gives the site management the opportunity to investigate with the respective plant manager and to check for issues in operations and projects.

Figure 8, Distance from excellence, Site View

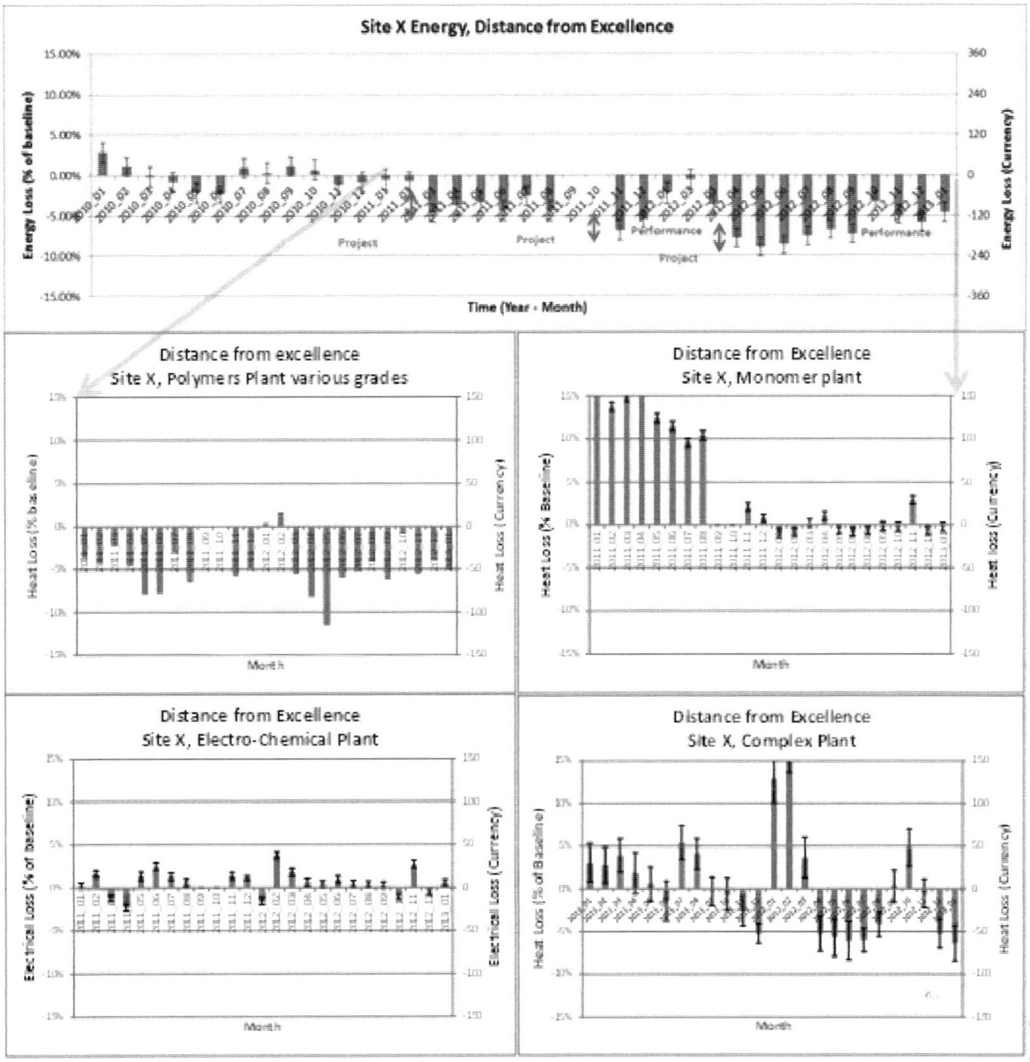

Figure 8 provides us with a good understanding of the site's performance with regards to primary energy usage performance and with key plant energy consumption performance. As a final step a performance indicator for energy conversion and distribution is required. The intuitive method is to do this in the "efficiency" form ie:

$$Energy\ Efficiency = \frac{\sum_i^n Energy\ Delivered\ to\ Customer}{\sum_i^n Energy\ Procured}$$

This certainly is not a bad way to evaluate your utilities operations. It will clearly show trends in performance. However, if the performance of the utilities operations is measured in loss, (and expressed in \$ or Euro), the size of the issue and, perhaps more important, the opportunities immediately becomes apparent.

$$Energy\ Loss = \sum_{i}^{n} Energy\ Delivered\ to\ Customers - \sum_{i}^{n} Energy\ Procured * Cost$$

Note that for cost we do not use the "real" or allocated cost, but say for an 80% efficient boiler we calculate the cost of the steam delivered as 80% of the energy inserted as 20% is lost in space. Similar like the Site and plant graphs discussed earlier, we've constructed similar models and loss graphs for key utilities equipment on site, the Cogen units, boilers, air compressors, etc.

Using Distance from excellence in the operations is also something that we have tried. It works, but the perspective needs adjusting. Our operations departments tend to become complacent when the bar goes below zero. (I'm doing good, move on to the next challenge). A key learning is that when changes are made that improve performance, the baselines need to be adjusted, so that the average is reflected at zero.

One shortfall of this method is that if fails to predict the effect of implementing projects. A successful project will show up, however, unsuccessful projects or projects which lack in control plans fail to show up.

On Site X, the Energy Leader facilitates a weekly discussion with the production engineer on the Energy and Utilities consumption. The consolidated issues and actions list provides a good starting point for small improvement projects. We are also working with the operators to implement a review cycle on a shift cycle 8 hours basis, embedding reporting on Energy Anomalies in the Asset Operations work streams. As of today, this remains an elusive process to implement.

Conclusion

As Energy cost and Sustainability have become important business drivers, it is paramount to be able to have robust discussions on Energy consumption and transformation performance. This allows management to monitor the improvement from operations and CAPEX improvements towards the sustainability goals. We believe that the basis for this discussion should be a form of normalized metric. The method of "Distance from Excellence" as it was provided to us proves to be an answer to this conundrum.

Removing influences like production volume, grade and seasonal variations enables robust performance discussions between operations, management and engineering. How good did you operate, did projects deliver and if not, where are interventions required?? Management wishes to review the unusual and to reward excellence. This is only possible by clearly defining the usual, and bringing laser focus to the unusual.

SABIC is a trademark of Saudi Basic Industries

References

[i] Plasschaert, Frank, Ken Somers, Gautam Swaroop, "Capturing the Lean Energy Opportunity in Chemical Manufacturing", McKinsey, Winter 2011

Project Management: The Good, the Bad, the Ugly

John H. Worstell
Fuel Project Engineer
ITSI Gilbane

Introduction

Whether a project is successful or not; i.e., meets schedule and budget, depends upon the onsite project manager or construction manager. This individual forms the focal point between owner, design engineer, and contractor. The onsite project manager translates engineering concepts into reality via the directed action taken by the contractor, thereby meeting the owner's goals. How well this individual performs this task depends upon communication and documentation, which are the foundation of every successful relationship. Communication and documentation with the owner focuses on budget and schedule; communication and documentation with the contractor focuses on translating drawings into results that meet specification. Understanding the relationships between owner, contractor, and design engineer is an understated concept in project management.

In this presentation, I discuss the communication and documentation tools required to develop and strengthen meaningful; i.e., functional, relationships between owner, contractor, and design engineer. I also discuss the proper techniques for reviewing construction budgets and schedules, thereby ensuring the owner value for the committed investment. My viewpoint throughout this presentation is that of a design engineer, construction manager and contractor on private and government sponsored projects.

The Players

The Owner funds the project based on a business or social need, depending whether the Owner is a commercial entity or a government entity. The Owner carries the burden for funding the project. The Owner empowers the onsite Project or Construction Manager[1] to ensure project quality and to protect the Owner from cost changing items; i.e., from design changes initiated by the Contractor at the project site.

The Design Engineer designs the project. All designs require approval by a registered Professional Engineer (PE). The Design Engineer may or may not be a PE; if not, then an independent PE will have to review and "stamp" the project design. Typically, the Design Engineer spends most of his/her[2] time prior to construction producing the Drawings and Specifications, which are used for bidding materials and construction. During construction, the Design Engineer will be "on call" if issues arise with regard to interpreting the design.

---

[1] Henceforth, I will refer to the Project Manger as the individual representing the Owner's investment and the Construction Manager as the individual who supervises construction. In some cases, one individual fills both roles.

[2] The current penchant for "he/she", "his/hers", and so forth is clumsy and repetitious. Its use suggests "his" alone is masculine, which it is not. "He" and "his" have traditionally been used in certain contexts when speaking and writing English; likewise, the same for "she" and "hers". Henceforth, I will use "he/his" and "she/hers" in the traditional English manner. (See M. Van Leunen, *Handbook for Scholars*, A. A. Knopf, Inc (1978).)

The Project Manager oversees and ensures project quality and manages the project budget. In some cases, the Project Manager will be onsite daily. In other instances, the Project Manager will oversee a Construction Manager identified by the Owner. The Project Manager represents the Owner; thus, the Project Manager manages the Owner's project funds, assesses cost changing proposals suggested by the Contractor, and serves as the communication conduit between Owner, Contractor, and Design Engineer, and Construction Manager, if one is utilized.

If a Construction Manager is utilized, he becomes the Project Manager's eyes and ear for project quality. The Construction Manager enforces Contractor adherence to the Drawings and Specifications issued for bid. The Construction Manager coordinates with third party inspectors to ensure that quality is professionally tested and documented. The Construction Manager also ensures that daily construction activities are documented. Of course, if a Construction Manager is not utilized, then the Project Manager performs these tasks.

The Contractor undertakes and performs the construction required by the project. The Contractor interacts daily with the Construction Manager and in most cases, weekly with the Project Manager, thus with the Owner. In an ideal world, the Contractor provides a quality product for the remuneration paid by the Owner. That quality product follows and fulfills the Design Engineer's specifications. But, the world is far from ideal; thus, the need for Project Managers, Construction Managers, and third party quality inspectors.

The third party inspector, henceforth "Tester", specializes in quality assurance, or control. The Tester can either be hired by the Contractor, Design Engineer, or Project Manager based on the written Specifications. The Tester ensures the quality of the Contractor's work. The Owner should think long and hard about allowing the Contractor to hire the Tester.

In large corporations, the Owner, Design Engineer, Project Manager and Construction Manager may all work for the same entity.

Project Phases

The project arises from a business or social need identified by the Owner. The Owner then incorporates the project in the annual budget and prioritizes it relative to other needs. Once the project funds are allocated, the Owner identifies a Project Manager who develops a Request for Proposal (RFP). The Project Manager then issues the RFP requesting companies to bid for Design Engineer. In large corporations, the RFP goes to the Design or Project Engineering group. Smaller companies and government agencies issue the RFP for bid. Typically, bids on the RFP are evaluated for best value rather than lowest price.

The Project Manager reviews the bid submissions and selects and contracts a Design Engineer, which may be an individual or a company. The Design Engineer then undertakes the design phase of the project. During this phase, the Design Engineer evaluates the Owner's needs and, working closely with the Project Manager, develops a solution that meets those needs.

This phase of the project may involve Other parties, who may be:

- Government agencies --- all projects require permitting and approvals;

- Property owners --- easements may be required from other property owners;
- Specialized engineers --- If the design is outside of the Design Engineer's expertise, then a contract is let to a specialized engineer to design that portion of the project;
- Lawyers --- a feature of modern life; they should negotiate easements with other property owners and be available for council if other issues that may arise.

The Design Engineer produces a final product consisting of Drawings and Specifications. The Drawings visually describe the project. Drawings are by trade: Civil, Architectural, Mechanical, Electrical, etc. The Specifications provide information about the materials and work quality required by the design. They are generally issued as a "book", segmented by trade. The Contractor must adhere to the Specifications.

The Project Manager joins the Drawings and Specifications with Contract Documents to form the Contract. The Contract also contains general and supplementary conditions to which the Contractor must adhere.

Following Owner approval of the design, the Project Manager advertises for bids to construct the project. This phase of the project, the advertisement phase, must be long enough to provide time for any and all bidding Contractors to review the contract and submit questions to the Design Engineer as well as provide enough time for bidding Contractors to obtain valid price quotes from material and equipment suppliers. If a question induces a design change, an Addendum to the Specifications must be issued to all interested Contractors. During this time, the Contractor assembles a bid and submits it during the Bid Opening.

Bid Opening is the period when all the Contractors submit their bids and learn the amounts bid on the project by each Contractor. During Bid Open, the Project Manager reviews each Contractor's bonding and insurance, and verifies that the Specifications and all Addendums have been met and acknowledged by the competing Contractors. If an issue arises with any of these items, then that Contractor's bid can be eliminated.

Bid Review generally lasts a week. In this phase; the Owner, Design Engineer and Project Manager review each Contractor's bid and evaluate the Contractor that offers the best value for the project. Picking the correct Contractor is crucial. The Design Engineer and Project Manager expend considerable energy making this choice. Note that most government funded projects are awarded to the lowest bidder.

Following the Bid Review, the Project Manager awards a contract. The Contractor will be given a time period to sign the Contract and return it to the Project Manager and Owner. The Project Manager must document justification of the contract award in case a disappointed Contractor submits a claim against the Owner for unjust selection. Thus, utilizing Lawyers during this phase is a must.

The Notice to Proceed period occurs when the Owner grants the awarded Contractor permission to begin the project. During this phase, the Owner may or may not assign a Construction Manager to the project.

Submittal/Shop Drawing Review occurs upon initiation of the project. During this phase of the project, the Contractor submits shop drawings for materials and equipment to the Project Manager. The Design Engineer must review all Contractor submittals and drawings. The Design Engineer return all submittals and drawings to the Contractor with one of the four criteria selected:

- Approved --- the submittal meets Drawings and Specifications and the product is approved for use;
- Approved as noted --- the submittal has minor issues that need to be changed. The Design Engineer will add comments to the submittal identifying the changes. It is the responsibility of the Project Manager to ensure that these changes are made when the materials or equipment are brought on site;
- Revise and resubmit --- the submittal has major issues that require change. If this is selected, the Design Engineer will typically make comments.
- Rejected – the submittal does not meet the Drawings and Specifications and another product must be submitted.

Work Commences, as a project phase, upon approval of all submittals and drawings. The Contractor mobilizes his crew onsite and has material and equipment delivered. The work commences and continues for the duration of the project. The Project Manager documents all budgeting activities while the Construction Manager, if one is appointed, oversees and documents construction activities. The Construction Manager informs the Project Manager of any issues arising with regard to the project schedule. The Project Manager should issue an open invitation to the Design Engineer to visit the project site and to inspect the integrity of the construction. The same invitation should be issued to any and all government agencies that issued permits for the project. The Project Manager or Construction Manager coordinates testing to maintain project quality.

The Project Manager and Contractor punch-list each portion of the project as it reaches completion. The punch-list arises from a walk through, detailed inspection by all concerned, with drawings "in hand". The result is an itemized list made by the Project Manager of the items that the Contractor needs to finish before the project can be commissioned.

Commissioning is the final part of construction or integration of the project into the Owner's existing facility. Upon approval by the Owner, the project moves to a point of Substantial Completion.

During Substantial Completion, the Owner agrees that the Contractor is ninety-five percent complete with respect to the contract. At this point, the Contractor will receive a fair portion of the retainage held throughout the project. Punch-list items and demobilization typically comprise the last five percent of the project.

Project Closeout is the final phase of the project. The Owner announces and advertises his satisfaction with the project and the Contractor. The Contractor then fully demobilizes his work force. Activity with regard to warranty work may occur during this period. The Contractor submits As-Built Drawings to the Owner, who then maintains them on file.

Typically, the Contract has a warranty period during which the Owner can request the Contractor to return to fix defective work. The Owner will rely on the end users or operators to communicate any such issues. In most cases, project warranties are effective for one year.

KEYS TO EFFECTIVE PROJECT MANAGEMENT

Understanding the Contract

A successful Project Manager knows more about the project than anyone else; that includes the Contract. The Contract documents the technical and budgetary aspects of the project. The technical portion of the Contract includes the Drawings and Specifications; the budgetary portion of the Contract includes the Contract Documents and General and Supplementary Conditions. To acquire technical competency with regard to the project, meet weekly with the Design Engineer. Have the Design Engineer thoroughly explain every feature and nuance of the design. Such knowledge helps communicate to the Design Engineer any issues that arise during construction. It accelerates the submittal review if the Project Manager can reject submittals that do not meet the Contract.

The Contract is a control tool to force the Contractor to perform quality work as well as stop work if there are major problems. The Project Manager needs to ensure that the Construction Manager understands the Contract as well, which can be accomplished through Contract review meetings. The better the Construction Manager understands the Contract, especially on the technical portion, the more likely he will identify an issue before it becomes a major problem. Expert knowledge of the Contract sets the tone of the project with respect to the Contractor.

Getting the Right Contractor

Choosing a reliable Contractor alleviates considerable problems for the Project Manager. Making this choice occurs during the Bid Review. Typically, the Project Manager investigates the credentials of the three lowest bidders and decides best value from their references. Of course, the reference check is only as good as the questions asked.

First, review the referenced projects and identify those most similar to your project. Second, interview the Owner, Project Manager and Construction Manager of those identified projects. Ask the below questions when investigating a Contractor's references.

- Did the Contractor meet schedule?
- Did the Contractor meet budget?
- If the Contractor went over-budget, why?
- Did the Contractor propose any change orders?
- How many change orders were there?
- Were the change orders warranted?
- Were the change orders priced fairly?
- How well did the Contractor understand the intent of the design?
- Were there any deficiencies noticed? And why?
- Did the Contractor assign a responsible person to site to ensure project quality?

Meeting schedule and meeting budget are the major indicators. If the Contractor does not meet schedule, it needs to be justified. Not meeting schedule indicates a mismanaged project and, most likely, the Contractor was not willing to add additional personnel to finish the project on time. Usually, liquidated damages are incurred for not meeting schedule, which covers the added time for the Project Manager and Construction Manager to oversee the construction past the finish date. If this is not properly calculated, the Owner loses money if the project extends beyond schedule.

There are many reasons for not meeting budget. First, the Contractor underbid the project to win the Contract. Second, the Contractor or the Project Manager or both mismanaged the project, thereby causing a budget overrun. Third, unnaturally inclement weather occurred during the Construction phase. Fourth, labor issues arise within the Contractor's workforce. Fifth, the Contractor issues and the Project Manager approves an inordinate number of change orders.

Change orders increase the cost the project. Change orders come in two types: those that are Owner requested; and, those that are Contractor requested. Change orders requested by the Owner typically indicates Owner

- satisfaction with the Contractor; therefore, the Owner increases the scope of the project;
- identification of a Design Engineer oversight.

Contractor requested change orders are usually formalized by value engineering on the part of the Contractor. A good Contractor will find cost saving changes as well as cost incurring changes. An inordinate number of Contractor initiated change orders indicates

- a poor design;
- an effort to gain additional remuneration on the project.

Additional questions to ask during Contractor investigation are:

- Did the personnel listed on the Contract appear and work at the site?
- How well did the Contractor work with the Construction Manager?
- How well did the Contractor work with the Project Manager?
- Did the Contractor effectively manage the construction?
- Was the Contractor fair?
- Any additional thoughts about the Contractor?

The last references that should be checked are with the bonding and insurance companies. Contractors typically dislike all contact with these entities and sometimes these entities will refuse to answer any questions without first consulting with the Contractor. There are typical questions that should be asked.

- Has there ever been a claim against the Contractor?
- Has the Contractor defaulted on a Bid?
- Has the Contractor defaulted on Performance?

If they answer yes to any of these questions, seek a different Contractor. A claim against the Contractor's insurance company indicates a dispute between a previous Owner and the Contractor or a claim from an ex-employee of the Contractor. Defaulting on the bid bond would indicate the Contractor signed the Contract but could not start the project. Defaulting on a performance bond indicates the Contractor did not fulfill the Contract. Both are signs the Contractor has funding issues.

Documentation

The Project Manager needs to document all activities so if the project enters litigation, evidence of events, meetings, discussions exists. Daily observation reports need to be filed. These reports should include the following.

- Date
- Project Name/Number
- Temperature
- Conditions
- Contractor Personnel and Trade
- Contractor Personnel Hours Worked
- Subcontractors Onsite
- Subcontractor Hours Worked
- Anyone Else that Visited the Site
- Equipment with Date Arrived and Date Departed
- Permits Required
- Description of Work
- Safety Issues
- Construction Issues
- Rework Items
- Testing Performed and Results
- Materials Delivered
- Signature of CM

Scheduled work cancelled due to inclement weather needs to be documented. Scheduled work delayed due to Contractor mismanagement requires documentation as well. These documents become evidence during litigation between the Owner and the Contractor. Thus, daily reporting is important.

Three critical items of the project require documenting by the Project Manager. These are the

- submittal log;
- RFI log;
- change order log.

The submittal log tracks the submittals the Contractor makes and notes how each submittal is returned to the Contractor: Approved, Approved as Noted, Revise and Resubmit, and Rejected. The submittals are key to the project because all materials and equipment need to be approved before they are installed. If a product is installed without a submittal approval, then work can be stopped until it is resolved, which causes future scheduling problems. Submittals are also important for long lead time items, such as customized items. The longer it takes the submittal to be approved, the longer it takes initiate manufacture of the item. Such a delay can also cause a future schedule issue. The typical submittal log contains

- Item Number
- Specification Section the Submittal Relates to
- Desciption of Submittal
- Required Submission Date
- Submitted Date
- Date Submitted to Engineer
- Date Returned from Engineer
- Date Returned to Contractor
- Submittal Approval/Rejection Status
- Additional Comments

It important during the submittal review to efficiently deliver the submittals to the Design Engineer and Contractor. Submittals should be held in binders or file drawers depending on the project size. The Construction Manager should always have a set in the field with him, as the Project Manager has in his office.

The second item that requires tracking is RFI's. RFI's are clarifications the Contractor will send to the Project Manager for potential design issues or product substitutions. These are usually items that require resolution by the Design Engineer. The log for this is quite similar to the submittal log and includes:

- RFI Number
- Specification or Drawing Sheet the RFI Relates to
- Description of RFI
- Submitted Date
- Date Submitted to Engineer
- Date Returned from Engineer
- Date Returned to Contractor
- Comments from the Engineer

In some cases, RFI's evolve into change orders if the design requires modification, which always increase project costs. If the RFI results in a change order, the Design Engineer typically reissues a Drawing or Specifications sheet for clarity. These sheets need to be added the Contract that the Project Manager uses. The old sheet should be clearly marked, so nobody is uses it.

The last item logged is changes to the project. Change proposals to the project are usually summarized in the Contract as to documentation requirements. Change orders consist of:

- Field Order – This is a non-cost incurring change. The form typically will document the change in writing with a supplementary Drawing if necessary. It is important that this change is agreed upon and signed by the Contractor so it will not return later in the project.
- Change Order Proposal (COP) – This is a proposal from the Contractor where a cost incurring change is proposed, which usually arises from a value engineering effort on their part or if any unforeseen field condition is encountered. It will usually include an estimate of the cost change and is submitted to the Project Manager for review by the Design Engineer.
- Work Change Directive (WCD) – This is a directive signed by the Owner where a cost incurring change is proposed. The Contractor will submit an estimate for the work that will be included in the document. A Work Change Directive either indicates that there was an oversight in the design or the Owner is satisfied with the work of the Contractor and added scope to the original Contract.

Once the latter two items are approved by all parties, the change will be documented by a change order, which adds to the project bid total, thereby increasing the overall cost of the project.

These items are not necessarily logged, but are retained in a secure filing system. Such documents may also result in Specification and Drawing sheet changes. Needless to say, it is important that all parties are using the same set of Specification and Drawing sheets.

Meetings

Developing a relationship between the Project Manager, Construction Manager, Design Engineer, Owner, Tester and Contractor is important for an efficient stream of communication during the project. Meetings, while time consuming, provide a relatively effective method for establishing an efficient communication network. Typical meetings for all parties include but are not limited to the following.

- Kick Off Meeting – This is a meeting where all the parties are introduced to each other and general questions are asked.
- Critical Construction Meeting – This is a meeting where an important event is about to happen requiring that all parties meet to ensure the work goes smoothly. Large concrete pours, punch-listings, commissioning, etc qualify for such a meeting.
- 50% Completion Meeting – This is a meeting where all parties assemble and organize the last half of the project.
- Project Closeout Meeting – This is a meeting where all parties meet and finalize the last few items that need to be completed to finish the project.

Agendas for these meetings vary based on the activity.

In addition to the above meetings, the Project Manager must organize a weekly meeting between himself, the Construction Manager, and the Contractor. These meetings typically last an hour and track schedule issues, submittals, change orders and any other issues with the project. A good format for such a meeting is:

- 4-Week Look Ahead Schedule – This item is supplied by the Contractor and shows the activities for the next four weeks. It is a way to coordinate inspections, testing and additional meetings.
- Safety Issues – This will be a review of any accidents from the prior week as well as potential safety issues from the items in the look ahead schedule
- Outstanding RFI's – This is a list of RFI's that still require resolution.
- Outstanding Submittals – This is a review of the submittal log with the Contractor.
- Change Order Status – This is a review of any future COP's and WCD's.
- Action Items – This is a list of activities that are typically Owner driven that the Contractor needs to expedite. Typical items include cleaning the jobsite, getting quotes for a WCD, etc.

Draft minutes from each meeting and distribute them to all attendees within the week. I prefer to just redline the agenda with any notes and add a section listing the attendees and their affiliations.

The Project Manager should schedule a monthly meeting with the Accounting Department to ensure that the Contractor is submitting invoices per schedule and procedure and that the Owner is paying per schedule and procedure.

Payment Application Management

The Contractor submits payment applications, a contractual invoice, monthly for payment. The Construction Manager and Project Manager ensure the numbers match what was accomplished the previous week. Getting the Contractor to segment costs is a major part of the payment application review. Such segmentation of costs provides a means for monitoring the productivity and efficiency of the Contractor. If the Contract is a unit bid contract, the project will be itemized by activities with a unit price associated with each. The Project Manager can easily monitor this type of Contract. If the Contract is lump sum, the Project Manager needs to have the Contractor segment the total project cost into activities that can be monitored easily. If a Contractor declines or resists this request, then the Project Manager must ask for monthly invoices describing the activities performed by the Contractor during the each month, which is painful to Contractors. Thus, the Contractor will more than likely prefer to segment the lump sum into job activities. If this is the case, the Project Manager must review the segmented costs and ensure their accuracy. To substantiate their accuracy, the Project Manager needs to understand the Contractor cash flow and material and equipment costs. In most cases, the Contractor will be in the negative for the first twenty percent of the project because the Contractor buys the materials and equipment but is unable to bill for these items until they arrive on site or are installed. If a lump sum is segmented and considerable money is mobilized, then early invoicing may be underway. The big money items generally enter the accounting process midway through a project.

Comparison of such activity with RS Means, a book that construction estimators typically use, will generally identify such activities.

CONCLUSION

Successful Project Management depends upon communication, relationship building, and careful scrutiny of cash flows.

(108f) Reliability Management Framework

Corey Evans, Shell, Deer Park, TX

The presentation will share an overview of the Reliability Management Framework (RMF) utilized by Shell Manufacturing sites to drive reliable operation. The RMF is intended to provide the framework of reliability elements that must be consistent and managed at each of Shell's Manufacturing sites. The RMF consists of the 16 most influential elements – many are core work processes, some are organizational behaviors, and a few are specific activities. Each of these elements is important on its own in driving site reliability, yet how these elements interface is also very important.

(109b) Reactive Distillation Using External Side Reactors for Synthesis of Bio-Renewable Organic Acid Esters

Aspi K. Kolah[1], Carl T. Lira[1], Dennis J. Miller[1], John C. Prindle[2] and Chandrakant B. Panchal[2], (1)Chemical Engineering and Materials Science, Michigan State University, East Lansing, MI, (2)E3Tec Service LLC, Clarksville, MD

Organic acid esters, produced from bio-based organic acids and alcohols, are truly bio-renewable, "green" chemicals that are promising replacements for halogenated and other petroleum-based solvents, lubricants, and plasticizers. With current uncertainty regarding the future of ethanol and other alcohols as biofuels, and with the availability of organic acids via fermentation, there exists strong motivation to develop esters as viable commodity bio-chemicals to support and expand the existing bio-renewable infrastructure.

Esters are advantageously produced using reactive distillation, where simultaneous reaction and separation take place in the same process unit. The hardware design of standard reactive distillation columns exhibits challenges for slow chemical reactions where the reaction requirements of high liquid holdup and high catalyst mass must be present inside the reactive distillation column. A similar situation arises when desirable reaction temperature inside the reactive distillation column is not in consonance with the distillation temperature or when side reactions occur.

In order to make the synthesis of esters economically competitive with their petroleum-based counterparts, in this study we have evaluated an alternative to the classical reactive distillation concept by using a conventional distillation column integrated with external side reactors.

In this SBIR Phase I Project, we have developed and demonstrated advanced reactive distillation concepts of heat-integration with side reactors for esterification of citric acid with ethyl alcohol to produce tri-ethyl citrate (TEC), which is a good candidate for replacing phthalate-based plasticizers. Phthalates have potential carcinogeneity and overall health concerns to the extent that they have been banned in Europe from use in children's toys and human contact applications. The potential of TEC is currently limited by the lack of large-scale, energy efficient and economic production.

AspenPlus process analysis is performed based on the available kinetic data for esterification of citric acid with ethanol for an integration of pre-reactor as well as side reactors with fractionation column. Heat integration is examined for energy efficiency and maintaining temperatures in side reactors for maximizing the catalyst utilization. The concept of integrated reaction and distillation is validated by experiments using the reactive distillation facility at Michigan State University with a pre-reactor and one side reactor coupled to a distillation column.

Pilot Plant Testing – The Key to Solving Liquid-Liquid Extraction Challenges

Authors:

Donald J. Glatz
Thomas D. Lightfoot

Koch Modular Process Systems, LLC.

Prepared for Presentation at

AIChE Spring Meeting

San Antonio, TX
April – May 2013

INTRODUCTION

The complexity of the process taking place in a liquid-liquid extraction (LLE) column (e.g. droplet breakup and coalescing, mass transfer, and axial & radial mixing), and the fact that small amounts of impurities can dramatically affect column performance, make it difficult to design a production extraction column based upon computer models and simulations alone [1]. It remains problematic today to design LLE equipment with any reasonable degree of confidence and accuracy without some type of pilot plant testing. This paper will focus on how pilot plant testing (1) identified potential scale up challenges, and (2) how the testing led to solutions. The scale-up designs could only be guaranteed using the solutions found during the pilot plant testing. Based upon many years of experience with pilot testing LLE processes, specific challenges/solutions will be presented including unfavorable density gradients, changing hydraulic behavior throughout the column, negative heat of mixing, and transitional phase in the process. The fact that most of these items could only be identified and solved during testing will clearly support the importance of pilot plant testing new LLE processes.

APPLICATION 1 – Unfavorable Density Gradient

An unfavorable density gradient results when the density of the continuous phase within an extraction column is significantly higher at the top of the column than at the bottom of the column. This phenomenon will result in greater tendency for back-mixing and loss of efficiency as the column diameter increases. This is particularly problematic when attempting to scale up from a small diameter column in the pilot plant to a large diameter production column.

A client tested a 25 mm diameter KARR® Column for a process in which the aqueous feed contained >60% solute (product to be extracted) and a maximum amount of organic solvent was specified in order to make the process economical. Initial tests in the KARR® Column were successful, however very unfavorable density gradients were observed. From bottom to top of the column the specific gravity of the light phase changed from 0.867 to 0.924 and the specific gravity of the heavy phase changed from 1.01 to 1.04. In our experience the KARR® Column is not the optimal design for these types of processes due to the large open area and reciprocating action. Stationary baffling that restricts back-flow is better and as such we recommended the SCHEIBEL® Column, for additional testing and scale-up.

A summary of the optimal test results for this process in a 3" diameter x 60-stage SCHEIBEL® Column are shown here:

Run No.	Capacity GPH/ft^2	S/F Ratio Mass Ratio	Total % Recovery
1	600	1.0	98.9
2	600	1.3	99.2
3	600	1.5	99.5
4	450	1.3	97.8
5	650	1.3	99.4

Some key points from these results that were utilized in scale up to the production column:

- Excellent extraction efficiency (>99% recovery) was achieved using 60 agitated stages, a S/F ratio of 1.3, and design capacity of 600-650 GPH/ft^2.
- When the column was operated at a lower capacity (450 GPH/ft^2), recovery dropped to below 98%, whereas at higher capacity (650 GPH/ft^2) the recovery remained >99%. This is expected for systems with unfavorable density gradient as higher velocities lead to better plug flow tendency, and thus less back mixing. As such the production column must be designed and operated at the higher capacity.
- For the production column design, a 20% safety factor was applied. Thus, the production column was built with a total of 72 agitated stages.

APPLICATION 2 – Changing Hydraulic Behavior throughout the Column

Very often, the hydraulic behavior is significantly different as an extraction progresses through a counter current extraction column. This can be caused by several factors including changing flux rates, changing capacity, changing density difference between the phases and changing interfacial tension. This phenomenon is evident when observing the interaction between the phases in a glass shell column. Compensation can be made by varying the mixing intensity throughout the column. For a column with rotating internals, this is often achieved by changing the diameter of the impellers. For a KARR® Column this is often achieved by varying the spacing between the perforated plates. Systems with a high amount of mass transfer (greater than 10% mass transfer), are obvious candidates for implementing variations prior to testing. For these types of systems, experience can lead to an "educated" guess as to the variation that will work. However, for some systems, it is not obvious that modified agitation intensity will be required.

An example of this was a system for extraction of methanol from an acrylate feed stream using water as the solvent. The organic feed contained approximately 2.5% methanol and the target raffinate concentration was less than 0.1% methanol. Liquid-liquid equilibrium data showed that approximately 6 theoretical stages would be required to achieve this result at the desired S/F of 0.3. For testing, a 25 mm KARR® Column was set up with 6' plate stack height. A summary of the results from testing in this column is shown here:

Run No.	Plate Stack	Feed Rate (ml/min)	Solvent Rate (ml/min)	Agitation SPM	Raffinate % MeOH
1	Standard plate spacing	150	45	75	0.165
2	Standard plate spacing	150	45	100	0.124
3	Standard plate spacing	150	45	110	Flooded
4	Variable plate spacing	180	54	200	0.078
5	Variable plate spacing	210	63	175	0.084

For standard plate spacing (2" spacing throughout the column), the minimum raffinate concentration was 0.124% at an agitation speed of 100 strokes per minute (SPM). Attempts to run at a higher agitation speed resulted in flooding. It was observed that the dispersed water phase droplets would fall very quickly through the first 4' of the column with very little hold-up. At >100 SPM, these droplets would start to back up and flood about 1-2 feet above the organic feed inlet. Thus, although there was only 2.5% mass transfer, the area immediately above the feed inlet was obviously the controlling region in the column. Prior to Run 4, the following plate stack was installed into the column; from the bottom, 1' of 6" spacing, 1' of 4" spacing and then 4' of the standard 2" spacing. As can be seen from the additional runs, this revised plate spacing allowed for both increased agitation speed and increased capacity. The higher agitation speed was required in order to achieve the desired < 0.1% methanol in the raffinate.

Several production columns have been designed from this pilot test work, and these are operating successfully today.

APPLICATION 3 – Negative Heat of Mixing

A process was tested in which an organic salt solution was fed to the bottom of an extraction column and chloroform was added at the top. Based upon the process, the client could control the feed such that the amount of organic product was in the range of 20-45 wt% and the salt content was in the range of 10-30 wt%.

The first feed tested was 40% organic product and 15% salt. The column was a 3" diameter SCHEIBEL® Column and the target operating temperature was 40 °C. For many runs, this column gave excellent extraction results, and based upon the hydraulic behavior inside the column, it was the best column design for this process. As testing progressed, capacity was pushed and lower solvent to feed ratios were attempted. Eventually, a cold spot became evident in the middle of the column (approximately 20 °C). This cold spot resulted in salt dropping out of solution and caused the column to flood. It was determined, that the transfer of the organic product from the aqueous phase to the chloroform phase was accomplished with a negative heat-of-solution, causing the temperature reduction in the middle of the column.

An experimental design was set up to evaluate the following variables; organic product concentration in the feed, salt concentration and solvent to feed ratio. Based upon the results from this testing, an allowable operating window was determined. Included in this window were the conditions that would produce the desired extraction result defined as the maximum allowable amount of product in the raffinate stream (lost product). From an extraction efficiency standpoint, lower solvent to feed ratios could produce the desired product recovery. However, minimum ratios were identified (based upon the feed composition) in order to prevent the "salting out" phenomenon and subsequent flooding.

APPLICATION 4 – Transition Phase in the Process

Extraction processes for recovery of metals are often accomplished via a multi-step process in which the metal ion is first recovered from an acidic "leach" solution using an organic solvent mixture following by back-extraction or stripping the metal from the organic phase using an aqueous "strip" solution. In these processes, the metal ions are essentially removed from the impurities in the original leach solution, and then put back into the "clean" aqueous phase from which they can be utilized as highly pure materials.

In one particular process, a metal ion was extracted from the leach solution very successfully using a SCHEIBEL® Column. This column worked efficiently due to a high density difference and high interfacial tension, for which the high agitation and internal baffling of the SCHEIBEL® Column provided the drop size and hold-up necessary to achieve the metal ion extraction. The stripping step was another matter. Upon start up for the stripping step, the concentrations of components in the process were such that side reactions occurred, resulting in a large amount of by-product formation. Insoluble solids were generated, having an appearance of white and fluffy material. Attempts to test this in columns with rotating internals resulted in plugging and flooding. Even mixer-settlers proved to be problematic for this process. Although not tested, it was determined that a packed or trayed column would not be able to handle the transition phase due to plugging.

Next the KARR® Column was tested. The benefits of the KARR® Column for this type of process were that the plate stack design is very open (55-60% open area) and the fact that the entire plate stack is reciprocating and thus there are no "dead" zones where solids can build up. Upon start up in the KARR® Column, the transition phase was once again observed by fine solids dropping out of solution. As hoped, these solids did not find any place to hold up in the column and flowed out of the column with the aqueous phase. After several minutes of operation, the system moved past the transition phase and no more solids were observed to form in the process. As the system reached steady state operation in the column, the entire column cleared up and clear solutions were observed to exit at both ends of the column. The extraction efficiency was excellent (complete stripping accomplished) and the KARR® Column was obviously the best choice for this process step.

APPLICATION 5 – Emulsion Formation

Emulsion formation in an extraction process can be a difficult problem for testing and design. The KARR® Column been demonstrated to be an excellent choice for these types of systems including fermentation and algae broths. A previous AIChE paper provides a good reference for this [3].

APPLICATION 6 – Agitation vs No Agitation

In some applications, agitation has been required in order to allow the two phases to pass counter currently through an extraction column, whereas flooding and interface control problems occur without agitation. This phenomenon was also covered in a paper presented at a previous AIChE meeting [4].

CONCLUSION

It is clear from the observations and results presented above, that many potential problems that can be encountered in extraction processes can only be identified during pilot plant testing. As shown in the examples presented, it was the observations and results during pilot plant testing that allowed for the accurate design of the production columns.

LITERATURE CITED

1. Cusack, R., Glatz, D., "Apply Liquid-Liquid Extraction to Today's Problems", Chemical Engineering, July 1996.

2. Cusack, R., Fremeaux, P., and Glatz, D., "A Fresh Look at Liquid-Liquid Extraction, Extractor Design and Specification", Chemical Engineering, February 1991.

3. Glatz, D., and Lightfoot, T., "Using Extraction Columns for Recovery of Valuable Products from Biomass", AIChE Fall National Meeting, October – November 2012.

4. Glatz, D., and Mason, L., "Extraction of Methanol and Glycerin from Crude Biodiesel", AIChE Fall National Meeting, November 5-9, 2007.

(109d) Development of a Standardized Method for Mass Transfer Measurements in Absorption and Desorption

Anna-Katharina Kunze[1], Philip Lutze[1], Manuela Kopatschek[2], Stephan Mueller[2], Marcus Gruenewald[2] and Andrzej Górak[3], (1)Department of Biochemical and Chemical Engineering, Laboratory of Fluid Separations, TU Dortmund University, D-44227 Dortmund, Germany, (2)Mechanical Engineering Department, University of Bochum, Bochum, Germany, (3)Department of Biochemical and Chemical Engineering, Laboratory of Fluid Separations, TU Dortmund University, Dortmund, Germany

Development of a standardized method for mass transfer measurements in absorption and desorption

A. Kunze[1], P. Lutze[1], M. Kopatschek[2], S. Müller[2], M. Grünewald[2], A. Górak[1]

[1] TU Dortmund University, Laboratory of Fluid Separations

Emil-Figge-Strasse 70, D-44227 Dortmund, Germany

Phone: +49 (0) 231 755-3034, E-Mail: anna-katharina.kunze@bci.tu-dortmund.de

[2] University of Bochum, Chair of Fluid Separations

Universitaetsstrasse 150, D-44780 Bochum, Germany

Due to the rising use of rigorous models for designing absorption and desorption processes a coherent method for mass transfer parameter determination must be developed. Currently, a comparability of mass transfer parameters investigated experimentally at different sights or different columns is not existent. Additionally, there is hardly one method to evaluate mass transfer parameters determined in the past, because often important information about the measurements and calculation is missing. That leads to the effort of a standardized experimental and calculation method for determination of mass transfer parameters in absorption and desorption. The main focus is to increase transparency and comparability for mass transfer parameters on gas and liquid side β_G und β_L and the effective interfacial area a_{eff} from different authors[1,2]. This will lead to a decrease in necessary safety factors resulting in a cost and resource optimized column design.

A consortium of 14 industrial partners, including manufacturers and users of packings, as well as two academic partners are working on this standard method in Europe. The first step of the standard method is to figure out an experimental method to achieve coherent results on pilot plant scale. Therefore, concentration measurements on different plants are carried out to calculate the volumetric mass transfer parameters $(\beta_G \cdot a_{eff})$ and $(\beta_L \cdot a_{eff})$. The method is developed for structured and random packings. During the development of an experimental method the main focus is on the configuration of the setup. Detailed investigations concerning the influence of e.g. gas and liquid distributors, packing height, and taking samples is carried out. The experiments for the experimental method are carried out in two columns with a diameter of 288 mm, which have exactly the same configuration. The experimental method will then be adapted additionally to a column with a higher diameter to carry out further experiments.

In addition a mathematical tool is necessary, which guarantees a consistent calculation of mass transfer parameters for rigorous modeling. The stepwise data processing from the experimental data to gain parameters for rigorous modeling is described in Figure 1.

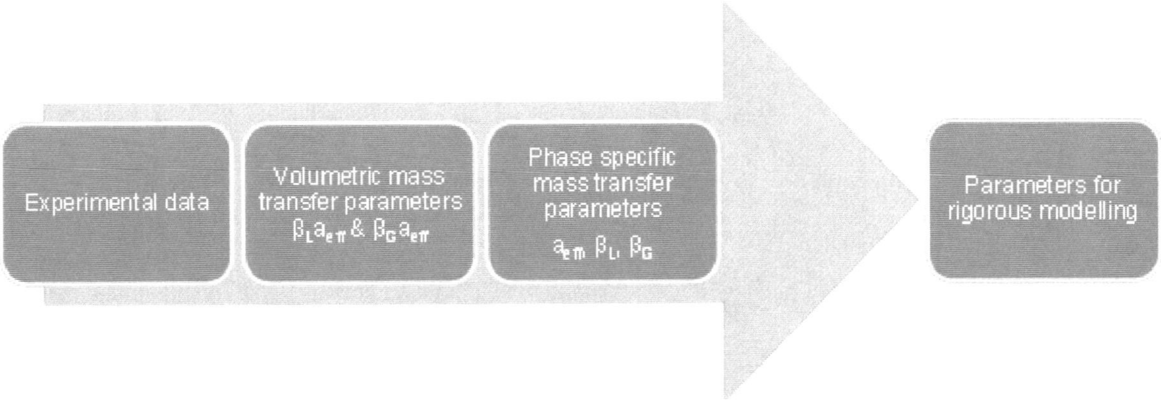

Figure 1: Pathway for the determination of model parameters of rigorous absorption models.

Based on the experimental data the first step is to define the volumetric mass transfer parameters **($\beta_G \cdot a_{eff}$)** and **($\beta_L \cdot a_{eff}$)**. To take temperature and concentration dependence of the chemical system data into account, a detailed analysis of possible calculation methods is carried out. To define the temperature and concentration dependence of the phase equilibrium, at least seven basic system data (such as vapor pressure as well as density, viscosity and diffusion coefficients in gas and liquid phase) have to be calculated. To minimize the calculation effort, a global sensitivity analysis is carried out. For the most sensitive parameters, different calculation methods are compared. In addition, also the experimental data are implemented in the sensitivity analysis to determine the influence of deviations in each measured value.

A combination of three chemical systems is necessary to determine **β_L, β_G**and **a_{eff}**, one of the following properties has to be fulfilled by one of the systems:

- mass transfer resistance on liquid side for **β_L**
- mass transfer resistance on gas side for **β_G**
- fast reaction pseudo first order for interfacial area a_{eff}

Via correlation of interfacial area between the three different chemical systems the phase specific mass transfer parameters can be determined.

Within this presentation, the detailed results of the analysis of equipment parameters, the method in general as well as the mathematical tool will be shown.

This work is part of GVT funded project "Standardization of mass transfer measurements in absorption and desorption".

Literature

[1] A. Hoffmann et al., "Standardization of mass transfer measurements: A basis for the description of absorption processes", Chemical Engineering Research and Design 2007, 85 (1), 40-49.

[2] J.F. Rejl et al., "Methods standardization in the measurement of mass-transfer characteristics in packed absorption columns", Chemical Engineering Research and Design 2009, 87 (5), 695-704.

A REVIEW OF THE ROTARY PRESSURE FILTER FOR THE FILTRATION OF A LIQUEFIED GAS SLURRY

Barry A. Perlmutter
BHS-Filtration Inc.
Charlotte, North Carolina 28270

Biography

Barry A. Perlmutter is President and Managing Director of BHS-Filtration Inc., a subsidiary of BHS-Sonthofen GmbH. Barry has over 30 years of technical and marketing experience in the field of solid-liquid separation, filtration, centrifugation and drying. He has published and lectured worldwide and has been responsible for introducing and creating growth for many European companies into the marketplace. He has a BS in Chemistry from Albany State, NY, MS from the School of Engineering at Washington University, St. Louis and an MBA from the University of Illinois.

ABSTRACT

In the manufacturing of this specialty chemical, the pilot plant examined the process to determine the most effective method for filtration to develop a production process. A study was undertaken, by plant engineering and development, to examine the use of a rotary pressure filter to develop the overall process operations for product quality and production rates.

This paper reviews the basics of two alternatives: rotary pressure filter and pressurized drum filters. The technologies are reviewed for the filtration of a liquefied gas slurry.

The article continues with the discussion of the process testing in the laboratory and in the field to evaluate the Rotary Pressure Filter. The lab testing used a BHS pressurized pocket leaf filter with 20 cm^2 of filter area while the fieldwork was conducted on a Rotary Pressure Filter with a filter area of 0.18 m^2.

The paper discusses test procedures, data collection, filter media selection and other process parameters. Scale-up data includes process conditions and production rates.

The paper conclusion is two-fold: a review of the unique specifications of the rotary pressure filter required for the process to meet production rates, product quality and site environmental and safety concerns. Finally, the paper demonstrates the creative process solution that is developed with close collaboration between the operating company and the vendor.

Acknowledgements: The author would like to thank Brian Moore, BHS Application Sales Engineer and Bruce Glines, BHS Senior Process Engineer for their help and assistance with this article.

LIQUEFIED GAS SLURRY

A liquefied gas is a clean, colorless gas that is easy to liquefy and transport. It is gaseous at normal temperature and pressure, but changes to a liquid when subjected to modest pressure or cooling. This easy liquefaction makes it easy to transport and store. This and other properties of certain liquefied gases, including a high oxygen content, lack of sulfur or other noxious compounds, and ultra clean make these versatile and promising process solvents. The most important point is to keep these gases, under pressure, so they behave as a liquid during filtration, cake washing and drying. The plant engineers reviewed the Rotary Pressure Filter and the Pressurized Drum Filter as two possible alternative solutions.

ROTARY PRESSURE FILTER (RPF)

The rotary pressure filter, shown in Figure 1, is a continuous pressure filter designed for thin cake to deep cake filtration with cake depths from 6 mm to 150 mm. The slowly rotating drum (6 – 60 rph), in Figure 2, is divided into segments (called cells) each with their own filter media (synthetic cloth or single or multilayer metal) and outlet for filtrate or gas. The outlets are manifolded internally to a service/control head where each stream can be directed to a specific plant piping scheme or collection tank. In this way, the mother liquor can be kept separate from the subsequent washing filtrates and drying gases. This allows for better process control as well as reuse and recovery of solvents and the gases. The service/control head, for this application, is pressure rated so the liquefied gas can be kept under pressure, acting as a liquid.

Figure 3 shows a typical operation as follows. The feed suspension enters each cell, under constant pressure, to form a filter cake. Internal divisions of the housing allow the cake to be processed in completely separate zones. Each zone can operate under different pressures depending upon the compressibility of the cake at each stage of filtration, washing and drying.

For example, pressure filtration is conducted up to a maximum differential pressure of 6 bar. Pressure filtration has the added benefit of eliminating post-precipitation of the solids if the process solvent is prone to flashing under vacuum filtration. In this case, the 6 bar pressure differential allows for a back-pressure of 2-bar or more, such that filtration can be at 4 bar and the gas can be a liquid.

A second benefit of the slow-rotating pressure filtration is the reproducible cake depth and the ability to control the residence time (by the speed of the drum). The slow rotation also results in much lower maintenance and energy costs.

Positive displacement washing or counter-current washing follows filtration. The design of the pressure cells, such that they have no free-space and are completely filled, eliminates the possibility of cake cracking or bypass, providing maximum washing efficiency. Multiple washing steps can be accomplished along with solvent exchanges, steaming and extraction.

The cake then undergoes a drying step by blowing ambient temperature or hot gas through the cake. The control of this drying step is influenced by gas flow, pressure, temperature as well as the cake depth and residence time.

Finally, the pre-dried cake is continuously discharged. The cake chute is pressurized which allows the cake to be fed directly into the downstream dryer without the use of rotary locks. This is an important benefit to keep the entire system pressurized. The cake discharge is assisted by a gas pulse.

After cake discharge, there is a cloth rinse step. This is controllable depending upon the product discharge, filter media and potential of cloth blinding. With the drum rotating, after the cloth rinse step, the cells are ready to be filled for filtration. A typical process flow is shown graphically in Figure 3.

PRESSURIZED DRUM FILTER (P-DF)

The other alternative to the RPF is the pressurized drum filter (P-DF) which is a rotating drum inside a pressure vessel. The unit consists of a filter drum, slurry trough, agitator, wash bars and a "pressure let-down rotary valve. The process begins by closing the pressure vessel manually with bolts and nuts, pressurizing the vessel with compressed gas, the rotary valve is also pressurized for sealing and the filter trough is filled via the suspension feed pipe and the agitator is started to keep the solids in suspension.

During the evaluation stage, the plant engineers reviewed the P-DF operation, as described above, and determined that the gas usage for the pressure vessel and rotary valve was very excessive. This is not required in the RPF.

Secondly, the RPF housing does not require the manual process of closing and sealing which saves man-hours and time as well as eliminates a possible leak path of the liquefied gas.

Finally, in the RPF, the solids fill the cell via a centrifugal feed pump which eliminates the need for a filter trough, agitator and feed pipe. It was for these points as well as the additional safety relief valves required, the back pressure to maintain the gas as a liquid and other points that the decision was to begin the test work for the RPF.

ON-SITE PROCESS TESTWORK: POCKET LEAF FILTER (PLF) TESTWORK

Process testing was conducted at the site's laboratory and in the plant. For the bench-top lab testing, the BHS pressurized pocket-leaf filter (PLF) with 20 cm^2 of filter area is used as shown in Figure 4. For the continuous pilot testing, a pilot RPF with 0.18 m^2 of filter area is installed, as shown in Figure 5.

The objectives of the PLF testing are as follows:

- Filtration time vs. filter media
- Filtration time vs. slurry feed mass
- Filtration time vs. differential pressure
- Filtrate quality vs. filter media
- Cake solids wash time and quality
- Cake solids discharge characteristics
- Production Scale-Up and Process Guarantee

The lab testing proved to be uniquely challenging both to feed the PLF as well as to maintain a pressure to keep the gas as a liquefied solvent. The plant engineers and BHS developed a confidential method to meet these challenges.

CONCLUSION AND PATH FORWARD STEPS FROM THE PLF TESTING

The PLF tests demonstrated that acceptable filtration and solids wash rates could be obtained for this product and acceptable solids levels were observed for the mother liquor filtrate. Washing results and drying quality parameters were also achieved.

Additional pilot plant tests with the BHS continuous pilot unit, RPF 0.18, are recommended to confirm the PLF lab tests. In these tests, BHS would be able to identify the necessary slurry solids percentage, cake solids thickness, solids wash time, solids drying time as well as cake discharge. Finally, the pilot testing will be the basis for the mechanical design of the RPF to ensure that the RPF can be designed for the process with a liquefied gas slurry.

DISCUSSION OF THE RPF 0.18 M^2 PILOT TESTWORK AND SCALE-UP

While the actual data is confidential, the plant engineers and BHS process engineers gathered the following parameters:

Process Parameters:

- Slurry Feed Pressure:
- Slurry Feed Flow:
- Wash Pressure:
- Wash Flow:
- Dry Pressure:
- Drying Air Flow:

RPF Parameters:

- Drum Speed:
- Separating elements:
- Cake blow back:
- Cloth blow back:
- Backpressure:
- Cake Thickness:
- Filter Cloth:

To fully evaluate the RPF performance, the site also compiled the following:

- Slurry solids concentration
- Filtrate quantity (mother liquor, wash, blow down, etc.)
- Filtrate yield
- Cake Moisture
- Total Cake quantity

Scale-Up From RPF 0.18 M^2 Pilot Data

- Calculate Specific Filter Performance from Pilot Testing
  = kg of dry solids/m^2/hour
- Calculate Production Area Required from Filter Performance and Client Required Production Rate
- Using the drum speed, time for filtration, washing and drying and several other RPF factors, the specific filter area is calculated

TYPICAL SCALE-UP CALCULATION - EXAMPLE ONLY

The scale-up is based on 224 g slurry with 1:1 composition = 112 g dry solids:

Drum revolutions:

$$n = \frac{3600\frac{s}{h}}{(4+8+15)s} \cdot \frac{270°}{360°} \cdot 0{,}85 = 85\frac{\text{revolutions}}{h}$$

270°:	active angle
0.85:	factor for separating elements

Specific filter performance:

$$Q = 500\frac{1}{m^2} \cdot 85\frac{1}{h} \cdot 112g = 4760\frac{kg\ dry\ solids}{m^2 \cdot h}$$

Required filter area:

$$A = \frac{20000\frac{kgdrysolids}{h}}{4760\frac{kgdrysolids}{m^2h}} = 4{,}2\ m^2$$

Selected filter: BHS-Rotary Pressure Filter, type B16 with 5.4 m² is sufficient to operate 20,000 kg dry solids per hour.

DESCRIPTION AND SPECIFICATIONS OF THE RPF

The RPF design has several unique features for this application to ensure that the liquifeid gas remains as a liquid and secondly to ensure the safe operation of the unit.

- Direct-coupled drive for completely stable drum rotation, shown Figure 6.
- Closed and pressurized housing and bearing shields, shown in Figure 7.
- Secondary containment of the control head
- Pressurized control head
- Pressurized discharge housing

SUMMARY

As operating companies develop new and unique chemical processes, there are many choices for filtration, cake washing and drying. Engineers must evaluate all outcomes to make an informed and successful decision. Technical evaluation and, as shown, laboratory and pilot testing are critical for a successful decision and project. The take-away is that close collaboration between the operating company and the vendor will allow for creative problem-solving and process filtration solutions to achieve the desired quality and production requirements.

Figure 1: RPF Filter

FIGURE 2: RPF DRUM AND CELLS

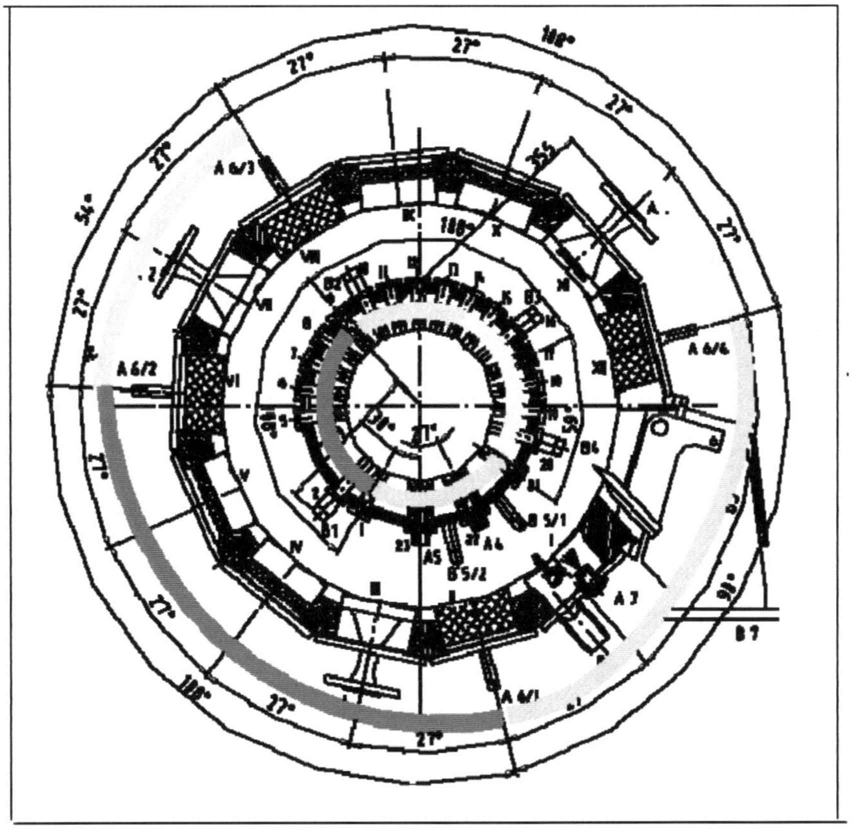

FIGURE 3: PROCESS FLOW

FILTRATION (RED)
CAKE WASHING (BLUE)
DRYING (YELLOW)
CAKE DISCHARGE & CLOTH RINSE (GREEN)

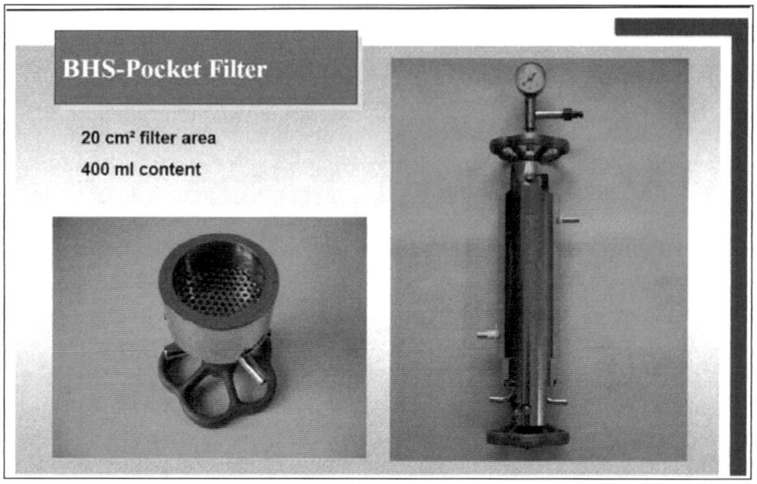

FIGURE 4: BHS POCKET LEAF FILTER (PLF)

FIGURE 5: RPF 0.18 M^2 PILOT FILTER

FIGURE 6: RPF DIRECT-COUPLED DRIVE

FIGURE 7: RPF PRESSURIZED HOUSING AND BEARING SHIELDS

IMPROVED BOILER SYSTEM OPERATION WITH REAL-TIME CHEMICAL CONTROL

Paper No. 295944

Deborah Bloom, Principal Consultant, NALCO, An Ecolab Company, Naperville, IL
David Puchan, Senior Industry Technical Consultant, NALCO, An Ecolab Company, Sugar Land, TX

ABSTRACT

The steam boiler system is a critical component of most manufacturing processes. Steam production reliability is often a key component in assuring product quality and overall production efficiency. Hourly steam load demands and varying system stresses make responsive water treatment of the boiler system difficult. A challenging production environment is made even more so by the volatile economic forces in today's world.

New technologies have been developed that provide real time monitor & control capabilities to help steam operations staff achieve more consistent, proactive boiler treatment by detecting system variability, determining the correct chemical or operational action, and delivering measurable environmental return on investment. These technologies are capable of communicating via web or DCS for monitoring and alarming existing conditions.

The new technologies will be described and case histories presented.

INTRODUCTION

Maintaining reliable and efficient steam operations in most manufacturing facilities requires good corrosion and scale control in the pre-boiler and boiler water system. Chemical treatments must be monitored and controlled within recommended ranges, an achievement that is often difficult with traditional methods. New state of the art systems provide real time monitor and control capabilities to proactively respond to system stresses and maintain tight control – increasing system reliability and asset protection, as well as typically providing chemical, water and energy savings.

As the U.S. marketplace has become more competitive, industry's environmental awareness has also increased. The competitiveness of the marketplace has resulted in a drive to extend equipment life (equipment such as a boiler system), reduce fuel and water costs, and optimize operational labor costs. The increased environmental awareness has resulted in major corporate initiatives to reduce greenhouse gas emissions, and fuel and water consumption.

SCALE AND CORROSION

There are many water-related challenges encountered when operating a boiler. Two of the most common problems are scale and corrosion.

Scale is the result of dissolved minerals, such as calcium and magnesium, exceeding their solubilities and forming deposits (1). Boiler heat transfer efficiency can be negatively impacted by scale. In addition scale can also create serious under-deposit corrosion problems. Scaling conditions, when not addressed in a timely fashion or when improperly dealt with, can result in ruptured boiler tubes or other boiler damage. This can cause boiler shutdown and lost or greatly reduced production.

For presentation at the AIChE 2013 Spring Meeting; San Antonio, TX April 28-May 2, 2013

Some of the more common ways to prevent scale from occurring in boilers include ion exchange (softening or demineralization) and reverse osmosis. Residual dissolved minerals are prevented from forming scale by the addition of chemical treatment.

Corrosion is a serious concern in the pre-boiler and boiler system for two reasons. First, it shortens the asset life of the boiler and pre-boiler system. Second, the corrosion by-products resulting from unchecked corrosion can deposit in high heat transfer areas of the boiler as scale. Oxygen corrosion is a specific and particularly devastating type of corrosion (Figure 1). It is a common problem that adversely affects the life of the key system components. Pitting attack from oxygen corrosion can cause premature metal failure.

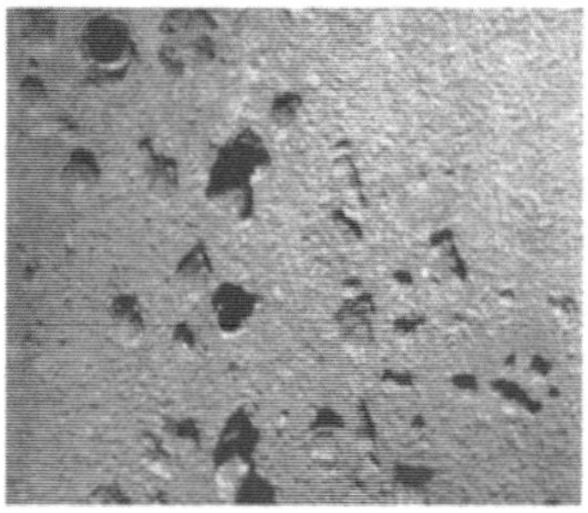

Figure 1. Oxygen attack, characterized by its classic pitting appearance

Iron oxide scale, just as the earlier described calcium and magnesium scales, reduces heat transfer efficiency. In uncontrolled scenarios, blown tubes (or other extensive boiler damage), boiler shutdowns and production losses can result.

Historically, the best practice for preventing oxygen corrosion in boiler and pre-boiler systems has been mechanical deaeration for primary oxygen removal, followed by chemical oxygen scavengers to remove remaining trace quantities.

For decades, the traditional practice employed to control many boiler water scale and oxygen control chemistries has been by sampling the blowdown. One primary reason for this was that the boiler concentrated the treatment chemicals sufficiently to enable them to be measured readily. Although modernization and improvement in test methods have enabled feedwater testing (e.g., cycling up is not required to run many of the tests), many operations hold onto the old practices of the 1930s through 1970s and test only in the blowdown. The major drawbacks resulting from this approach are related to the lag time between changes in feedwater conditions and those changes showing up in the blowdown.

In a blowdown based control scenario, the operator would sample the blowdown, perform an array of chemical tests, and adjust the feed rate of the scale inhibitor or the oxygen scavenger. This constant "add chemical - sample – test – adjust" control loop was the standard. This meant that if a test was missed, an opportunity for adjustment was also missed.

In addition to the challenge of the above testing loop, the physical capacity of the boiler often meant that a negative change in feedwater conditions might go completely undetected or not have a notable impact on the boiler blowdown composition for hours or even days. Feedwater conditions that were potentially adverse to system integrity (e.g., an oxygen excursion, a reduction in feedwater pH, or a badly contaminated condensate return stream) would only be seen after they had become history, if they were detected at all, because only the boiler water was being tested. What was missing was a proactive approach to protect the boiler as opposed to a reactive approach that adjusts after damage has occurred.

Ideally, one would want to measure and control boiler chemistry in the feedwater, because that would enable one to make adjustments to changing conditions in real-time, rather than chasing events or missing them entirely and suffering the outcome.

NEW CONTROL TECHNOLOGY

The new boiler feedwater treatment automation system is a combination of two different Nalco technologies. First is the patented solid state fluorometer, which is used to control boiler internal treatment (2, 3, 4). The second is the Nalco Corrosion Stress Monitor™ technology. This instrument works by measuring the oxidation reduction potential at the temperature and pressure of the boiler feedwater. The At Temperature Oxidation Reduction Potential, or AT ORP, detects changes in the corrosivity of the feedwater and can be used to control the oxygen scavenger feed. The AT ORP has a design rating up to 500°F and 2,800 psig. A uniquely designed controller controls both the fluorometer and AT ORP. Monitoring, On / Off Control, and PID (Proportional-Integral-Derivative) Control are all possible with this controller. These technologies are capable of communicating via web or DCS for monitoring and alarming existing conditions.

Scale Control

The use of fluorescent tracers for system diagnostics and dosage control of treatment chemicals is a technology that has been in practice for many years. This method involves adding an inert fluorescent molecule to a water treatment product in a known proportion to the active ingredients. The fluorescent molecules are inert and are unchanged by the conditions in the water being treated. They are also easily detectable in the water with an instrument called a fluorometer, and measurement of the tracer gives a quick and highly accurate value for the amount of treatment that has been added to the system. Typically, tracer measurements are used to control the feed of the scale inhibitor treatment.

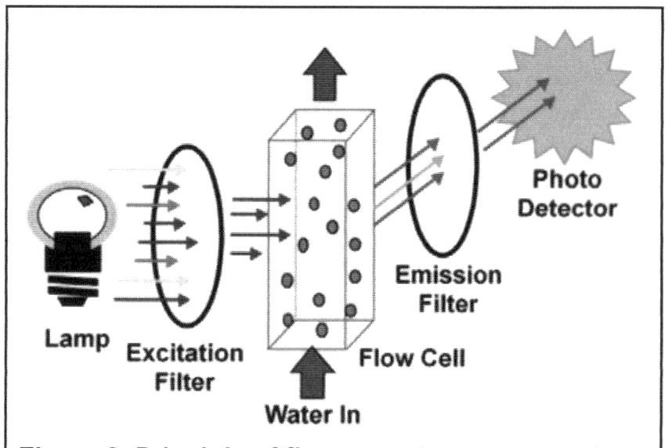

Figure 2. Principle of fluorescent measurement

Fluorometer

The fluorometer used with this technology can measure the presence of these fluorescent materials at the part per billion level of detection. Today's fluorometers are based on solid state electronic technology that makes these devices much simpler, more compact, less costly and more reliable, Figure 2.

The fluorometer enables us to monitor and control traced products either with a handheld measurement device using grab samples (Figure 3) or with on-line feed and control monitoring technology. In addition the fluorescent materials can be used by themselves as a system diagnostic tool for flow measurement, determination of boiler cycles of concentration, holding time index, boiler leak indication, or detection of boiler water carry over.

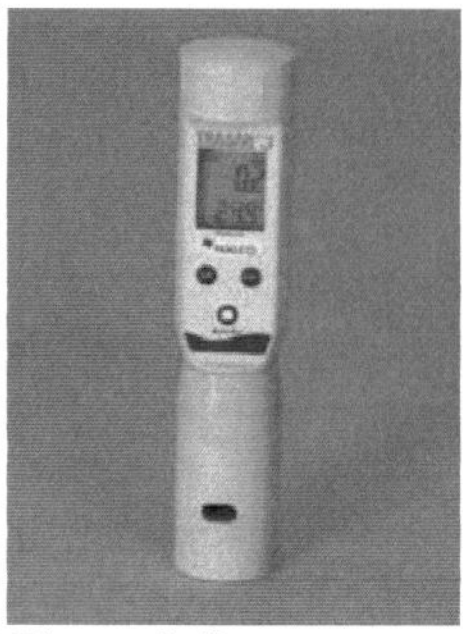
Figure 3. Pen fluorometer used with grab samples

This unique tracer technology makes it possible to determine the mass balance of a boiler system with greater accuracy, Figure 4. It also makes it possible to accurately and automatically control and monitor system chemistry and operating conditions, respond to boiler system changes, and maintain optimum treatment levels of highly effective scale and deposit inhibitors to keep boilers clean. As an example, this technology has been used in conjunction with soluble iron (Fe^{+2}) monitoring in a boiler system to quantify flow-assisted corrosion rates.

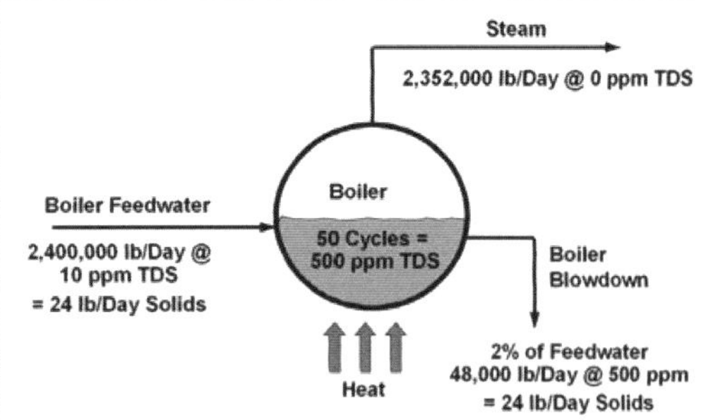

Figure 4. Boiler cycles of concentration

Corrosion Control

Corrosion of metals involves electrochemical reactions with charge transfer occurring at interfaces. The rate at which this transfer of charge occurs, or current, depends on the driving force, or voltage applied across the local system. Typically, as dissolved oxygen increases in a system, the ORP increases. When oxygen scavenger, or reductant, is added to the system, ORP decreases. The ORP reflects the composition of the system with respect to electroactive species. It is important to note the impact these species have on corrosion. While the ORP is not a direct measurement of corrosion potential, it is a useful indicator of the feedwater corrosivity (5).

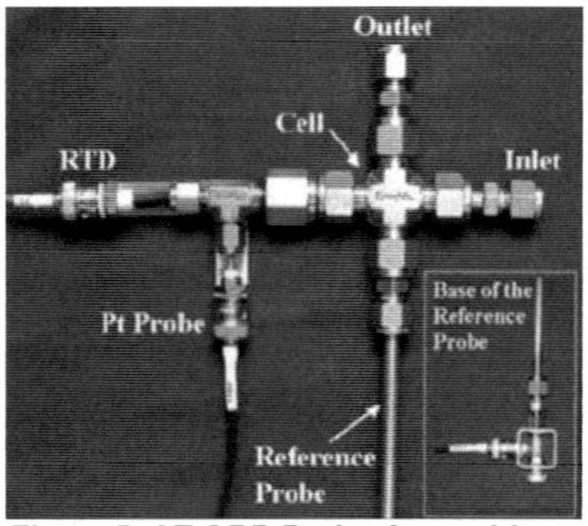

Figure 5. AT ORP Probe Assembly

At Temperature ORP

Figure 5 shows a commercial AT ORP probe. The probe assembly consists of a system water sample inlet and outlet, a connection for an integral RTD (resistance temperature detector) and platinum electrode (Pt Probe), and a separate external pressure-balanced reference electrode (Reference Probe). The reference electrode is filled with 0.1 N KCl in this case. All of the active portions of the three attached probes reside in close proximity within the cell itself.

The cell arrangement also consists of a high-pressure cell and high-pressure fittings that house the electrodes. The potential of the Pt Probe is measured against the external, pressure- balanced silver/silver chloride reference electrode (0.1 N KCl internal filling solution). The temperature of the water flowing through the cell is also recorded as a function of time.

There are a number of reasons why it is important to measure the feedwater ORP value at the boiler feedwater temperature and pressure:

- AT ORP is much more sensitive. It can see effects at high temperature that cannot be seen at room temperature.
- Corrosion rates vary with temperature and pressure. Corrosion is occurring at the operating temperature, so we need this correlation for proper corrosion control.
- AT ORP has a fast response time due to quick probe response coupled with sampling earlier in the stream.
- Scavengers/passivators work better at higher temperatures and pressures so control is more sensitive and realistic with AT ORP.

When comparing ORP readings, it is important to understand the differences between reference electrodes. Relative scales must also be standardized. Since both ORP and corrosion rates are very temperature dependent, AT ORP is a much more useful reading than Room Temperature ORP (RT ORP).

OPPORTUNITIES FOR ENERGY SAVINGS

Controlling boiler scale can have real energy savings benefits. There are many different sources of boiler scale – two of the most common are feedwater hardness and iron entering the boiler feedwater.

The new control technologies can have profound impact on minimizing the potential for scale, as well as mitigating its impact. Since the scale inhibitor is measured in the boiler feedwater and adjusted in real time, regardless of changes in steam load, underdosing and overdosing can be almost completely eliminated. Boiler feedwater corrosivity is likewise monitored and controlled in real-time, so feedwater corrosion excursions, and the resultant scale-forming iron can also be greatly reduced.

Other opportunities for energy savings spring from these technologies. It is important to remember that sulfite, one of the most common oxygen scavengers, contributes dissolved solids to the boiler feedwater. These solids increase boiler water conductivity. In a scenario where excessive levels of sulfite oxygen scavenger are being fed, improved control of oxygen scavenger feed can result in reduced boiler feedwater dissolved solids, higher cycles of concentration in the boiler, and reduced blowdown, all with associated water and fuel savings.

This technology also has the capability of monitoring and controlling feedwater pH, so energy and water savings from reduced feedwater dissolved solids, higher boiler cycles, and reduced blowdown can also be obtained by facilities that use caustic (sodium hydroxide) for controlling feedwater and boiler water alkalinity.

CASE HISTORY: GULF COAST REFINERY

A Gulf Coast refinery experienced several boiler tube failures on their 600 psig steam generators resulting in poor steam system reliability. The refinery management team wanted to improve the overall reliability and total system cost of its steam system. An Engineering Approach Analysis was done to understand system gaps and discover opportunities for improvement.

The 600 psig boiler system receives 100% sodium zeolite softened water. Operations ran samples to test for boiler feed water total hardness. The results are shown below in Figure 6.

Typical response to elevated boiler feed water hardness levels would include:

- Review of the sodium zeolite operation and corrective action where needed to bring the total hardness levels back into target
- Increased dosage of scale inhibitor
- Increased blowdown

Operations attempted to take corrective action with the tools and knowledge available.

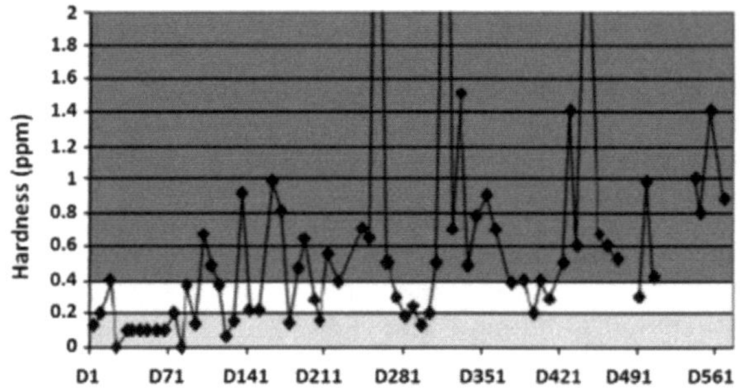

Figure 6. Boiler feed water hardness was in the target zone 45% of the time.

Boiler Cycles

Our analysis of the operational data found that boiler cycles were rather erratic and there was significant room for optimization as shown in Figure 7.

Refinery Operations was taking daily measurements and attempting to make blowdown adjustments using a manual valve. Achieving good manual control on a highly variable system is very difficult.

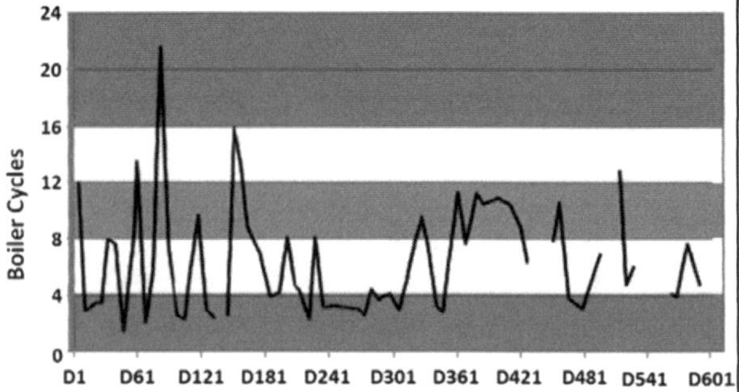

Figure 7. Boiler Cycles - In the Zone Analysis

Recommendations

After careful review of the customer's system, chemical treatment program, and daily operations using the Engineering Approach Analysis, several recommendations were made.

- Develop a plan to reduce hardness levels in boiler feedwater through additional operator training and implementation of an excursion reaction plan so that, when there is a hardness event, any negative impact can be minimized.
- Replace the current internal treatment product with the most advanced all polymer boiler treatment chemistry.
- Utilize boiler automation to continuously monitor and control the chemical treatment and other critical steam system parameters based on system stress and demand.

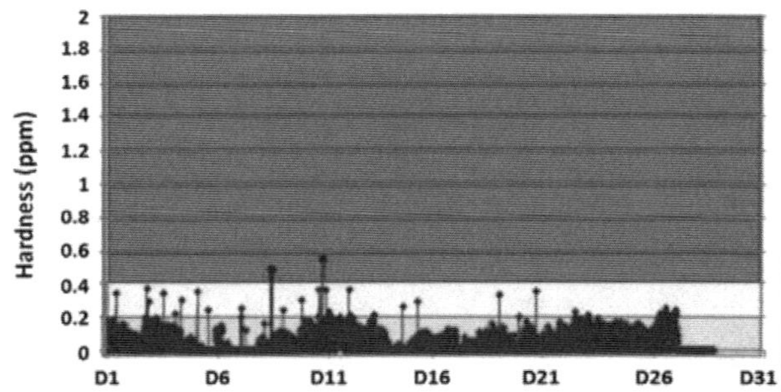

Figure 8. Boiler feed water hardness was in the target zone 89% of the time

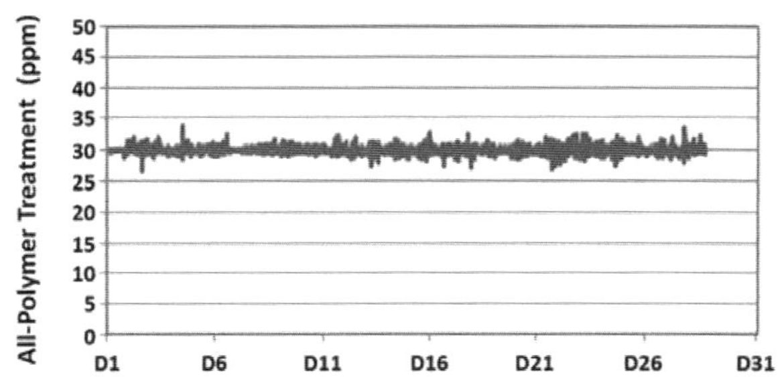

Figure 9. All-polymer internal treatment feedwater dosage

Boiler Feed Water Quality Results

The trend in Figure 8 shows the continuous boiler feed water total hardness measurement. Although there were occasional spikes above the best practice target, the off-spec measurements were quickly brought back into control. Plus, any time the hardness went out of target, the unit went into alarm mode and action was quickly taken to bring the hardness back into specification or control.

In the event of a significant hardness event, the controller would not only alarm, but would also automatically increase the all-polymer treatment dosage to account for the additional stress on the system. When the system stress was relieved, the product dosage automatically returned to its normal dosage.

Given the system design, an all-polymer/fluorescent tracer set point of 30 ppm was selected. The all-polymer dosage control was achieved by the automation system as shown in Figure 9.

The boiler automation controller is able to maintain a precise dosage even when there are system upsets. The trend shown in Figure 10 shows how the all-polymer set point of 30 ppm was maintained even with a sudden load increase on the #9 boiler due to another boiler going down unexpectedly. Although the boiler feed water flow rate increased by almost 30% in a matter of minutes, the new automation control system was able to maintain good control, assuring boiler reliability even under stressed conditions.

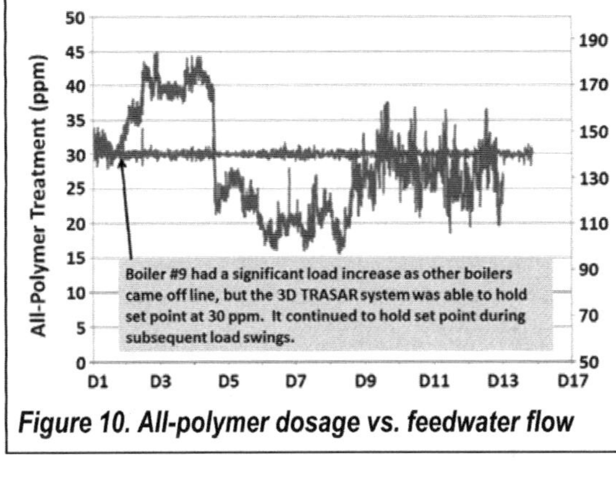

Figure 10. All-polymer dosage vs. feedwater flow

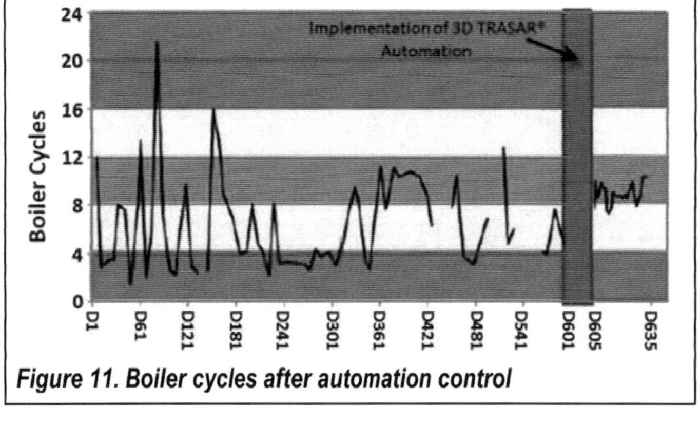

Figure 11. Boiler cycles after automation control

In fact, the boiler automation controller was able to maintain a six sigma control range of 29 to 31 ppm for the all-polymer internal treatment.

Table 1. Energy and Water Costs and Credits

		With Initial Conditions	Under New Conditions	Savings
Blowdown Energy Cost	\$/year	669,614	334,807	334,807
Blowdown Sewer Cost	\$/year	99,749	49,875	49,875
Makeup Water Cost	\$/year	235,893	214,448	21,445
Sub Total (Costs)	**\$/year**	**1,005,257**	**599,130**	**406,127**
Returned Condensate Fuel (Credit)	\$/year	0	0	0
NET SAVINGS or (COSTS)	**\$/year**			**406,127**
Calculated Cost of Steam	\$/1000 lb.	9.83	9.39	0.44
NET CO_2 EMISSION SAVINGS (INCR)	Fuel is natural gas		Tons CO_2/yr.	3,236

Boiler Cycles Results

Prior to implementation of the new automation control technology, plant operations had to manually control boiler cycles. As a result, the cycles ranged anywhere from 3 to 22 cycles. This variability created stress on the boiler system and resulted in poor reliability as well as wasted water and energy.

The new technology can be used to optimize and automatically control boiler blowdown. With boiler cycles control, the boiler cycles were automatically maintained between 8 and 10 cycles, Figure 11.

This boiler cycle control will result in significant annualized financial and sustainability savings of over \$400K as shown in Table 1. In summary, implementing the new automation technology resulted in tremendous improvement in operational control leading to improved boiler reliability, lower total cost of operations, and a significant eROI™ value.

CASE HISTORY: EAST COAST CHEMICAL PLANT

After a turnaround, the deaerator supplying water to the 900 psig Utility boilers was unable to maintain dissolved oxygen levels within normal ranges (<7 ppb). In fact, dissolved oxygen readings were consistently in the 200 – 300 ppb range as shown in Figure 12.

The chemical plant determined that major modifications were needed to the deaerator. The challenge was to maintain boiler operation with high feedwater dissolved oxygen concentrations until the deaerator repairs could be completed. Since boiler feedwater was used for steam attemperation, the use of sulfite oxygen scavenger was not an option. The chemical plant was using Nalco ELIMIN-OX® oxygen scavenger, which is both an oxygen scavenger and metal passivator.

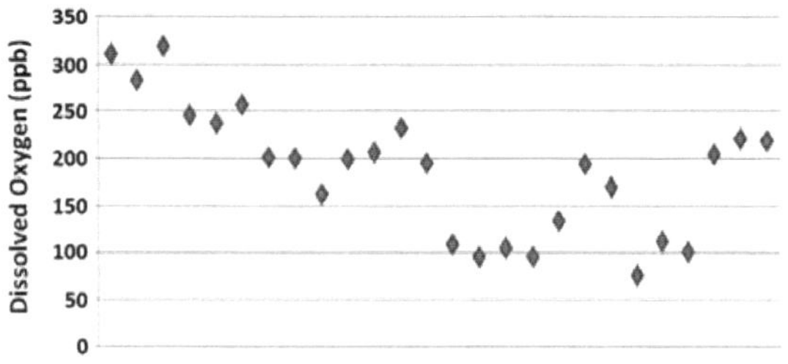

Figure 12. Manual Boiler Feedwater Dissolved Oxygen Test Results beyond normal ranges of less than 7 ppb.

All of the Utility boilers had an economizer, which is the area that is most susceptible to oxygen corrosion in a boiler system. One of the economizers was two years old; the other economizers were 20 years old. All economizers had outlet sample points that could be utilized to monitor iron-related corrosion in the economizers. The challenge was to utilize the passivating properties of the ELIMIN-OX oxygen scavenger and control the feed to a level that minimized the pre-boiler corrosion, especially in the economizers.

Solution

Initially, only daily grab samples were tested for dissolved oxygen, ELIMIN-OX oxygen scavenger residual, and total iron before and after the economizer. An on-line dissolved oxygen analyzer was installed on the boiler feed water to continuously monitor the actual dissolved oxygen levels. The results of the manual and on line monitoring are shown in Figure 13.

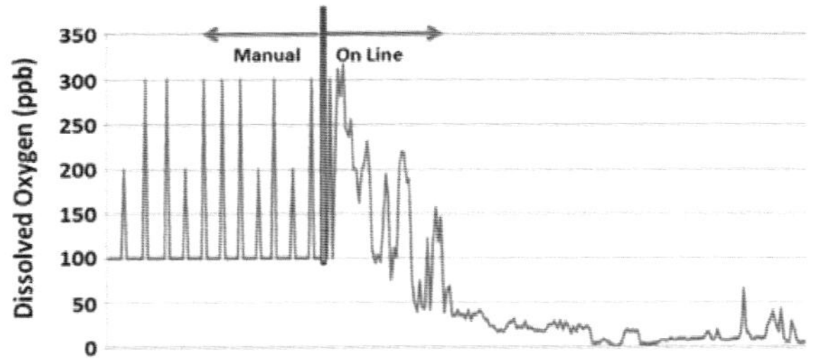

Figure 13. Dissolved Oxygen measurement was steady with on line monitoring.

To determine the amount of potential corrosion taking place with the high dissolved oxygen concentrations, total iron was monitored on the outlet of the economizers. As shown in Figure 14, economizer outlet iron levels remained extremely low. On-site total iron values were cross-checked with inductively coupled plasma (ICP) analysis. ELIMIN-OX feed rate was maximized and controlled to minimize the economizer outlet iron values.

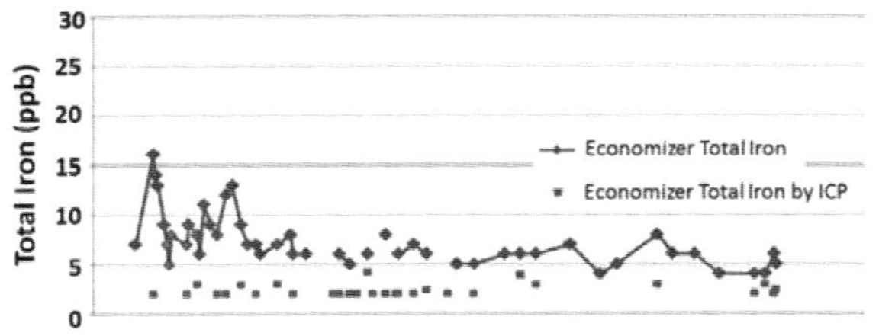

Figure 14. Economizer outlet total iron values on-site and by ICP.

The next step was to install the AT ORP to determine the overall pre-boiler corrosion potential using at temperature ORP. As shown in Figure 15, the AT ORP is far more responsive than the dissolved oxygen analyzer in detecting corrosion stress.

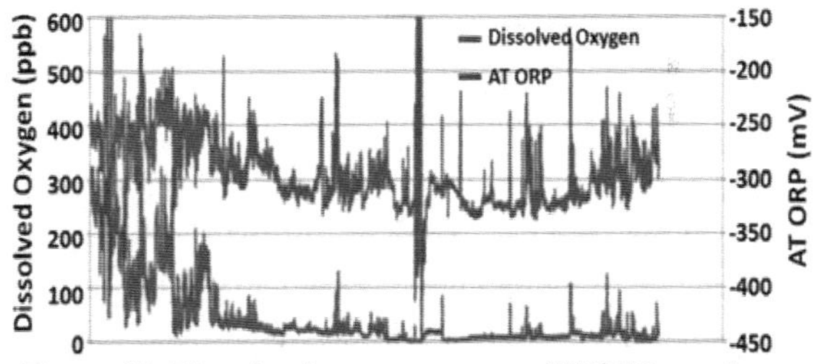

Figure 15. Dissolved oxygen versus AT ORP readings.

The AT ORP was used to optimize the ELIMIN-OX feed rate. As shown in Figure 16, ELIMIN-OX feed was reduced from 20 ppm to 13 ppm (35% reduction) with no measurable increase in corrosion potential as shown by the AT ORP readings. These results were also correlated to dissolved oxygen, as shown in Figure 17.

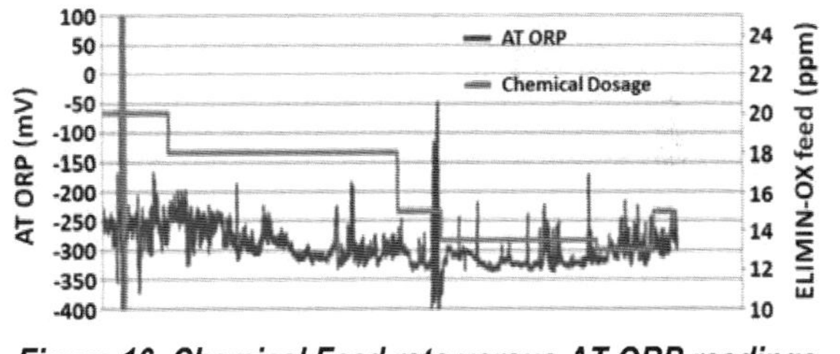

Figure 16. Chemical Feed rate versus AT ORP readings, showing the ability to reduce ELIMIN-OX feed while maintaining corrosion potential.

The final monitoring step was to correlate the economizer outlet total iron reading with each step-change in ELIMIN-OX feed and AT ORP reading to validate that no economizer corrosion was occurring. Figure 18 shows that, even with the reduction in the ELIMIN-OX feed rate, economizer outlet total iron values remained low.

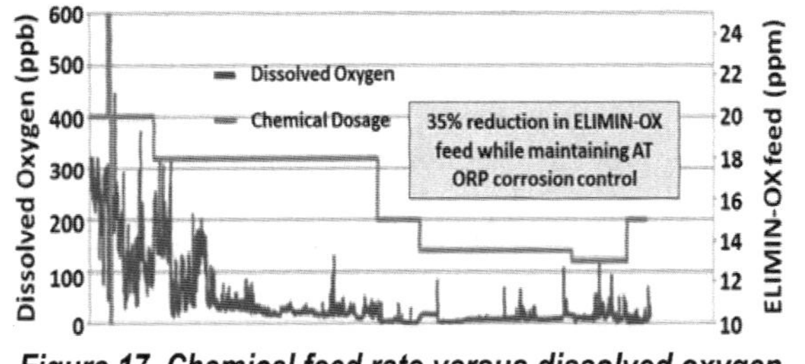

Figure 17. Chemical feed rate versus dissolved oxygen.

Results

All data showed that by utilizing the passivating properties of ELIMIN-OX technology and maintaining control with the AT ORP, the plant was able to operate their boilers confident that the long term reliability was not compromised. The plant easily made their goal of safely operating the boilers until the deaerator could be repaired.

However, for final verification, a tube sample from the newest economizer was extracted and sent to an external lab for metallographic analysis. The analytical report stated: "Internal corrosion is not extensive, and some areas are completely unaffected. The deepest pits were less than 9 mils deep (<0.009"). Overall tube thickness meets or exceeds 0.165" nominal."

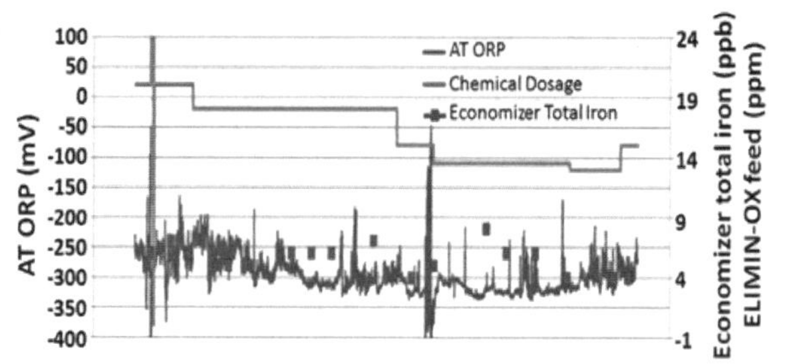

Figure 18. Dissolved oxygen, AT ORP, and chemical dosage.

The conclusion from the metallographic analysis was that the economizer tube that had been in service for two years showed no significant internal corrosion, thereby confirming the success of mitigating the potential corrosion effects of an improperly operating deaerator.

The eROI values achieved were:

- Asset Protection and Reliability – Allowed the refinery to operate for more than a year with high dissolved oxygen levels without damage to downstream equipment until a replacement deaerator could be procured. By minimizing oxygen corrosion to the economizer, the economizer life was extended.
- Chemical Savings – Allowed optimization of the oxygen scavenger feed while maintaining asset protection (35% reduction in chemical feed).
- Energy Saving – By minimizing incoming iron values, optimum boiler efficiency was maintained.

CONCLUSIONS

Even the best RT ORP probes are not as sensitive as the AT ORP probes. With this new boiler feedwater treatment automation, chemical is now fed based on a true at-temperature corrosion demand. Production events can occur quickly, and this new automation has shown that it can respond to the events quickly and with the appropriate response. The solid state fluorometer and AT ORP are both valuable instruments that can protect the boiler and feedwater systems from scaling and corrosion events. In addition, this technology can be used as a troubleshooting tool to help diagnose issues and correct them.

Given the recent economy, it is becoming even more important for all boiler house managers, engineers, and operators to take a fresh look at ways to reduce operating costs, protect asset life, and improve productivity. This technology has clearly shown that it can help find issues that previous technologies cannot see, and allows users to protect their pre-boiler and boiler systems.

REFERENCES:

1. Nalco Water Handbook, 2nd Edition, 1988, McGraw Hill Publishing
2. United States Patent #6,255,118 "Method for Using an All Solid-State Fluorometer in Industrial Water Applications"
3. United States Patent #5,736,405 "Monitoring Boiler Internal Treatment with Fluorescent-tagged Polymers"
4. "Monitoring and Control in Boilers With All Solid-State Fluorometers", Martin Godfrey, Corrosion / 2002, Paper No. 02012, NACE International Conference, Houston, TX (2002).
5. "A New Approach to Feedwater Corrosion Control: A Detailed Nalco AT ORP Study at a Paper Mill", Peter Hicks, David Grattan, and Paul Desch, Nalco ADV-1105.

SOLID-LIQUID FILTRATION TRENDS FOR PARTICLE FINES REMOVAL IN MANUFACTURING ENVIRONMENTS

Barry A. Perlmutter
BHS-Filtration Inc.
Charlotte, North Carolina 28270

Biography

Barry A. Perlmutter is President and Managing Director of BHS-Filtration Inc., a subsidiary of BHS-Sonthofen GmbH. Barry has over 30 years of technical and marketing experience in the field of solid-liquid separation, filtration, centrifugation and drying. He has published and lectured worldwide and has been responsible for introducing and creating growth for many European companies into the marketplace. He has a BS in Chemistry from Albany State, NY, MS from the School of Engineering at Washington University, St. Louis and an MBA from the University of Illinois.

ABSTRACT

In the manufacturing of chemicals, petrochemicals, pharmaceuticals, bioenergy etc., after the initial mixing-reactions, there is always the necessity to remove the mother liquor from the process slurry. In some cases, the liquid is the valuable component, in some cases it is the solids and in other cases it is both compounds. The nature of the requirements will determine the type of equipment needed for the separation process.

Generally, after the initial mixing - reactions, there will be bulk separation to remove the larger product solids or unwanted solids. This is the easy step. As processes have become more sophisticated and quality requirements increased, there is a trend to be able to remove the residual particle fines from the slurries. These fines are very small, typically in the 1 – 5 micron range and smaller and are very low in concentration down to ppm levels.

This paper provides a brief overview of crude separation equipment and then continues with a detailed view of technologies that can remove – recover small particle fines. The general term for this is "clarification" technologies and this will cover alternatives to traditional bag and cartridge filtration systems.

The paper concludes with a review of the installations in chemical, refinery, bioenergy and pharmaceutical plants to provide engineers with new and creative ways of thinking to evaluate process filtration problems and solutions.

SOLID / LIQUID SEPARATION FOR PROCESS SYSTEMS

Filtration, pressure or vacuum, is the art of finding a filter media which allows the liquid to pass through while retaining the solids. The driving force may be gravity, vacuum, or pressure.

The ultimate solid phase is always drier than that from gravitational systems and a great deal of processing can often take place in the actual filtration stage, especially with regard to the conversion or exchange of the residual cake moisture.

The driving force, causing the liquid to exit while leaving the solids behind, can be divided into four groups:

1. Gravitational, which is really little more than draining, but can be very useful to reduce large quantities to more manageable proportions. They can be batch operated or, more usual, continuous.
2. Vacuum, this is in many ways mechanically the simplest driving force available. Vacuum filters can be batch operated but are normally continuous. In general, the solid (cake) thickness can be controlled within close limits; there are few limits to materials of construction and some of vacuum filters offer the best solids washing possibilities.
3. Pressure, with or without compression, which obviously involves mechanical constraints. All pressure filters are batch-operated units with one exception, the exception being the Rotary Pressure Filter. Cake washing can be excellent and the final cakes are usually as dry as can be expected without heat input.
4. Centrifugal, which can sometimes offer a compromise between vacuum and pressure filtration. Centrifugal filters can be continuous in operation or they can operate in an automated continuous batch-mode. The nature and behavior of the solids play a great part in the success or failure of centrifugal filters. Cake washing can be good depending on the type and the behaviour of the solids.

TYPES OF FILTRATION SYSTEMS

With the above in criteria in mind for chemical and pharmaceutical production operations, there are several basic choices for a process engineer. These are segmented by particle size removal as well as whether or not the filter will discharge the cake batch-wise or continuous.

BATCH FILTERS

Autopress: This horizontal pressure filter is tubular in construction and operates as a contained filter press. Circular plates with welded metal or synthetic media are contained in a pressurized housing. This allows for pressure filtration, cakes washing and vacuum or pressure drying. After the cycle is completed, the housing is moved and automatic cake discharge is via scraper knives that move between the plates

Bag filters: A bag with a connection for a high pressure inlet held captive in a metal cage. Filtrate runs through the bag and the solids stay inside.

Basket strainer: Similar to the bag filter, except that the strainer may be the filter medium in which case the solids have to be dug out, or the basket may be lined with a bag, for easy cake solids disposal.

Basket centrifuge: A metal basket rotating at centrifugal speeds and a cloth bag is fitted inside the basket. The drive mechanism runs the basket up to high speeds and feed is poured into the basket. Filtrate runs through the basket and is collected in an outer mantle. When sufficient solids have collected inside the bag, the machine is stopped and the solids removed or it is slowed down to allow a mechanical scraper to remove the bulk of the solids (a "heel" stays behind). In a horizontal version, the Peeler Centrifuge, the bulk of the cake can be removed without slowing the machine.

Cartridge filters: Pressure vessels inside which are fitted a number of filter cartridges. Almost without exception the flow is from the outside inwards. Cartridges are available in cleanable or (much more common) disposable form. A very wide variety of materials, textures, pore size and physical sizes are available.

Filter press: The original filter presses were of the "plate and frame" construction. Modern presses use almost exclusively "recessed plates". A very wide variety of sizes, configurations, plate supports and degrees of automation exist.

Leaf filter: Leaf filters consist of a pressure vessel inside which a number of filter leafs are vertically mounted. The leaves are normally metal and fitted with a synthetic media. For removal of the solids different arrangements exist, ranging from opening the bottom of the vessel to mechanically spinning or vibrating the cake into a chute.

Membrane presses: The membrane press is basically a filter press but instead of having drainage grooves in the plates, the plates are fitted with an elastomer sheet (which has the drainage grooves) and which sheet can be inflated. By inflating the sheet at the end of the filter cycle any residual moisture will be expelled and the cake itself will be squeezed, usually resulting in better cake moisture figures.

Nutsche filter: Circular or rectangular filters with a drainage bottom onto which a filter medium is fastened. If the drainage section is connected to a vacuum source the filters are often open top. If they are closed at the top, they can be pressurized and thus benefit from a higher driving force.

Tube press: These may be vertical or horizontal and consist of a metal tube and inflatable membrane. Inside this membrane is mounted a central filter core consisting of a drainage tube with a filter medium. The feed is introduced into the tube under pressure, filtrate exits from the central filter core. After filtration and / or washing the membrane is inflated to squeeze the cake solids to extremely low residual moisture values. To discharge the cake, the membrane is relaxed, the bottom section of the tube opened and the tube slightly or totally withdrawn allowing the cake solids to drop off or to be scraped off.

CONTINUOUS FILTERS

Disc Filter: These consist of a flat disc mounted on a hollow shaft. The disc is made of metal and has on either side an open cloth support structure which connects with the hollow shaft which carries the filtrate. A filter cloth is fastened to the disc and the hollow shaft connects to a vacuum source. The disc, usually 30 % submerged, rotates slowly in a feed trough where it picks up the solids, which can later be scraped off just prior to re-entry into the feed trough.

Rotary Pressure Filter: The Rotary Pressure Filter is a continuously operating unit for pressure filtration, cake washing and drying of slurries up to 50% solids. The filter has a rotating drum inside a housing. The drum has cells within which are cell inserts fitted with the filter media. Filtration is conducted via pressure of up to 6 bar. Positive displacement washing or counter-current washing follows filtration. Multiple washing steps as well as solvent exchanges, steaming and extraction are accomplished. Finally, the cake is dried by blowing hot or ambient gas through the cake. The Filter has a uniquely designed discharge system, which provides for atmospheric discharge from pressure filtration. After automatic cake discharge, the filter cloth is washed; the clean filter cloth the re-enters the feeding / filtration zone thereby continuing the process. All solvent and gas streams can be recovered separately and reused in the process to minimize their consumption.

Pusher Centrifuge Horizontally mounted filtering centrifuge with a filter cage made of metal bars the smallest opening of which is about 100 micron. Inside the cage is a "pusher" arrangement, basically a sturdy ring which lies very close to the filter medium. The ring is connected to a central shaft which rotates at the same speed as the filter cage but reciprocates (approx 1 – 3 cm) causing the collected solids to be pushed out of the machine. Multi stage machines exist. In these the solids are pushed onto the second or third stage which is each time of a larger diameter. The higher linear velocity as well as the tumbling action can be beneficial for dewatering and / or cake washing, although it may cause the crystals to fracture.

Rotary Vacuum Drum Filter: The original rotary drum filter consisted of a cylindrical drum, fitted with a drainage grid on the outside over which a filter cloth was stretched. The drum had perforations and was connected to a vacuum source. The drum itself was mounted on a horizontal shaft and submerged for about 30% in a feed trough. The vacuum caused the liquor phase to be sucked through the cloth, leaving the solids behind as a filter cake. The cake was scraped off just before re-entering the feed trough.

Vacuum Belt Filter: The two main types are carrier belt and tray (fixed or indexing). These filters have a continuous belt as the filter medium which is subjected to a vacuum source. Feed is introduced at one end, where it is normally allowed to settle under gravity for a few seconds, causing the coarser fraction to form its own pre-coat. The suspension is then subjected to full vacuum. Since the cake thickness can be controlled by adjusting the speed of travel or indexing time, the optimum cake thickness to obtain the required residual moisture and / or cake washing efficiency is achieved. The horizontal configuration makes this filter one of the most efficient cake washing filters.

CLARIFICATION OF SLURRIES & RECOVERY OF SOLIDS

In the previous section, batch and continuous filtration systems were discussed. As processes have become more sophisticated and quality requirements increased, there is a trend to be able to remove the residual particle fines. These fines are very small, typically, smaller than 1 – 5 micron range and are very low in concentration down to ppm levels.

Candle Filters and Pressure Plate Filters are installed for clarification and recovery applications from liquids with low solids content. The candle filters are vertical candles while the pressure plate filters are horizontal plates. The cake structure as well as the process parameters determines the optimum thin-cake technology.

Description and Operation of the Candle Filter

Candle Filters provide for thin-cake pressure filtration, cake washing, drying, reslurry and automatic discharge as well as heel filtration in an enclosed, pressure vessel. Units are available from 0.17 m^2 up to 200 m^2 of filter area per vessel.

Filter Candles & Media

The filter candles, as shown in Figure 1, consist of three components: single-piece dip pipe for filtrates and gas, perforated core with outer support tie rods and filter sock. The filtrate pipe is the full length of the candle and ensures high liquid flow as well as maximum distribution of the gas during cake discharge. The candle can be a synthetic, stainless steel or higher alloys. The outer support tie rods provide for an annular space between the media and the core for a low pressure drop operation and efficient gas expansion of the filter media sock for cake discharge. The filter media is synthetic with a clean removal efficiency to less than 1 micron (um). As the cake builds up, removal efficiencies improve to less than 0.5 um.

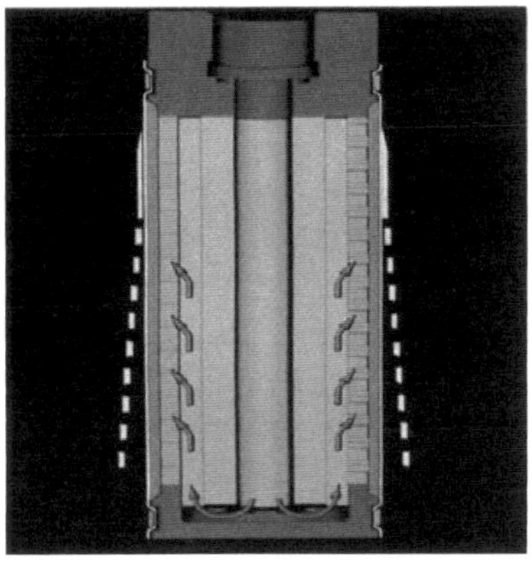

Figure 1: BHS Candle Showing Gas Flow to Expand the Filter Media Sock for Cake Discharge

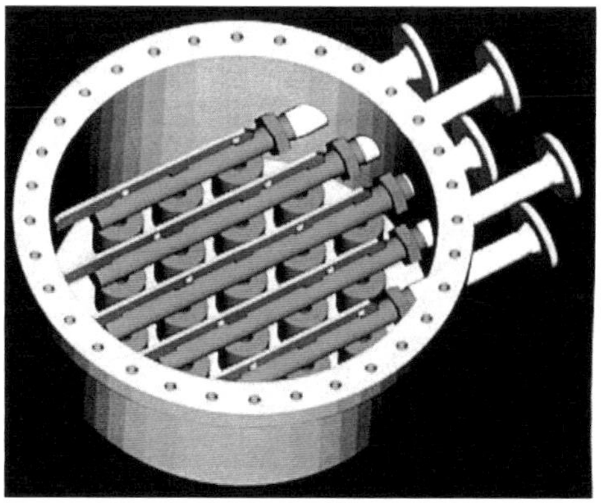

Figure 2: BHS Candle Filter

Filter Vessel & Candle Registers

The candle filter vessel is constructed of stainless steel or higher alloys. Within the vessel are horizontal manifolds called candle registers. Each candle is connected to a register with a positive seal to prevent bypass. Each register may contain from 1-20 candles depending upon the filter size. The registers convey the liquid filtrate in the forward direction as well as the pressure gas in the reverse direction for filter media sock expansion. Each register is controlled with automated valves to ensure optimum flow in both directions. Figure 2 illustrates the candle filter vessel.

Automatic Process Cycles

Filling: The slurry feed enters the bottom of the filter vessel.

Filtration: The slurry is either pumped or pressurized from the reactor into the vessel. Cake will deposit on the outside of the candle; the separated filtrate will flow through the filtrate pipe and the registers. This process continues until one of the following conditions is achieved: maximum pressure drop, maximum cake thickness, minimum flow or time.

Washing: Displacement washing or recirculation washing.

Drying: Blowing gas, steam or "shock" drying.

Heel (Falling-Film) Filtration: The liquid remaining in the vessel cone after filtration or washing is completely filtered.

Cake Discharge: Gas flows sequentially through each of the candle registers, down each of the filtrate pipes and then is distributed by the perforated core. The filter media sock gently expands by the gas flow and pressure allowing for cake discharge, as shown in Figure 1. Alternatively, the cake can be discharged as a slurry.

Description and Operation of the Pressure Plate Filter

The pressure plate filter has similar operating characteristics to the candle filter. The filter design is shown in Figure 3.

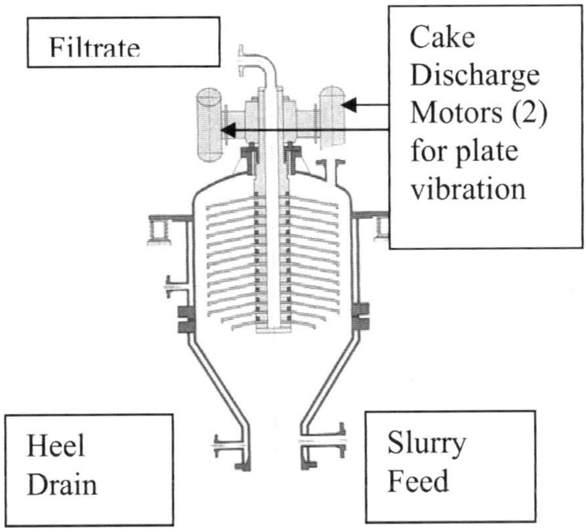

Figure 3: BHS Pressure Plate Filter

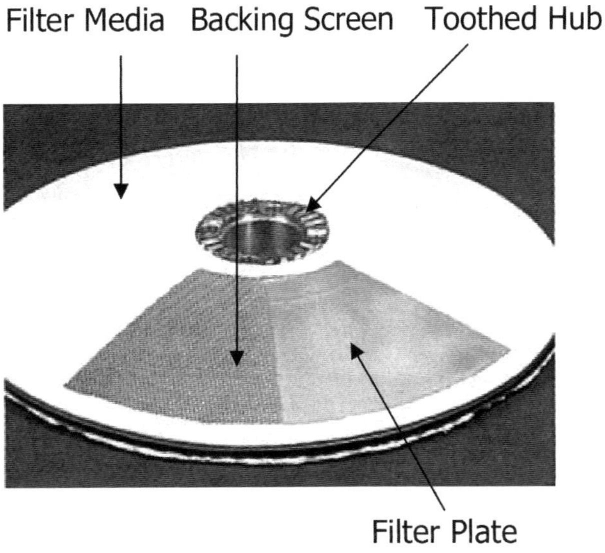

Figure 4: Filter Plate

Automatic Process Cycles

Filling: The slurry feed enters the bottom of the filter vessel.
Filtration: The slurry is pumped under pressure into the vessel or via gas pressure through the reactor. Cake will deposit on the top of the plates. The separated filtrate will flow through the plates to the center main filtrate outlet. This process continues until one of the following conditions is achieved: maximum pressure drop, maximum cake thickness, minimum flow or time.
Washing: Displacement washing or recirculation washing.
Drying: Blowing gas, steam or "shock" drying.
Heel Filtration: The liquid remaining in the vessel cone after filtration or washing is completely filtered.
Cake Discharge: As shown in Figure 5, the motors on the top of the filter operate at different frequencies and the plates gently vibrate for cake discharge. The plates vibrate in the vertical and horizontal planes and the solids are conveyed in an elliptical pattern to the outside of the vessel. Gas assist helps in the discharge process. There are no rotating plates, gears or bushings and mechanical seals are not required.

Description and Operation of the Pressure Plate Filter (Continued)

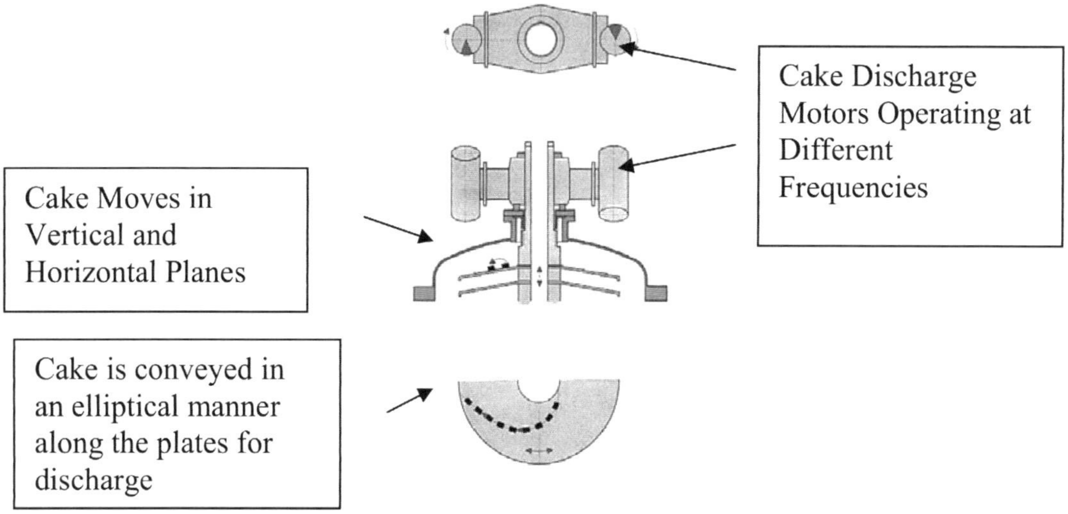

Figure 5: Cake Discharge in the Pressure Plate Filter

SELECTION OF CANDLE VERSUS PRESSURE PLATE FILTER TECHNOLOGIES: CAKE STRUCTURE AND PROCESS PARAMETERS

The major difference between the two technologies depends on the cake structure that is formed. Some cakes are better handled in the horizontal and some in the vertical.

Cake Thickness and Filtration: The candle filter is limited to cake structures that can be formed to about 5 – 20 mm. The pressure plate filter can handle cakes up to 75 mm. Both units can conduct filtration up to 150 psig.

Filter Media: The candle filter uses only synthetic media with a clean removal efficiency from 1 –3 micron range and finer down to 0.5 microns. The pressure plate filter can also use metal media. For the pressure plate filter, the clean micron range removal efficiency is also 1 – 3 microns and finer.

Cake Washing: If the process requires washing of the solids, then generally the pressure plate filter is a better alternative. If washing is not as critical, then the candle filter may be the optimum technology for clarification and recovery.

Heel Filtration: The remaining liquid in the vessel (liquid heel) after filtration or washing can be removed from the candle filter or pressure plate filter by circulation, heel filter in the cone of the vessel, or additional heel filter plates in the pressure plate filter.

Cake Drying: The candle filter can produce cakes with approximately 10% moisture. This moisture level depends upon the specific cake but the moisture lower limit is that moisture just above the cake cracking point. The pressure plate filter can produce bone dry cakes.

Cake Discharge: Both designs can easily discharge most cakes equally with no residual heel.

Clean-In-Place (CIP)/ Steam-In-Place (SIP): Both units conduct CIP / SIP operations in identical manners by filling and circulating cleaning fluids while blowing gas in the reverse direction to the filtration direction, which creates a turbulent mixture or a quasi-ultrasonic cleaning effect. The pressure plate filter further enhances this operation with plate vibration.

APPLICATION 1:
REPLACING A MANUAL PLATE FILTER AND BAG FILTER COMBINATION

This specialty chemicals manufacturer produces various resins that require filtration. Current production includes a neutralization step which yields metal salts. These salts are filtered out with a manual plate filter followed by a bag filter for polishing. Two solvent washes follow the filtration step to recover as much resin as possible. After washing, the filters are steamed and opened. The solids are disposed manually for each batch and the filter paper is replaced. The goals are to eliminate exposure to heptane, reduce the maintenance and operation on the two filters and to recover a dry, as possible, catalyst. Current production is 3000 gallons in 4 – 5 hours.

Results & Conclusions

The filtration flux rate from the BHS laboratory tests ranged between 10-30 L/m^2min at approximately 20 psi feed pressure. The filter cloth for the sock is polyester with an air permeability of 1.0 cfm/ft^2.

The tests showed that one BHS candle filter with 10 m^2 of filter area can complete the cycle in a time of 4.3 hours and replace the manual plate filter and bag filter.

The cycle time is as follows:

Cycle Times			
	Filling	5	min
	Filtration	10	min
	Wash	4	min
	Drain	10	min
	Dry	5	min
	Vent	2	min
	Discharge	5	min
	Reserve	9	min
	TOTAL	**50**	**min**

Figure 6: Dry Cake After Discharge

APPLICATION 2:
REPLACING A FILTER PRESS FOR A HOT SLURRY FILTRATION

This specialty chemicals manufacturer produces various grades of polyols from ethylene glycols that require filtration of small particles in the 1 – 2 um range. Current production includes a manually-operated filter press at 200 – 250 degrees F. The goals are to eliminate operator exposure, reduce the maintenance and operating costs for the filter press and to increase the polyols yield. Current batch sizes are 45000 pounds of slurry.

Results & Conclusions

BHS conducted the lab tests in the BHS laboratory in Charlotte, North Carolina. The slurry temperature was between 200 – 250 degrees F at a pressure of 90 psig. The filter cloth for the sock was successful using PTFE or PEEK material.

The tests showed that one BHS candle filter with 27 m^2 of filter area can complete the required batch is three (3) cycles, each of 6-hours. The cycle time is as follows:

Cycle Times			
	Filling	12	min
	Precoat	8	min
	Filtration	60	min
	Drain	15	min
	Dry	3	min
	Vent	2	min
	Discharge	15	min
	Reserve	5	min
	TOTAL	**120**	**min**

Figure 7: Jacketed Candle Filter with Register-Filtrate Piping & Junction Box

APPLICATION THREE:
REPLACING CENTRIFUGE & CARTRIDGE FOR TIN HYDROXIDE SLURRY

This process was previously done with a centrifuge which did not provide much separation at all and the fines were causing issues in the downstream operation at their sister plant.

Process Observations:

Precoating with Hyflo (5 lbs) and water – filtrate goes to sump
Precoating based on volume and then switch over to feed slurry (0.4 – 0.5% body feed)
Filtration - Pressure built to around 20 psi in about 30 minutes, batch size is approximately 3000 lbs. (10-15 gpm)
Washing with water- 40 psi until filtrate looks clean with 12 um PEEK filter sock
Draining- initially through candles first (25 psi) and then the heel drain
Drying – Blowing of gas for 5 minutes at 20 psi
Cake discharge – dry cake with a cake thickness of 15-20 mm

Results:

The BHS candle filter with 10 m^2 of filter area replaced the centrifuge and cartridge polishing filter. The results are below.

Cycle Times			
	Filling	5	min
	Precoat	10	min
	Filtration	45	min
	Wash	5	min
	Drain	10	min
	Dry	5	min
	Discharge	5	min
	Reserve	5	min
	TOTAL	**90**	**min**

APPLICATION FOUR:
REPLACING A FILTER PRESS USING PAPER MEDIA

This specialty lube oil process, for final clarification, is conducted in a filter press using a paper media at 170° C, and using carbon, lime or celite as a body feed. There are five (5) grades of lube oils and cleaning between each campaign is required. The material is dried with nitrogen and the solids are removed and burned in the power plant as fuel. The process has the manual problems associated with a filter press plus the safety concerns of being burned with hot oil. The average particle size distribution (PSD) is 0.7 um.

Results:

The pilot testing demonstrated that the oil slurries could be filtered to meet the customer's specifications, that the resultant cake could be dried to the customer's desired moisture content, and that the cake could be automatically discharged satisfactorily from the candles. The purchased system was a 10 m^2 jacketed BHS candle filter using PEEK filter socks for the current five (5) products with expansion possibilities.

APPLICATION FIVE: DUAL STAGE CANDLE AND PRESSURE PLATE FILTERS FOR FINES REMOVAL AT REFINERIES, GAS PLANTS AND GREY WATER AT COAL GASIFICATION PLANTS

The fines generated in grey water as well as by cat crackers and cokers are generally less than 1 um in size and very low concentration. For these large flows, upward to 800 gpm, simple bag or cartridge filters or filter presses cannot be used. For these applications, BHS has developed a dual-stage process using candle filters for concentrating and then followed by a pressure plate filter for the final filtration, cake washing and drying. The benefits include increased reliability, much lower consumption (25% less) of compressed air/gas for drying and cake wash water. A typical example and photo is below.

Process & Results:

Total Flow:	94 gpm
Design Solids Load:	28 lb/hr (600 ppm)

- Two (2) Candle Filters, each with 20 m^2 of filter area, for concentration
- One (1) Pressure Plate Filter, 4 m^2 of filter area, for final filtration, cake washing and drying

Figure 8: Complete Skid Package with Candle Filters, Plate Filter, Tanks, Pumps and PLC Controls

SUMMARY

There are many choices for the initial filtration steps in a process. The take-away is that one filtration system may not achieve those quality requirements. Engineers must evaluate all outcomes to make an informed and successful decision. A secondary filtration system for fines removal, while adding capital expense to the project may result in an overall more reliable and optimum process solution. While not discussed in this article, laboratory and pilot testing is critical for a successful decision and project.

Kneader Technology for the Direct Devolatilization of Temperature Sensitive Elastomers

Synthetic elastomers have been produced for over 50 years. Advances in catalyst systems and polymer formulations have been somewhat diminished by the continued use of the same processing technology. In particular, the use of coagulation, steam stripping, mechanical dewatering, and convective drying for the devolatilization of temperature sensitive elastomeric solutions can be replaced with direct devolatilization using kneader technology. A two-step, direct devolatilization process has demonstrated energy savings and advantages in environmental emissions and process/product flexibility when compared to the conventional steam stripping process.

Meeting: Spring AIChE National Meeting & 9th Global Congress on Process Safety

San Antonio, TX – April 28th through May 2nd, 2013

Topical 3: Chemicals Manufacturing for the 21st Century

Title: ***"Separation Technologies for the New Biochemical Applications"***

Authors: Glenn Shiveler, Sulzer Chemtech USA Inc. – Tulsa, OK

Dr. Antonio Garcia, Sulzer Chemtech USA Inc. – Houston, TX

Abstract:

Advances in the next generation of biofuels will include development of commercial process technologies for the global transportation markets. R&D has led to new biofuels from renewable feed stocks: grains, bagasse, switch grass, sugar cane, beets, turnips and wood pulp. Fermentation of these feed stocks produces alcohols along with contaminates that include acetates, aldehydes and acids. Separation of the valuable biofuel components from these contaminants is made challenging due to azeotropes, separation of the chemicals from the fermentation biomass, potential use of extraction solvents, efficient heat integration and dehydration of the biofuel product.

Additional challenges are: process economic feasibility and fast-track commercialization process.

For many years Sulzer has been active in providing mass transfer equipment in many existing 1st generation biofuels production facilities. Currently, Sulzer is supporting the development of 2nd generation biofuels with separation process technologies, process equipment, skid mounted systems for demonstration scale and basic engineering and key equipment for commercial plants. The new biochemicals plants involve various design considerations that are described in this paper.

(132e) In-Process Particle and Droplet Characterization for Process Design and Optimization in Manufacturing Operations

Anjan P. Pandey, METTLER TOLEDO, Columbia, MD

Particles and droplets are pervasive in the chemical, consumer products, and bio-pharmaceutical industries. Particles grow, agglomerate, break, dissolve, and change shape - all of which can have significant impact on process performance and product quality. Droplet distribution can impact emulsion stability, polymerization kinetics, and oil-water separation rate. In order to ensure repeatable product yield, purity, and bulk density at production scale, it is critical to understand and control particle/droplet distribution. However, manufacturing conditions (high pressure, high temperature etc.) often make it extremely difficult in obtaining samples for offline measurements. In addition, challenges of obtaining representative sample and extensive sample dilution could lead to erroneous information about particle/droplet distribution.

With FBRM® and PVM® probes the researcher can track the entire dynamic particle and droplet system in production scale at standard operating temperatures and pressures, without sampling or sample dilution. For over 20 years, scientists and engineers in chemical/pharma/oil/consumer industries have routinely used FBRM® and PVM® technologies for tracking particle/droplet distribution in real time to ensure less batch failures, batch to batch consistency, and greater production throughput. This presentation will focus on real case studies applying FBRM® and PVM® technologies at production scales to:

- Ensure batch-to-batch consistency during crystallization
- Target a desired crystal endpoint distribution
- Optimizing filtration rates and product throughput
- Optimize droplet distribution to control oil-water separation

(132f) Saving Agitator Power Costs by Using the "Off" Switch (and other techniques)

Gregory Benz, Benz Technology International, Inc., Clarksville, OH

Though agitator power consumption may not be one of the primary energy costs in most processing plants, it is still a contributor which may be minimized by judicious equipment selection as well as operating procedures, such as shutting the agitator off when agitation is not needed, or using variable speed drive when less agitation is needed. The right approach depends on the kind of agitation problem. In this article, we will focus on 3 major problem classifications: liquid motion, solids suspension and gas dispersion.

(133a) The Evolution of Chemicals R&D in the Information Age

Felipe Tavares, Jansley Pascoal and Bruno Maia, Intratec Solutions LLC, Houston, TX

Most strategic decisions regarding product and process development need to be made in the initial phase of a project lifecycle. In today's competitive environment, R&D teams are under constant pressure to produce results faster with fewer resources. Decisions must be made based on limited data obtained during a very early phase.

In the past, the difficulty in the access to quality information and the lack of suitable tools led to spend significant time and effort to produce realistic economic estimates, and get a sense of the opportunity attractiveness in early stages of the project workflow. The great dilemma experienced by R&D teams was: develop an estimate too quickly but inaccurate or develop a detailed estimate but spending too much time and effort.

In recent years, the world is experiencing constant technological advances reflected in the increasing flow of information through the internet and the exponential growth in the processing power of computers. The easy access to huge amount of information and innovative engineering tools have completely re-invented the workflow of processes and project development. The combined use of extensive public information sources, such as patents and articles databases; conceptual process and equipment design tools; rigorous process simulators; and consistent cost information are re-shaping how R&D teams are doing and managing research.

This session will cover how to make the most of all available resources to evaluate new products and process, in a fast and accurate way, during the conception stage. Also, it is disclosed how to define metrics to help R&D group to conduct their projects, either internally or through partnerships with research centers and universities.

Currently, there are research firms taking benefit from this approach, significantly reducing cycle time of their projects and increasing the number of alternatives assessed by their staff. Also, there are already companies relying on sale of services based on these concepts, which would be impossible decades ago.

Two cases are presented as illustrative examples: the acrylic acid production from sugar fermentation and ethylene production from ethanol dehydration. These processes technologies were quickly assessed from few public available information and hypothetical assumptions allied with engineering tools.

(133b) Software for Synthesizing Complete Set of Pareto Efficient Designs of Energy Supply Chains According to Cost, Ecological Footprint, and Emergy

Leisha Vance[1], **Botond Bertok**[2], Istvan Heckl[3], Heriberto Cabezas[1] and Ferenc Friedler[2], (1)Office of Research and Development, U.S. Environmental Protection Agency, Cincinnati, OH, (2)Department Computer Science and Systems Technology, University of Pannonia, Veszprem, Hungary, (3)Department Computer Science & Systems Technology, University of Pannonia, Veszprem, Hungary

Methodology and computer aid are proposed for designing sustainable supply chains in terms of sustainability metrics by utilizing the P-graph framework. The methodology is an outcome of the collaboration between the Office of Research and Development (ORD) of the U.S. EPA and the research group led by the creators of the P-graph framework at the University of Pannonia. The integration of supply chain design and sustainability is the main focus of this collaboration. A recent extension to the P-graph framework provides a mathematically rigorous procedure for synthesizing the complete set of Pareto optimal networks subject to multiple objectives and constraints, which include profitability and sustainability in the proposed methodology. Specifically, to evaluate the sustainability of a given process under construction including its supply chain, sustainability metrics are incorporated into the design procedure.

The proposed methodology and the software are demonstrated with the optimal design of a supply chain for providing heat and electric power to an agricultural region with relatively limited land area where agricultural wastes can potentially be recovered as renewable resources. Possible sources of heat and electricity included electricity from the Hungarian grid, and heat and electricity generated from natural gas, corn, corn silage, grass silage, or wood or some combination of these sources. Each supply chain was ranked according to cost, and assessed environmental impacts using the ecological footprint (representing land use burden), and emergy (representing energy resource burden). Decisively feasible supply chains were found with cost variations of +2% to -17% compared to "business as usual" scenarios, i.e., using only natural gas and electricity from the Hungarian grid. For these supply chains, the sustainability profile as represented by the ecological footprint varied from +8% to -78%, and the emergy results ranged from -54% to -93%. Most importantly, it appeared feasibly possible to design supply chains for heat and electricity generation which were both cheaper and more sustainable than the supply chain currently in use.

(133c) Multiobjective Optimization of Solar Energy Driven Desalination Systems

Eman Tora, Chemical Engineering and Pilot Plant Department, National Research Centre, Giza, Egypt and Tarek Moustafa, Chemical Engineering Department, Cairo University, El Giza, Egypt

Solar energy offers a sustainable energy source that can be used as thermal energy and electrical energy. Due to energy resources shortage and cost increase through the world, obvious attention is paid for using solar energy systems especially in the processes that require extensive energy quantities like desalination. Desalination of sea water can be divided into thermal desalination and electrical desalination units whereby solar energy can be used to drive both categories. However solar energy has much low operating cost, it suffers from other defects like dynamic profile availability, expensive initial capital cost, and need to large areas. These are interrelated parameters but conflicting as well. Hence, determining the optimum operating and design variables sound a combination of different conflict variables. Thus multiobjective optimization is introduced to tackle these contradictory parameters and to determine the optimum tradeoffs. The target is to reduce the cost and increase the system performance simultaneously. Different types of solar collectors are considered to provide the energy needed to drive the desalination process. Likewise, different desalination units are tackled. Performance throughout the year is addressed. Case study considering the Egyptian weather condition is given to illustrate the feasibility and applicability of the proposed multiobjective optimization approach.

Numerical Simulation in a Novel Polysilicon CVD reactor

HUANG Zheqing, LIU Chunjiang

Abstract: The paper presents a novel polysilicon CVD reactor that the flow pattern of the gas mixture is plug flow, which is fundamentally different from that of traditional CVD reactor. A three dimensional theoretical model of mass, momentum and heat transfer process have been proposed. The velocity and temperature profile of gas mixture in two different reactors have been simulated by using computational fluid dynamic method. The simulation result of the flow field shows that the flow pattern of the gas mixture was plug flow and the flow short circuit between the entrance and exit can be avoided in the novel reactor. The simulation result of the temperature field shows that the novel reactor temperature field can be controlled. By changing the operating parameters, the size of high temperature region which can lead to the generation of silicon powder can be diminished. Furthermore, the energy analysis for traditional reactor show that the total energy consumption is around 70kWh·kg-1 and the radiation loss is about 60% of the total energy loss. The energy consumption of the novel reactor is lower 8% at least than that of the traditional reactor.

Key words: Numerical Simulation; Polysilicon Product; CVD Reactor; Energy Consumption

1. Introduction

With the depletion of the petroleum resources, solar energy is cleaner and more abundant, which have been developed rapidly. The polysilicon as the main raw material of the solar cell will increase sharply in the following years [1-3]. Many processes to produce polysilicon have been tested and patented, but the most popular process is the improved Siemens process, which shares about 80% of total polysilicon manufactured world-wide [4-6]. In the process, more than 60% of the energy is consumed in the decomposition process[7-8]; this indicates that the polysilicon CVD reactor is the most important equipment during the process. The polysilicon CVD reactor, also called a Siemens reactor, consists of a gas-tight chamber where several high-purity silicon slim rods are heated by an electric current flowing through them. Polysilicon is deposited on the seed rods through the thermal decomposition of trichlorosilane in a hydrogen environment.

Polysilicon CVD reactor as the key equipment attracted a lot of research attention during the past few years. For instance, a detailed radiated heat transfer behavior in a 36 rods polysilicon reactor have been investigated[8], and the effect of enlarging the reactor capacity, the impact of thermal shields and the influence of the wall emissivity were studied. Kang et al[9].have investigated the gas nozzle effect on the deposition of polysilicon by monosilane Siemens reactor. Gum et al[10] have presented a patent in which the energy consumption was reduced by providing a thin layer of gold inside a reactor chamber. Pazzaglia et al[11] have presented a thermal shield in Siemens reactor to block thermal radiation emitted from the heated polysilicon rods. Liu et al[12] have developed a novel polysilicon CVD reactor, and the velocity profile and temperature field was simulated by computational fluid dynamic method.

It can be found that the research works mentioned above mainly devoted to describe the polysilicon deposition process and the energy saving strategies for polysilicon CVD reactor. In this paper, a more energy saving CVD reactor [13, 14] with a new flow pattern was developed and detailed analysis have been carried out using computational fluid dynamic.

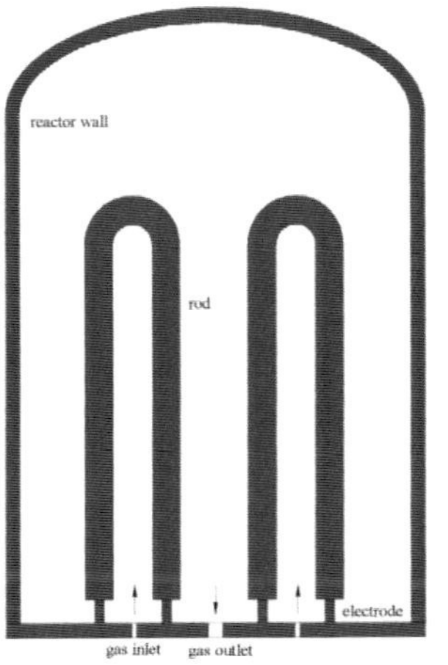

Fig1. Frame structure of traditional polysilicon CVD reactor.

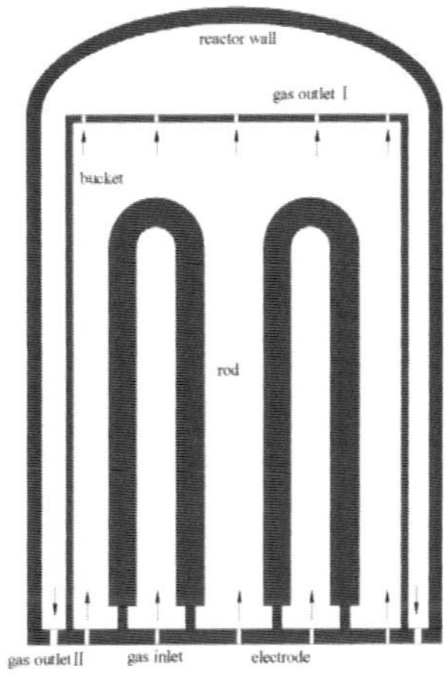

Fig2. Frame structure of novel polysilicon CVD reactor.

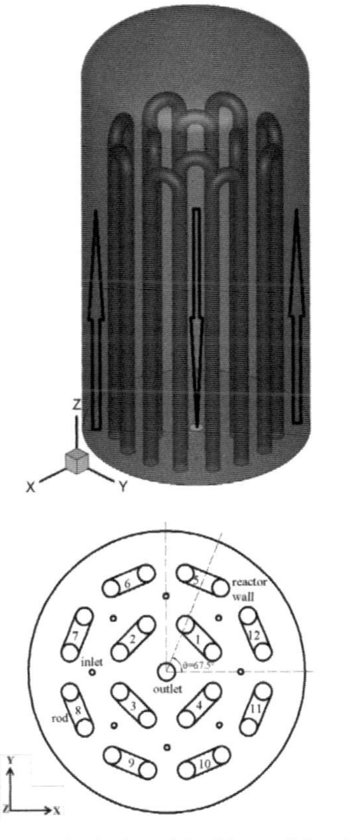

Fig.3 Physical model of the traditional polysilicon CVD reactor.

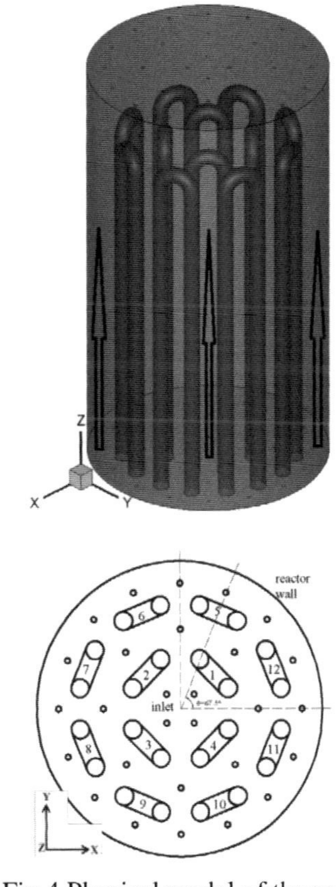

Fig.4 Physical model of the novel polysilicon CVD reactor.

2. Numerical method

2.1 *Physical model*

The frame description of the traditional polysilicon CVD reactor and the novel reactor was given in Figure 1 and Figure 2 respectively. For the traditional reactor, the gases are injected through eight inlets and exhausted through the outlet in the chassis. The difference for the novel reactor is that a bucket is added to change the flow pattern in the reactor, and the gases are exhausted through the outlets in the bucket top. In order to describe the transport phenomena by computational fluid dynamic method, the reactor have been simplified and the physical models with 12 rods have been presented in Figure 3 and Figure 4. Their structure parameters have been presented in Table1.

Table 1 Parameters of the physical model of the CVD reactors

	Traditional	Novel
H/mm	2800	2800
D/mm	1600	1600
h/mm	2300	2300
d_i/mm	20	20
d_o/mm	100	20

2.2*Governing equations*

The governing equations including continuity, momentum, energy and radiated transfer heat are presented in this section.

1） *Continuity equation*

$$\nabla(\rho \boldsymbol{U})=0 \tag{1}$$

2） *Momentum equations*

$$\nabla\cdot(\rho \boldsymbol{U}\boldsymbol{U}) = -\nabla P + \rho \boldsymbol{g} + \nabla\cdot\left[\nabla \boldsymbol{U} + (\nabla \boldsymbol{U})^{\mathrm{T}}\right] - \nabla\cdot\left(\rho \overline{u'u'}\right) \tag{2}$$

3） *Transport equations for the standard k-ε model*

$$\nabla\cdot(\rho \boldsymbol{U} k) = \nabla\cdot\left[\left(\mu + \frac{\mu_t}{\sigma_k}\right)\nabla k\right] + G_k + G_b - \rho\varepsilon \tag{3}$$

$$\nabla\cdot(\rho\varepsilon \boldsymbol{U}) = \nabla\cdot\left[\left(\mu + \frac{\mu_t}{\sigma_\varepsilon}\right)\nabla\varepsilon\right] + C_{1\varepsilon}\frac{\varepsilon}{k}(G_k + C_{3\varepsilon}G_b) - C_{2\varepsilon}\rho\frac{\varepsilon^2}{k} \tag{4}$$

4） *Energy equation*

$$\nabla\cdot((\rho E + p)\boldsymbol{U}) = \nabla\cdot(\lambda_{eff}\nabla T) + S_h \tag{5}$$

The source term, $S_h = \nabla\cdot(I(\boldsymbol{r},s)s)$, is the increment of radiation intensity in direction $\boldsymbol{s}$.

5） *Radiative transfer equation*

$$\nabla\cdot(I(\boldsymbol{r},\boldsymbol{s})\boldsymbol{s}) + aI(\boldsymbol{r},\boldsymbol{s}) = a\frac{\sigma T^4}{\pi} \tag{6}$$

In order to calculate the transport properties of gas mixture in the above equations, the properties of species should be obtained. The data for specific heat, viscosity, and thermal conductivity of relevant gas species as a function of temperature has been given in appendix A[15].

2.3 *Boundary conditions*

Walls: No slip boundary condition for the gas velocity has been used on the reactor walls. The isothermal boundary condition is employed. The reactor wall is made of stainless, and the wall surface is polished, and the value of the emissivity for the reactor wall is taken as 0.2[16]

$$U = 0,\quad T = 373\text{K} \tag{7}$$

$$U = 0,\quad T_{Si} = 1373\text{K} \tag{8}$$

Rods: No slip boundary condition for the gas velocity is specified, the temperature is 1373K. The emissivity of the polysilicon rods is considered to be 0.7[17].

Inlets: At the inlets, the inlet mass flow, temperature, and the composition of gas mixture are specified, the pressure is 6atm. The radiation absorption coefficient of gas mixture is 0.05[18].

$$Q = 0 \sim 1400\text{m}^3 \cdot \text{h}^{-1},\quad T_{inlet} = 300 \sim 600\text{K},\quad Y_{SiHCl_3} = 0.25 \tag{9}$$

Outlets: At the outlets, the flow is assumed to be fully developed.

$$\nabla U = 0,\quad \nabla T = 0 \tag{10}$$

2.4 *Simulation Scheme and Grid strategy*

The polysilicon deposition process in the CVD reactor is non-steady-state, so a series quasi-steady state simulations corresponding to different rod size have been carried out. The physical models and their mesh have been established by using the Gambit. The same grid strategy was adopted in different physical models; the grid with 0.01m was adopted on the polysilicon rods surface; the boundary layer mesh with 0.01m adjacent to the polysilicon rods was adopted; and the tetrahedral mesh with 0.02m was applied in the whole zone. The Fluent software package is used to calculate the governing equations, and the SIMPLE method is used for pressure-velocity coupling.

3. Results and Discussion

The velocity profile, temperature field and the energy consumption of the traditional reactor and novel reactor have been analyzed in the following sections. According to the industrial data (the initial diameter of polysilicon rod is 10mm, and finally diameter is 125mm after 120h reaction), the growth rate of 8μm·min-1 ($0.133\mu\text{m}\cdot\text{s}^{-1}$) is taken in this paper. The quasi-steady state models corresponding to the rods diameter with 29.2mm, 48.4mm, 67.6mm, 86.8mm, 106mm, 125.2mm corresponding to the reaction time from 20 hours to 120 hours have been simulated. Firstly the validation of the proposed model is investigated.

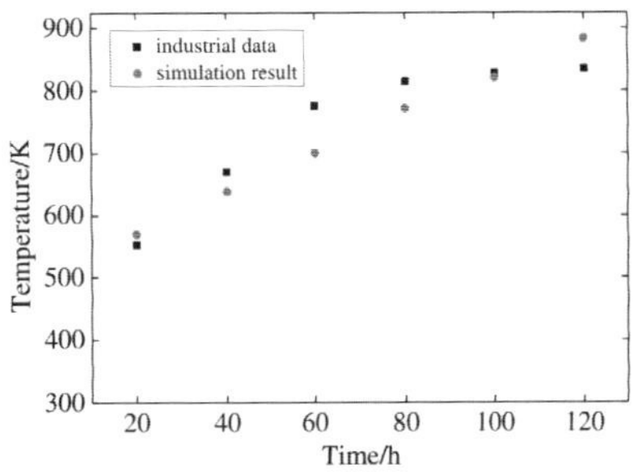

Fig.5 Outlet temperature of the traditional polysilicon CVD reactor.

3.1 *Validation of the CFD model*

In order to verify the validation of the model, the comparison of the outlet temperature between the simulated results and the industrial data obtained by a polysilicon factory in the traditional polysilicon CVD reactor has been done. Fig.5 presents the outlet temperature of the simulated results and the industrial data. It can be found that the deviation of the simulated results and the industrial data is less than 5%. This shows that the proposed model is reliable for describing the behavior of the process.

3.2 *Flow Behavior*

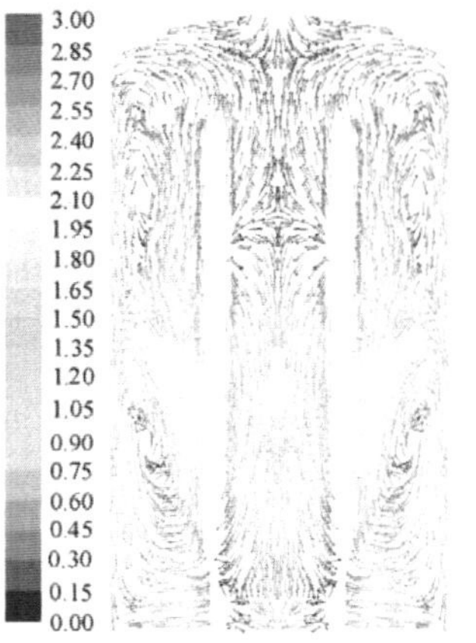

Fig.6 Velocity vector along θ=67.5° plane in traditional reactor/m·s^{-1}

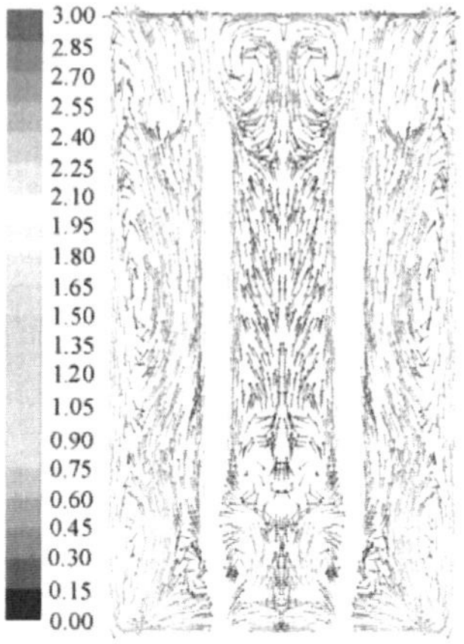

Fig.7 Velocity vector along θ=67.5° plane in novel reactor/m·s^{-1}

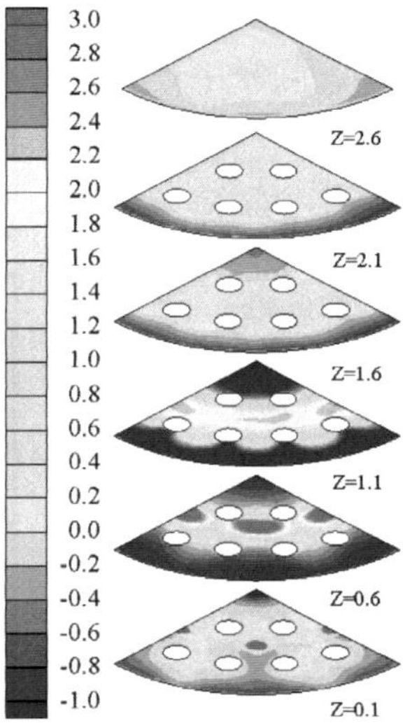

Fig.8 Z velocity contour along different plane in traditional reactor/m·s^{-1}.

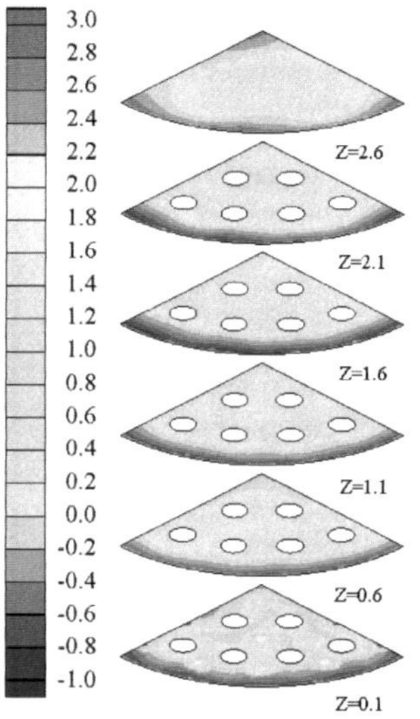

Fig.9 Z velocity contour along different plane in novel reactor/m·s^{-1}.

Fig.6 and Fig.7 give the velocity profile along θ=67.5° plane in different reactors when the rods diameter is 0.106m, and Fig.8 and Fig.9 present the z direction velocity contour along z=0.1, z=0.6, z=1.1, z=1.6, z=2.1, z=2.6 planes in different reactors. It can be found from these figures that the velocity in novel reactor is more uniform than that in traditional reactor. For traditional reactor, a dead zone can be easily formed in the reactor top and the gas mixture go short circuit between the entrance and exit. This problem can be attributed to that the inlets of reactant gases and outlets of exhaust gases in traditional reactor are both opened on the chassis, and the flow pattern of the gas mixture is almost a mixed flow. However, for novel reactor, the problem of the short circuit can be solved and the velocity of the whole zone is almost same, namely the gas flow pattern is almost plug flow (as seen in Fig.7), which is fundamentally different from that of traditional reactor (as seen in Fig.6), so the velocity profile can be easily to control in novel reactor. And to control the velocity profile is very important. This is due to the temperature field distribution greatly affected by the velocity profile.

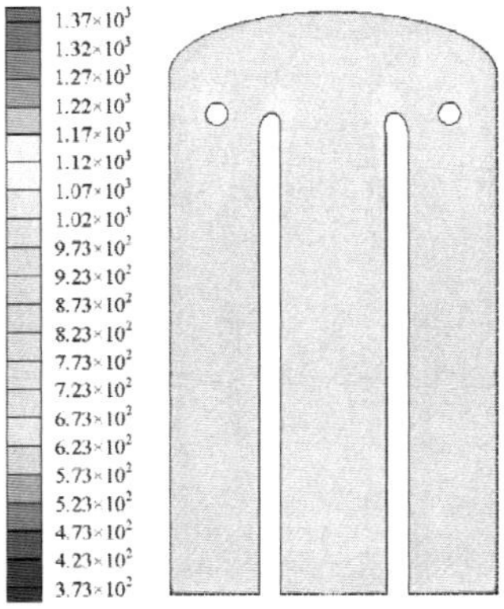

Fig.10 Temperature profile along θ=67.5° plane in traditional reactor/K

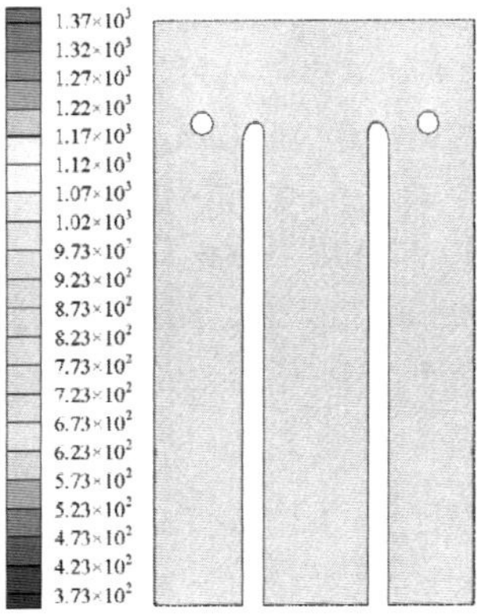

Fig.11 Temperature profile along θ=67.5° plane in novel reactor/K

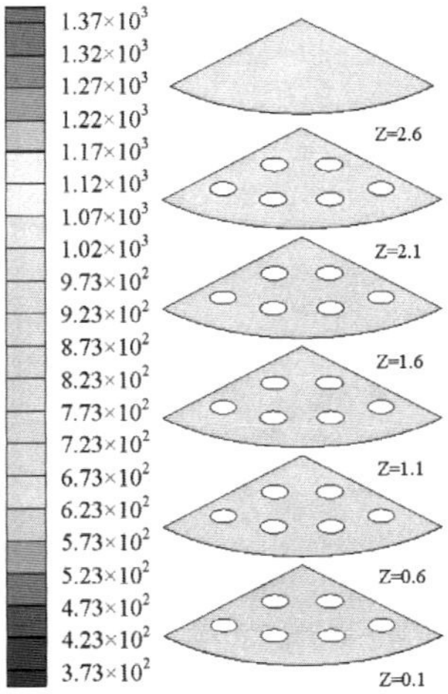

Fig.12 Temperature contour along different plane in traditional reactor/K.

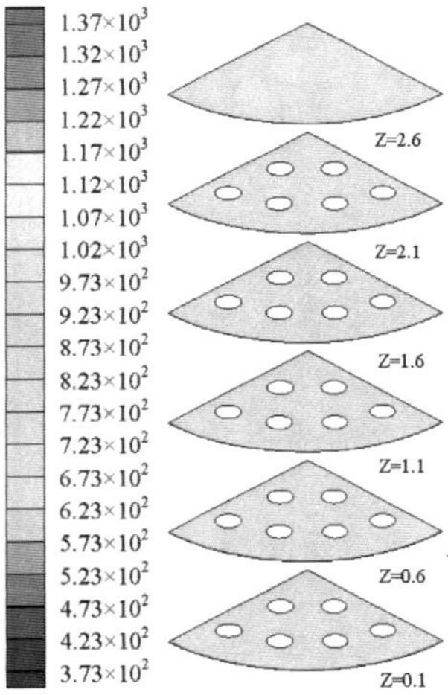

Fig.13 Temperature contour along different plane in novel reactor/K.

3.3 *Heat Transfer Behavior*

Fig.10 and Fig.11 show the temperature field along θ=67.5° plane in different reactors when the rods diameter is 0.106m, and Fig.12 and Fig.13 present the temperature field along z=0.1, z=0.6, z=1.1, z=1.6, z=2.1, z=2.6 planes in two reactors. It is clearly found that the temperature field in traditional reactor is more uniform than that in novel reactor, but the local temperature in traditional reactor is higher than that of novel reactor under the same process conditions. This is due to that the flow pattern in traditional reactor is a mixed flow, thus the temperature field is more uniform. But the simulated result of velocity profile show that the dead zone can be easily formed in the reactor top, so the local temperature is higher. For novel reactor, the flow pattern is almost plug flow, and the dead zone can be eliminated; then the problem of the local high temperature can be solved. It is important for polysilicon CVD reactor to eliminate the dead zone and the high temperature zone, this is due to that the local high temperature will lead to the silicon powder generation, the silicon powder can contaminate the reactor wall during the production process and the cleanness of the wall went dirty, finally the reflectivity of the wall will decrease and the heat loss due to radiation will increase dramatically.

3.4 *Comparison and analysis of energy consumption*

Fig.14 shows the energy loss in traditional reactor and novel reactor throughout the process. It can be found that the energy loss in different reactors decreased with the growth of the polysilicon rods. This is due to the fact that the greater of the polysilicon rods diameter, the more of the polysilicon production per unit time, this means that the average energy loss per unit time and production is lower. When the value of the emissivity in reactor walls are all 0.2, the energy loss is 70kwh·kg^{-1} and 65kwh·kg^{-1} in traditional reactor and novel reactor respectively. And the radiation loss is about 60% of the total energy loss in traditional reactor. The simulated result in traditional CVD reactor is consistent with the industrial data. However, when the value of the emissivity is changed from 0.2 to 0.05[19], for example, the reactor wall and bucket wall were platinum plating (ε=0.05), the energy consumption decrease from 65 kWh·kg^{-1} to 46 kWh·kg^{-1}, which means at least 35% of the energy loss can be saved. This demonstrates the fact that if the cleanness of reactor wall in traditional reactor is contaminated by the silicon powder, the energy loss will increase dramatically.

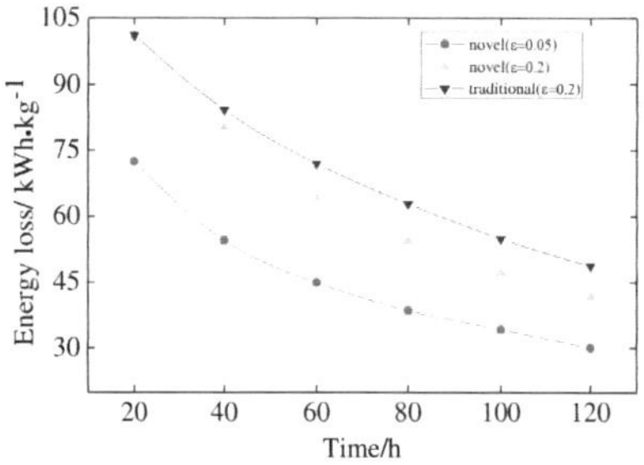

Fig.14 Energy loss for different polysilicon CVD reactors

4. Conclusion

A novel CVD reactor with more energy saving has been developed and a three dimensional theoretical model of mass, momentum and heat transfer process have been proposed. The velocity and temperature profile of gas mixture in two different reactors have been simulated by using computational fluid dynamic method. The simulation result of the flow field shows that the flow pattern of the gas mixture was plug flow and the problem of the flow short circuit between the entrance and exit can be solved in the novel reactor. The simulation result of the temperature field shows that the novel reactor temperature field can be controlled. Due to the temperature field in the novel reactor can be controlled; the high temperature zone in traditional reactor can be eliminated, and the silicon powder generation can be avoided, thus the problem of the inner reactor wall probably contaminated by silicon powder can be solved, and the cleanness of the wall can be remained unchanged. By keeping the cleanness of the inner wall, power loss of the reactor due to radiation can be greatly reduced. The simulation result shows that the total energy consumption is around 70kWh·kg-1 and the radiation loss is about 60% of the total energy loss. The energy consumption of the novel reactor is lower 8% at least than that of the traditional reactor.

Appendix A

The gas density obeys the ideal gas law and the operation pressure is 6 atm;

$$\rho = \frac{PM_{av}}{RT} \tag{A-1}$$

The specific heat of the gas mixture is calculated using the following relationship[20]:

Table2 Heat capacity as a function of temperature

Gases	$J \cdot kg^{-1} \cdot K^{-1}$
Hydrogen	C_p=13749+2.97T-3.84e-3T^2+2.09e-6T^3
Hydrogen chloride	C_p=817.95-0.149T+2.81e-4T^2-8.67e-8T^3
Trichlorosilane	C_p=287.3+1.21T-1.12e-3T^2+3.65e-7T^3
Dichlorosilane	C_p=220.8+1.65T-1.29e-3T^2+3.68e-7T^3
Tetrachloride	C_p=180.1+1.88T-3e-3T^2+2.08e-6T^3-5.27e-10T^3

$$C_p = Y_{SiHCl_3} C_{p,SiHCl_3} + Y_{H_2} C_{p,H_2} \tag{A-2}$$

The viscosity of the gas mixture is calculated using the following relationship[20]:

$$\mu_m = \sum_i^n \frac{y_i \mu_i}{\sum_j^n y_j \phi_{ij}} \tag{A-3}$$

$$\phi_{ij} = \frac{\left[1 + \left(\frac{\mu_i}{\mu_j}\right)^{1/2} \left(\frac{M_j}{M_i}\right)^{1/4}\right]^2}{\left[8\left(1 + \frac{M_i}{M_j}\right)\right]^{1/2}} \tag{A-4}$$

Table3 Viscosity as a function of temperature

Gases	Pa·S
Hydrogen	μ=3.08e-6+2.04e-8T-2.88e-12T^2
Hydrogen chloride	μ=-5.97e-7+5.47e-8T-1.04e-11T^2
Trichlorosilane	μ=4.88E-7+3.78e-8T-3e-12T^2
Dichlorosilane	μ=-1.91e-6+4.8e-8T-1.27e-11T^2
Tetrachloride	μ=2.54e-7+3.77e-8T-1.41e-11T^2

The thermal conductivity of the gas mixture is calculated using the following relationship[20]:

$$\lambda_m = \sum_i^n \frac{y_i \lambda_i}{\sum_j^n y_j \phi_{ij}} \tag{A-5}$$

Table4 Thermal conductivity as a function of temperature

Gas	$W \cdot m^{-1} \cdot K^{-1}$
Hydrogen	λ=0.03754+4.64e-4T-6.72e-8T^2
Hydrogen chloride	λ=1.47e-3+4.51e-5T-4.4e-10T^2
Trichlorosilane	λ=-6.78E-3+5.1e-5T-8.74e-9T^2
Dichlorosilane	λ=3.17e-3+1.08e-5T-4.33e-8T^2
Tetrachloride	λ=8.75e-3-1.59e-5T+5.22e-8T^2

References

[1] C. Del Canizo, G. Del Coso, W. C. Sinke. Crystalline silicon solar module technology: Towards the 1€ Per Watt-Peak Goal[J]. *Progress in Photovoltaics: Research and Applications*, 2009, 17(3):199-209

[2] P. D. Maycock. PV Market Update[J]. *Renewable Energy World*, 2003, 6(4):84-101

[3] S. Pizzini. Towards solar grade silicon: Challenges and benefits for low cost photovoltaics[J]. *Solar Energy Materials and Solar Cells*, 2010, 94(9):1528-1533

[4] J. W. Liang. The Production Technology of Electronic Grade Polycrystalline Silicon[J]. *Engineering Science*, 2000, 2(12):34

[5] A. Luque. Handbook of Photovoltaic Science and Engineering[M]. New York: John Wiley&Sons, 2010.

[6] W. O. Filtvedt, M. Javidi, A. Holt, M. C. Melaaen, E. Marstein, H. Tathgar, P. A. Ramachandran Development of fluidized bed reactors for silicon production[J]. *Solar Energy Materials and Solar Cells*, 2010, 94(12):1980-1995

[7] C. Del Canizo, G. Del Coso, A. Luque. Disclosing the polysilicon deposition process // The 25th European Photovoltaic Solar Energy Conference and Exhibition[C]. Valencia, 2010.

[8] G. Del Coso, C. Canizo, A. Luque. Radiative energy loss in a polysilicon CVD reactor[J]. *Solar Energy Materials and Solar Cells*, 2011, 95(4):1042-1049

[9] S. O. Kang, U. J. Lee, H. J. You, K. Park. Gas nozzle effect on the deposition of polysilicon by monosilane Siemens reactor[J]. *International Journal of Photoenergy*, 2012,

[10] J. C. Gum, C. Fero, D. Desrosier. Gold-coated polysilicon reactor system and method[P]: US, 20110159214. 2011-01-12.

[11] G. Pazzaglia, M. Fumagalli, M. Kulkani. Bell Jar for Siemens Reactor Including Thermal Radiation Shield[P]: US, 2011128729. 2011-10-20.

[12] Z. Q. Huang, C. J. Liu, X. G. Yuan. Numerical simulation of flow and heat transfer in a novel polysilicon reduction furnace[J]. *Chemical Industry and Engineering Society of China*, 2013, 64(2):484-489

[13] C. J. Liu, Z. Q. Huang, L. Duan, X.G. Yuan. A 18 rods novel polysilicon reduction furnace arrangement and connection method[P]: CN, 201110190062.X. 2013-01-02.

[14] X. J. Wang, Z. Q. Huang, L. Duan, C. J. Liu. A 42-rods polysilicon reduction furnace and connection mode[P]: CN, 201110173457.9. 2012-12-19.

[15] R. C. Reid, J. M. Prausnitz, B. E. Poling. The Properties of Gases and Liquids[M]. NewYork: McGraw-Hill, 1987.

[16] W. Bauer, H. Oertel, M. Rink. Spectral Emissivities of Bright and Oxidized Metals at High Temperatures // The Fifteenth Symposium on Thermophysical Properties[C]. Boulder, 2003.

[17] T. Sato. Spectral Emissivity of Silicon[J]. *Japanese Journal of Applied Physis*, 1967, 6(3):339-347

[18] J. L. Qian, Z. H. Yu, L. T. Wang. Tube Furnace[M]. Beijing: Hydrocarbon Processing Press, 1987.

[19] W. L. Wolfe, J. Z. George. Infrared Information and Analysis Center[M]. Washington, D.C.: Office of Naval Research, Department of Navy, 1985.

[20] L. Y. Carl. Matheson gas data book[M]. NewYork: McGraw-Hill, 2001.

(133e) Catalytic Transesterification of Waste Cooking Oil to Biofuel and Conversion of Glycerin with Supercritical Methanol

Chee Kai Tan and Maoqi Feng, Div. of Chemistry and Chemical Engineering, Southwest Research Institute, San Antonio, TX

In the conventional transesterification of fats/vegetable oils for biodiesel production, free fatty acids and water always present a problem and result in a low conversion. The object of this study was to investigate the yield of methyl esters treated by acid-free and alkaline-free with and without a heterogeneous catalyst in supercritical methanol (SCM) condition. Glyceride tributyrate was used as a model compound for waste cooking oil to study transesterification with methanol at supercritical condition (239.4°C and 1140 psi). The transesterification reaction was studied with an automatic controlled fixed bed tubular reactor in a continuous mode as shown in Figure 1. The reactor was constructed from a 3/8" OD x 4' long stainless tube with wall thickness of 0.049" (reactor volume = 47.4 mL). This system was continuously used for the biodiesel synthesis from and waste cooking oil and animal fats.

Figure 1. Continuous Process Flow Diagram for biodiesel production with supercritical methanol.

For glyceride tributyrate, the conversion was 100% under SCM condition. For waste cooking oil, the fatty acid methyl esters (FAMEs) yield was >90% at the following reaction conditions: reaction temperature = 300°C and 250°C, reaction pressures = 2000 psi, 2500 psi, and 3000 psi, residence time = 18 minutes, and molar ratio of methanol to oil = 40:1, flow rate = 2.5 mL/min. However, when feed flowrate increased to 7 mL/min. 250 °C and 2500 psi, the transesterifcation conversion dropped to <86 %. Heterogeneous catalyst, ETS-10, lowered the transestrification reaction to sub-supercritical conditions, the conversion was consistently achievable over 99% for a 24 hours study at a flow rate of 6mL/min (LHSV 4.47 g. min./mL). Other nanoporous solid catalysts were also studied in similar SCM conditions. The catalysts include hydrotalcite, Zeolite-X, Zeolite-Y, sulfate-$ZrSO_2$, ML-3, and ZSM-5. The findings demonstrated that, by a supercritical methanol approach with heterogeneous nanoporous catalysts, vegetable oil and its wastes could be readily used for biodiesel fuel production in a simple, continuous, and automatic process.

Glycerin is a main byproduct in triglyceride transesterification for biodiesel production as shown in Equation 1.

CH_2-OCO-R CH_2-OH

| |

CH-OCO-R + 3R'OH → 3R-COOR" + CH-OH Eq (1)

| Alcohol Alkyl Ester |

CH_2-OCO-R Biodiesel CH_2-OH

Triglyceride (byproduct) Product Glycerol

R = fatty acid chain R' = CH_3 : fatty acid methyl esters R" = CH_2CH_3 : fatty acid ethyl esters

Continued expansion of biodiesel capacity has led to a rapid increase in glycerin supply, the supply of glycerol to the market could affect the need for chemical derivatives. When the transesterification reaction was conducted at 350ºC with SCM, there was no phase separation of the byproduct glycerol and biodiesel; both components were dissolved in the methanol solvent. GC-MS analysis showed that there were some oxygenated low molecular weight compounds, which are miscible with biodiesel. After carefully analyzing these compounds, glycerol had reacted with methanol and produced isobutanol and 3-methoxy-1,2-propandiol at the higher temperature.

At lower temperature (300 ^oC) in SCM, spiked glycrol remained unconverted. Heterogeneous catalysts, ETS-10 and Zeolite-X, contain basic surface states had catalyzed the complete conversion of the glycerol bio-based byproduct to useful fuel additives with higher carbon/oxygen ratio (C/O) at sub-supercritical temperature and pressure.

Research on Isobaric Vapor-Liquid Equilibrium for 1-Propanol-Water-Mixed Solvent System

Xiao-hua LI, Zhi-chao XU, Jian-bing JI

(College of Chemical Engineering and Materials Science, Zhejiang University of Technology,

Zhejiang Province Key Laboratory of Biofue, Hangzhou, Zhejiang, China 310032)

Abstract: Isobaric vapor-liquid equilibrium(VLE) data for 1-propanol-water system containing mixed solvent were measured in a novel ebulliometer at 101.3kPa. The experiments showed mixed solvent(ethanediol and postassium acetate) lead to eliminate the azeotropic point of 1-proanol-water system and rise the relative volatility of 1-propanol to water prominently. The relativity of 1-propanol to water increased with more addition of mixed solvent, so mixed solvent can be considered as an extraction solvent for extraction distillation of 1-propanol and water. The suitable mass fraction of mixed solvent was about 60%. The VLE data were correlated by NRTL model, and comparing the correlated results with the experimental ones, the standard deviations of bubble point temperatures and vapor compositions were 0.45K and 0.00872, respectively.

Key words: vapor-liquid equilibrium ,1-Propanol, water , ethanediol , postassium acetate

1 Introduction

1-Propanol is an important chemistry material and organic solvents. It widely used effective solvent for chemical, pharmaceutical and pesticide industry[1]. In fact, 1-propanol and water has azeotrope. The binary mixtures exhibits a minimum boiling temperature azeotrope with 0.40 of 1-propanol molar composition at 87.67℃under atmospheric pressur.VLE data are useful for the study of solution properties of non-ideal mixtures as well as for the design of the separation equipment. It is relatively difficult to separate the 1-propanol from water by traditional rectification. The extractive distillation method suggested to solve this problem by using an additional agent to alter the relative volatility.

Extractive distillation for dehydrating various alcohols has been studied [2-8]. Many solvents were suggested. Duan Zhanting [9], extractive distillation with salt for anhydrous ethanol, Chen Xiaoping[10],separation of azeotrope of methyl ethyl ketone-water system using extractive distillation with salt, the results were very effective. Mixed solvent（80% ethanediol and 20% postassium acetate）was chosen as the entrainer in this paper Then the VLE data at 101.3 kPa for 1-propanol-water-mixed solvent system were measured and correlated.

2 Experimental

2.1 Materials

Water was double distilled. 1-propanol and ethanediol were analytical grade reagents and were used without further purification. Purities of 1-propanol and ethanediol were checked by GC, and the results were>99.5 %.The postassium acetate were dried for 6hr under vacuum at 400K to remove volatile impurities.

2.2 Apparatus and Procedure

VLE data were measured by a novel ebulliometer made of quartz glass featuring vapor-phase sampling and synthetic liquid composition. Experimental device and method of calculating the liquid composition refer the literature see [11]. The modified ebulliometer was shown in Fig 1.The pressure was measured by a U-tube filled with water, with one end connected to a pressure buffer and another to the atmosphere. Atmospheric pressure was measured by a standard barometer(p). The temperature was measured by a standard platinum thermometer and a 6-1/2-digit multimeter(t). The VLE data measurements for the systems were performed in a way of continuous synthesis. At the beginning of the measurement, certain amounts of 1-propanol and mixed solvent were introduced into the ebulliometer to form the first synthesis. The water contents of the samples in the first synthesis were determined by Karl Fischer analysis, added in or taken out of the ebulliometer was weighed by an electronic balance (Mettler-Toledo Bs224s), Vapor-phase compositions were measured by the Karl Fischer method, when the water mole fraction was less than 0.1,or by gas chromatography (GC-17A,shimazu), when the water mole fraction was greater than 0.1.

In order to verify the reliability of the novel ebulliometer, the VLE data for 1-propanol-water system were measured in this ebulliometer at 101.3kPa, the results agreed with literature[12] values, The thermodynamic consistency test was passed for 1-propanol + water binary system by Herrington. It showed that the novel ebulliometer measured VLE data was reliable.

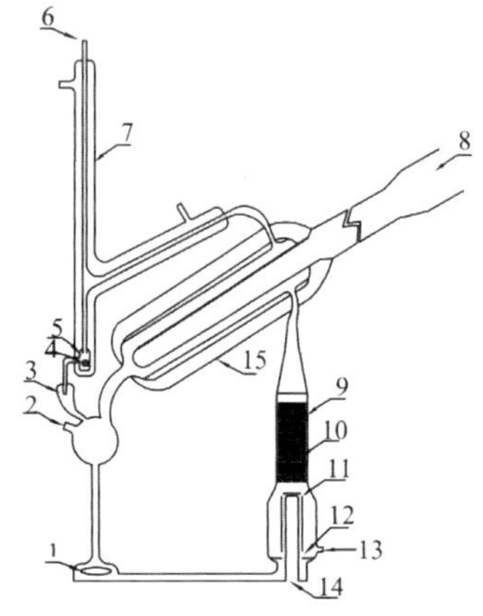

Fig.1 Double circulation ebulliometer

1--Magnetic stirrer functioned as a centrifugal pump; 2-Sample injection port; 3-Drop counter; 4-Vapor phase sampling port; 5-Vapor phase sampling chamber; 6-Pressure measuring port; 7-Vapor condenser; 8-thermometric instrument; 9-Boiler; 10-Column packed by glass coils; 11- Holes on the top of the jacket; 12-Holes on the lower part of the jacket inside the boiler; 13-Bulk material sampling port; 14-heating rod; 15-Equilibrium cell;

3 Results and discussion

VLE of the system 1-propanol (1) + water (2) + mixed solvent (3) were measured at different mass fractions of mixed solvent, data were measured at w_3=0.400, 0.500, 0.600, 0.700and 0.800. Results were shown in Tables 1. Relative volatility were also calculated. Here $x_1'$was the liquid-phase mole fraction of component 1, namely, 1-propanol, calculated on mixed solvent free basis. y_1 was the vapor-phase mole fraction of component 1,namely, 1-propanol.

Table 1 Vapor-Liquid Equilibrium Data for the System of 1-Propanol (1) - Water (2) - mixed solvent (3) at101.3kPa

w_3	T/K	x_1'	y_1	a_{12}	w_3	T/K	x_1'	y_1	a_{12}
		w_3≈0.8					w_3≈0.7		
0.8055	393.70	0.8958	0.9418	1.988	0.7016	386.02	0.8964	0.9293	1.651
0.8048	393.48	0.797	0.9013	2.150	0.6982	385.45	0.8001	0.8672	1.864
0.8017	391.67	0.6955	0.8670	2.362	0.7008	384.01	0.6995	0.8320	2.001
0.8015	390.80	0.6036	0.8212	2.691	0.7014	382.64	0.5982	0.7717	2.305
0.8011	390.12	0.5012	0.7480	3.100	0.7023	381.95	0.4975	0.7227	2.690
0.8022	389.21	0.4037	0.7072	3.537	0.7001	380.45	0.3997	0.6508	3.098
0.8003	388.65	0.3021	0.6220	4.120	0.6994	379.60	0.3012	0.5910	3.705
0.8014	388.48	0.1954	0.5394	4.686	0.6989	378.99	0.1998	0.5107	4.297
0.8009	387.29	0.1005	0.4581	5.317	0.6983	378.13	0.1013	0.4406	5.102
		w_3≈0.6					w_3≈0.5		
0.6053	381.73	0.8997	0.9096	1.415	0.5068	378.62	0.8993	0.9018	1.091
0.5994	380.3	0.7975	0.8431	1.531	0.5011	376.78	0.7968	0.8311	1.281
0.6044	379.16	0.6984	0.7998	1.701	0.5002	376.87	0.7013	0.7570	1.394
0.6029	377.84	0.6011	0.7403	1.981	0.5046	375.15	0.5997	0.6906	1.596
0.6012	376.48	0.4997	0.6822	2.419	0.5033	372.60	0.5028	0.6428	2.000
0.6064	375.30	0.3993	0.6248	2.903	0.5037	371.40	0.4016	0.5891	2.401

0.6031	374.47	0.3002	0.5645	3.301	0.4993	370.71	0.2996	0.5293	3.091
0.5996	373.19	0.2029	0.5003	4.095	0.5052	370.58	0.2001	0.4792	3.697
0.6001	372.23	0.1002	0.4300	4.789	0.5002	369.78	0.0991	0.4154	4.688
		w_3≈0.4					w_3≈0.4		
0.4042	375.86	0.8988	0.8849	0.931	0.4032	368.36	0.4027	0.5511	2.094
0.4032	373.79	0.7998	0.8001	0.965	0.4012	367.67	0.2998	0.5087	2.568
0.4056	372.11	0.7012	0.7221	1.091	0.4028	366.65	0.2059	0.4539	3.367
0.4043	370.99	0.5994	0.6582	1.313	0.3988	365.97	0.1002	0.4017	4.257
0.4042	369.61	0.5052	0.6089	1.651					

The experimental VLE data were correlated with the NRTL equation[13]. In the correlations, the binary parameters for water and 1-propanol were taken from the literature 12. For the simplicity of application, the binary parameters were regarded as temperature independent and the values of a_{12},$a_{13}$and a_{23} were set as 0.3. The remaining four binary parameters were fitted by optimization of the following objective function(1).Results for the optimized parameters and deviations in the correlations were summarized in Table 2

$$F=\sum_{j=1}^{m}[(T_{cal}-T_{\exp})_j^2+(y_{1,cal}+y_{1,\exp})_j^2] \qquad (1)$$

Table2 Interaction parameters and standard deviations for NRTL

i	1-Propanol	1-Propanol	mixed solvent	Standard deviation	
j	water	mixed solvent	water	$\sigma_T(k)$	σ_y
$g_{ij}-g_{jj}$	1648.8	7245.7	-2430.2		
$g_{ji}-g_{ii}$	7896.7	389.8	4668	0.45	0.0087
a_{ij}	0.3	0.3	0.3		

The effect of the mixed solvent on the phase behavior of the volatile components can be observed in Figure 1. At a given value of x_1', the vapor-phase mole fraction of 1- propanol y_1, increases with the increase of mixed solvent mass fraction. This indicates the enhancement of the relative volatilities of 1- propanol to water. For the mixture of water+1-propanol, the azeotrope was broken with addition of mixed solvent at w_3=0.500.

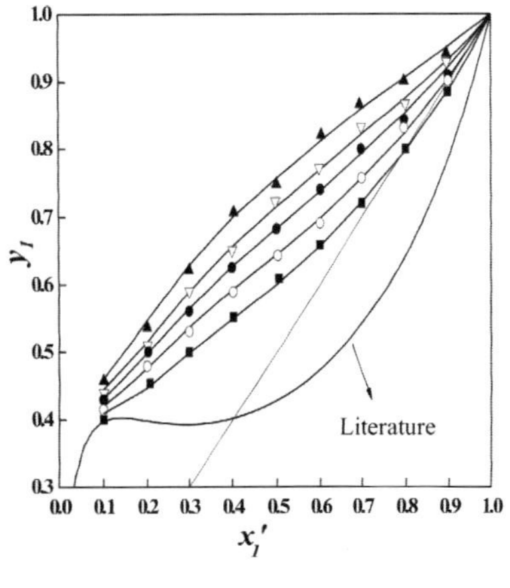

Fig. 2 Vapor-liquid equilibrium curve of 1-propanol-water-mixed solvent system at 101.3KPa

▲$w3$≈0.8; ▽$w3$≈0.7; ●$w3$≈0.6; ○$w3$≈0.5; ■$w3$≈0.4

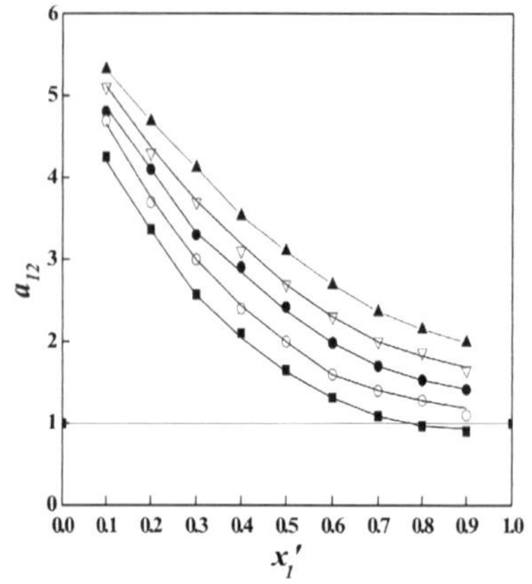

Fig.3 Relative volatility of 1-propanol(1) to water(2) at different mass fraction of mixed solvent

▲$w3$ ≈0.8; ▽$w3$ ≈0.7; ●$w3$ ≈0.6; ○$w3$ ≈0.5; ■$w3$ ≈0.4

The effect of the mixed solvent on the phase behavior of relative volatility can be observed in Figure 2. At a given value of $x'_{1,}$ the relative volatility a_{12},increases with the increase of mixed solvent mass fraction. This indicates the enhancement of the relative volatilities of 1- propanol to water. At the same time , when the mixed solvent remain unchanged and 1-propanol mole fraction x'_1 increases, the relative volatility of 1-propanol to water decreases. The effects of vapor-liquid equilibrium can be attributed to its different molecular interaction between these molecules, thus changing the activity coefficient of each component to improve the the relative volatility; That is because the solvent and water are strong polar molecules and they would be priority interaction.Then the activity coefficient of water decreases and 1-propanol increase. That lead to enhancement of the relative volatilities of 1-propanol to water; When the water decreases in system, a few water molecules were priority interaction with the solvent molecules, At the same time, same free solvent molecules will be interaction with 1-propanol.Compared to the former, the relative volatility of 1-propanol to water are decrease. The relative volatility of 1-propanol to water $a_{12}>1$ at $w_3>0.500$. Account into the re-use of solvents and the relative volatility of the separation costs, the option w_3= 0.6~0.7 for the extractive distillation separation of 1-propanol - water azeotrope.

4 Conclusion

1 VLE data for 1-propanol + water + mixed solvent systems were measured by a novel ebulliometer. The vapor composition was obtained by sampling and analysis, while the liquid composition was based on synthetic quantities and was calculated with regard to vapor phase composition, quantity of vapor material, and vapor-to-liquid circulation ratio.

2 The effect of the mixed solvent on the 1-propanol + water azeotrope was obtained by their relative volatility. The experiments showed that mixed solvent(w3>0.5) can eliminate the azeotropic point of 1-proanol-water system and rise the relative volatility of 1-propanol to water prominently. The more mixed solvent adds, the more salting-out effect becomes. The relativity of 1-propanol to water increase with more addition of mixed solvent, so mixed solvent can be considered as an extraction solvent for extraction distillation of 1-propanol and water.

3 The thermodynamic consistency test was passed for 1-propanol + water binary system. The parameters of NRTL models for the ternary system were obtained. The average deviations in temperature and vapor mole fraction were 0.45℃ and 0.00872, respectively.

References

1 LIANG Cheng, Analysis on Development Tendency of Domestic and Global n-propanol Market[J].Economic Analysis of Petroleum and Chemical Industry,2004,10,49-53

2 VERCHER E MUNOZ R. Isobaric vapor-liquid equilibrium data for the ethanol-water-potassium acetate and ethanol-water-(potassium acetate/sodium acetate) systems[J].J Chem Eng Data,1991,36(3):274-277.

3 LIN C L,LEE L S,TSENG H C. Phase equilibria for 1-propanol+water+sodium chloride and +potassium chloride and 2-propanol+water+lithium chloride and +lithium bromide[J].J Chem Eng Data,1993,38(2):306-309.

4 ILLIUTA M C,THYRION F C,LANDEUER O M. Effect of calcium chloride on the isobaric vapor-liquid equilibrium of 1-propanol+water[J].J Chem Eng Data,1996,41(3):402～408.

5 VERCHER E,ROJO F J,ANTONI M A. Isobaric vapor-liquid equilibria for 1-propanol+water+calciu -mnitrate[J].J Chem Eng Data,1999,44(6):1216-1221.

6 WU Wenliang,ZHANG Yaming,Wang Yanru etal. Vapor-liquid Equilibrium Data for System 1-Propanol-water-potassium Acetate[J]. Journal of Chemical Engineering of Chinese Universities，1997,11(4):413-416.

7 BAO Jing,ZHANG Yaming,ZHANG Lei. Vapour-liquid equilibrium for the ethanol-water-mixed solvent systems[J] Journal of Nanjing University of Technology, Vol. 26 No. 5 Sep.2004

8 CHEN Wei miao,ZHANG Ya ming, Recent progress in vapor-liquid equilibrium for alcohol-water-salt systems[J], Journal of Nanjing University of Technology, Vol. 24 No. 6 Sep.2002

9 Duan Zhanting,LEI Liangheng,Zhou Rongqi, Research Advances in Extractive Distillation with Salt[J]. Petrochemical Technology, 1980,9(6):350～353

10 CHEN Xiaoping,ZHANG Zhu, Study on the Separation of Azeotrope of Methyl Ethyl Ketone-water System Using Extractive Distillation with Salt, Journal of Yantai University(Natural Science and Engineering Edition),Vol. 15 No.2 Apr. 2002

11 ZHANG Lianzhong,HAN Jianzhi,DENG Dongshun,etal. Selection of ionic liquids as entrainers for separation of water and 2-propanol[J].Fluid Phase Equilibria,2007,255(2):179-185

12 J GMEHLING,U ONKEN,W ARLT. Vapor-liquid equilibrium data collection[J].Frandfurt dechema chemistry data series,1977-1984,1/6a,122.

13 John M.Prausnitz,Rudiger N.Lichtenthaler,Edmundo Gomes de Azevedo. Molecular Therodynamics of Fluid-Phase Equilibria[M]. BeiJing, Chemical Industry Press,2005,149

14 SANDER B,FRWDENSULUND A,RASMUSSEN P. Calculation of vapour - liquid equalibria in mixed solvent salt systems using an extended UNIQUAC equation[J]. Chem Eng Sci, 1986 ,41(5):1171-1183.

(138a) Synthesis of 2-Acetoxy Propanoic Acid Via Reactive Distillation an Intermediate in Synthesis of Biobased Acrylic Acid From Lactic Acid

Aspi K. Kolah, Lars Peereboom, Carl T. Lira and Dennis J. Miller, Chemical Engineering and Materials Science, Michigan State University, East Lansing, MI

The selective formation of 2-acetoxypropanoic acid (2-APA) by reaction of lactic acid with acetic acid has been carried out in continuous reactive distillation. The reaction takes place in a six-meter tall, 50 mm ID Pyrex column operated at atmospheric pressure and is catalyzed by Amberlyst-15 cation exchange resin supported in Katapak-SP11 structured packing in the middle four meters of the column. Lactic acid solutions in water were used as feedstock in the experimental study which was reacted with glacial acetic acid in the reactive distillation column. High conversions of lactic acid and its oligomers were achieved along with high selectivity to monomeric 2-APA. Experimental studies are complemented by process analysis using AspenPlus simulation software. By implementing a kinetic model and phase equilibrium data generated in our laboratories, we are able to predict the observed pilot-scale column behavior and simulate commercial-scale facilities.

The formation of 2-APA represents the first step of a process to produce acrylic acid from lactic acid. In the second step of the process, 2-APA is cracked under high temperature to form acrylic acid and liberate acetic acid that can be recycled.

Acrylic acid, its esters, and acrylate polymers are critical chemical building blocks used in a wide gamut of applications including superabsorbents, hard plastics, paints, detergents, and adhesives. The global acrylic acid market is estimated at 3.5 million tons and $8 billion annually, with a growth of 4% per year. Acrylic acid synthesized from biobased lactic acid is a drop-in replacement for the petroleum-based product; lactic acid is a mature biobased feed stock with a proven track record of high yield (>95%) from glucose and low cost.

(138b) Revolutionizing the Production of Transportation Fuels and Chemicals

David St. Angelo, Engineering, Joule Unlimited, Bedford, MA

Many technological alternatives to petroleum-based fuels have been explored, and with good reason. Limited energy supplies, pricing volatility and environmental concerns have converged to create one of the greatest challenges of our time. Joule® has an answer that surpasses biomass-derived fuels, channeling the sun's energy into a new, industry-changing category of direct solar fuels. The company's ***Helioculture™*** platform incorporates proprietary, engineered photosynthetic microorganisms to directly produce infrastructure-ready diesel, ethanol and multiple chemicals with no dependence on biomass feedstocks, agricultural land or fresh water. In parallel, Joule has developed a novel ***SolarConverter®*** system to enable the direct, continuous process with productivities greater than biomass-dependent methods, which require numerous energy-intensive steps and downstream processing to achieve an end product. In contrast, using sunlight, non-potable water and waste CO2 from industrial emitters or pipelines, Joule ultimately targets productivities of up to 15,000 gallons of diesel and 25,000 gallons of ethanol per acre annually, at stable costs as low as $50/bble and $1.28/gallon respectively (without subsidies).

(138c) Production and Upgrading Of Renewable Bio-Crude From Large Scale Algae Ponds

Neil Osterwalder, Dan Sajkowski and Ben Saydah, Sapphire Energy Inc., San Diego, CA

Renewable fuels from algae are moving closer to commercialization as Sapphire Energy showed with their Integrated Algae Bio-Refinery (IABR). The IABR in New Mexico was commissioned in 2012 and is producing algae derived bio crude oil on a continuous basis using Sapphire's proprietary conversion and extraction process. This algae bio-crude contains a wide range of hydrocarbons, fatty acids, aromatics, and nitrogen compounds and other heteroatoms. This chemical profile suggests that algal bio-crude can be economically processed by blending with existing refinery streams and co-processing the combined streams as opposed to the more costly development of algal-specific refineries. Different options of co-feeding the algae crude to a refinery can produce a full suite of refinery products from petrochemicals to gasoline, diesel and other refinery products. This study describes the products that can be produced from algae bio-crude and its implementation into existing refining infrastructure.

(138d) Membrane-Assisted Downstream Processing for Biobutanol Purification

Sebastian Heitmann, Daniel Welz, Philip Lutze and Andrzej Górak, Department of Biochemical and Chemical Engineering, Laboratory of Fluid Separations, TU Dortmund University, Dortmund, Germany

One of the main challenges for chemical industry is the change from fossil resources to a sustainable production of chemicals. The change to biobased raw materials and biotechnological processes often leads to product inhibitions of microorganisms and diluted system, resulting in an expensive downstream processing. One bulk chemical that can be produced in a sustainable way is biobutanol. One method of producing biobutanol utilizes the so-called acetone–***n***-butanol–ethanol (ABE) fermentation that can be carried out on basis of e.g. cellulosic and lignocellulosic biomass or agricultural wastes. However, the productivity of this fermentation is still limited, because ***n***-butanol is toxic towards the microorganisms used in ABE fermentation. That is why the concentration of n-butanol in the fermentation broth is often limited to values lower than 2 wt.-percent [1]. For n-butanol recovery mainly distillation processes are used which - as result of the low concentrations - have a high energy demand. About 80 % of the total energy demand of the downstream process is required for the separation of acetone, n-butanol and ethanol from the fermentation broth. Alternatively to distillation the fermentation can be coupled with a pervaporation unit, thus n-butanol can be separated from the broth continuously to prevent the n-butanol concentration from reaching the toxicity limit. Because of the mild operating conditions, pervaporation enables an energy-efficient n-butanol separation and facilitates the recycling of the broth with lower n-butanol content back to the fermenter.

Therefore, the aim of this work was to evaluate if it is more efficient to separate n-butanol by means of pervaporation or distillation or a combination of both. For this purpose extensive experimental studies were performed to characterize the permeation behavior of a commercial available poly(dimethylsiloxane) (PDMS) membrane (Sulzer Pervap^{TM} 4060). These studies provide a wide data basis for the downstream process analysis and are used to identify possible operating windows for the application of a pervaporation. An experimental comparison between the pervaporation of a binary mixture of n-butanol and water, the pervaporation of a synthetic medium and the pervaporation of real fermentation broth was carried out. The synthetic medium contained acetone, n-butanol and ethanol in mass ratios of 3:6:1 together with acetic and butyric acid, which are the solvent precursors. Additionally process parameters like feed composition, pervaporation temperature and permeate pressure were varied. The detailed experimental procedure is described elsewhere [2]. Next to the PDMS membrane, composite membrane materials containing ionic liquids are promising for separation of n-butanol from an aqueous solution. These membrane materials show advantageous permeation properties compared to conventional polymer membranes [2]. Therefore different supported ionic liquid membranes (SILMs) have been tested to evaluate their potential in n-butanol pervaporation.

Based on the experimental results a correlation has been developed describing partial fluxes of acetone, n-butanol, ethanol and water. Fluxes of acetic and butyric acid were neglected. With the help of this correlation modular-based process studies have been carried out using Aspen Custom Modeler^{TM} to find an optimal process configuration for n-butanol separation from the fermentation broth. It was found that pervaporation as stand-alone process as well as distillation is comparatively expensive. Because n-butanol should be almost completely separated from the broth, the concentration at the outlet of the membrane module becomes small, leading to a large required membrane area and a small permeate concentration of n-butanol.

In contrast to the application of distillation and pervaporation as stand-alone processes, a combination of pervaporation and distillation was found to be best suited for efficient n-butanol recovery. A possible membrane-assisted process is shown in Fig. 1. Pervaporation is used for continuous n-butanol removal in order to extend the fermentation cycles. Because the mass fraction of n-butanol $w_{B,Rec}$ in the recycle stream is only slightly smaller compared to the mass fraction of n-butanol in the fermentation $w_{B,Ferm}$, the required membrane area is comparatively small and a higher permeate concentration can be obtained. The complete removal of n-butanol from the broth is then carried out in distillation columns. In this case, the extension of the fermentation periods and the reduction of the stream to be processed in the energy-intensive distillation lead to a reduced energy demand and lowered downstream costs for n-butanol. The permeate stream and the solvent phase from the distillation can be processed further in a conventional distillation sequence. As existing biobutanol production plants mainly rely on distillation sequences, retrofit of a pervaporation thus enables an increase in the biobutanol production efficiency, while existing paid-off plants can still be used.

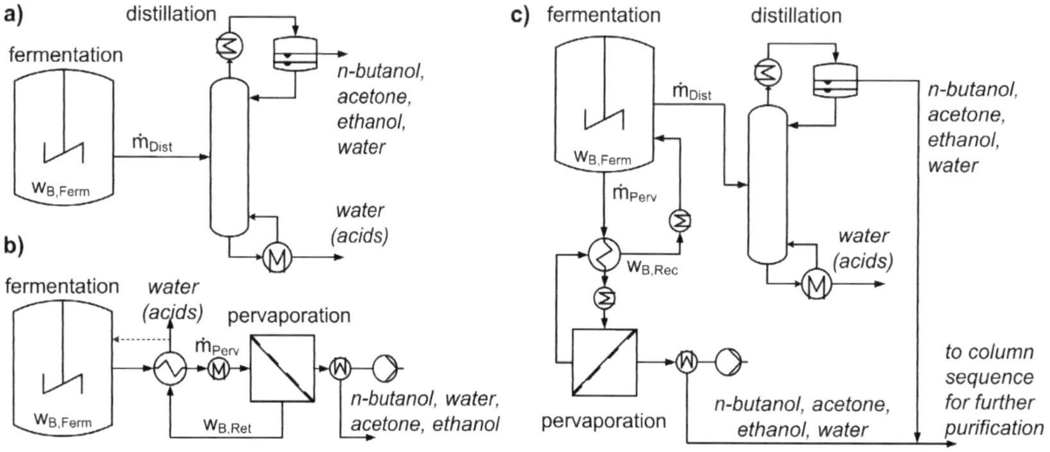

Fig. 1: Solvent removal from fermentation broth by distilliation (a), pervaporation (b) and a membrane-assisted process (c).

Acknowledgement:

The research leading to these results has received funding from the European Union Seventh Framework Programme (FP7/2007-2013) under grant agreement n° 241718 EuroBioRef. We acknowledge Sulzer Chemtech AG and the Laboratory of Biochemical Engineering, TU Dortmund University for providing PDMS membranes and fermentation broth.

[1] T.C. Ezeji, N. Qureshi, H.P. Blaschek, Bioproduction of butanol from biomass: from genes to bioreactors, Curr. Opin. Biotechnol. 3 (2007) 220-227.

[2] S. Heitmann, J. Krings, P. Kreis, A. Lennert, W.R. Pitner, A. Górak, M.M. Schulte, Recovery of n-butanol using ionic liquid-based pervaporation membranes, Sep. Purif. Technol. 97 (2012), 108-114.

Cool Planet Energy Systems and the Production of Carbon Negative Fuels

Mike Rocke, Vice President Marketing
Cool Planet Energy Systems, Inc.
mr@coolplanet.com

Abstract: Cool Planet Energy Systems is a 3-year old start-up producing gasoline blendstock from renewable non-food biomass feedstock. We have devised an economical biomass-to-liquid thermo-chemical conversion process that simultaneously produces liquid fuels and sequesterable biochar useful as a soil amendment. Fuel produced from this process is termed negative-carbon fuel since more carbon dioxide is removed from the atmosphere than released from combustion. The process for producing this fuel relies in part on a unique thermochemical decomposition of biomass that ramps the biomass temperature while the biomass is being subjected to pressure shocks. Co-produced biochar is converted into a soil amendment. We have built a 200,000 gallon per year pilot plant using this process and have demonstrated in laboratory studies that biochar generated using this process can exhibit excellent soil amendment properties. Fuel produced from this pilot plant has been fleet tested in a blend formulation in several standard configuration automobiles. The process is widely tunable in the produced biochar/fuel ratio. We also describe our plans to mass produce trailer-sized modular refinery units using this technology. We are backed by some of the world's largest energy companies and top Silicon Valley venture capitalists. We will discuss the company's trajectory from infancy to present status, and its fit in the bio-economy.

Overview

Cool Planet Energy Systems is a CA based start-up that has developed ground breaking, disruptive technology that economically converts non-food biomass into a sustainable, high-octane gasoline blend stock. Cool Planet's technology is based on twelve awarded, US patents. Hemi cellulosic or lignocellulosic biomass is processed through a mechanical biomass fractionating system that uses pressure and heat to create a series of useful volatile components that then go through proprietary catalytic systems to make high octane fuel. The process also captures the residual carbon to create a byproduct, called biochar, which is an inert, high surface area carbon that enhances soil quality through improved water and fertilizer retention and thereby increasing crop production on marginal or arid farmland. This carbon negative process removes atmospheric carbon to reverse the effects of fossil fuel consumption resulting in up to 150% carbon footprint reduction for every gallon of fuel used.

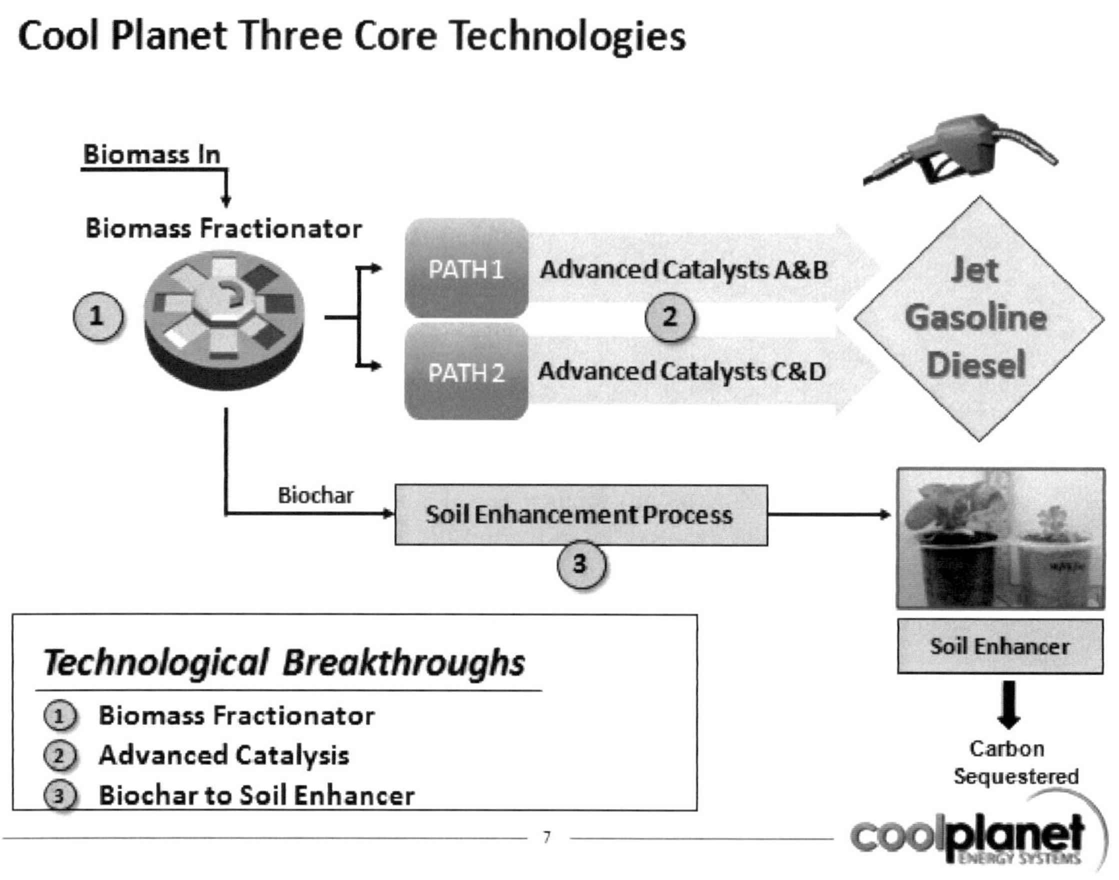

Technology of the Biomass fractionator

Ground biomass is transported by a commercial, mechanical transport system to a load hopper that pre-stages the biomass for the fractionator. Methods of introducing biomass into a processing station can include introduction via conveyor belts, hoppers, and/or pulverizers. The biomass is fed to the top of a sliding gate valve above a load station, and a precise amount of biomass is placed in a moving tray that holds the biomass for the next stage of processing. The biomass moves to the 1st heating and processing stage where the function is to drive off entrapped water, and to perform 1st step of de-hydration of the biomass while mechanically compacting it for more effective heat transfer at later stages of processing. The processing stages rotate or use a shuttle drive from step to step where the biomass is progressively heated and mechanically compacted from an initial stage temperature of 150 degrees C. As each of the following stages heat and compress the biomass, it will outgas commercially viable bio-intermediary compounds that are then taken off to catalytic reaction chambers. At each successive processing stage, the biomass is heated to a higher temperature profile and the material is compacted to a thinner layer with increasing level of elemental carbon in the remaining biomass. The processing stations thus produce volatile and non-volatile components when heated. Because the material can be heated within a relatively narrow temperature range, the non-volatile component (in the form of partially formed char) has additional volatile components embedded therein, which are extractable and processed in subsequent processing stations. The output of these processing stations is connected to an array of catalysts as described in the next section.
The final step or stage does not compact but controls the final biochar product output quantity and quality.

Advanced Catalytic Reaction Arrays

The Cool Planet system has a programmable number of processing stations, connected to an array of catalysts; a means for subjecting biomass within these stations to programmable temperatures and increments to produce a volatile and non-volatile component; a means for incrementing the non-volatile component temperature to further produce additional volatile and non-volatile components; and a means for subjecting the volatile components generated in each station through a series of catalysts. The catalysts involved are selected from the group of dehydration catalysts, aromatization catalysts, and gas-upgrading catalysts to produce different types of renewable fuel. The chemical compounds from biomass decomposition are routed to these distinct catalyst chains to produce products containing renewable, pure HC fuels and other value added products. Biomass is typically comprised of a wide array of compounds classified within the categories of cellulose, hemicelluloses, lignin, starches, and lipids. These compounds go through multiple steps of decomposition when subject to the biofractionation "pyrolysis" process. For example, hemicelluloses comprise C5 sugars such as fructose and xylose, which yield furfural and hydroxymethylfurfurals upon thermolysis. The latter compounds can be further converted to fuel intermediates furan and tetrahydrofuran. The narrow temperature windows experienced within a processing station allows for the collection of these useful intermediates.

The processing stations are connected to a series of catalyst components comprising a dehydration catalyst, an aromatization catalyst, and a gas-upgrading catalyst. The main catalysts are proprietary modified ZSM-5 catalysts. The output from each catalytic array when cooled is comprised of volatile gases, renewable fuel and water. The volatile gases are then programmed to pass through subsequent catalytic columns. Depending on the temperature (300C to 600C) of the processing stations and the biomass composition, the volatile components from any given processing station can be channeled through one or more catalysts using program selector switches.

Biochar

The last step as heat energy transfer is completed to the biomass is the production of biochar (BMF char). The BMF char remains after the previous processing steps, and the physical characteristics of the char will differ depending on the starting biomass material, which can include any of the above-identified cellulosic materials such as wood, grasses, energy crops, and agricultural waste. Different biomass feedstocks are expected to produce different types of BMF chars, varying in porosity and other physical characteristics. The biomass feedstocks can be fed individually or as mixtures of different feedstocks to produce chars containing different physical characteristics. After the BMF char is formed, a last processing step is to transfer the BMF char out of the reaction chamber for a subsequent storage or filling into a bio-char reactor for subsequent optional activation. The transfer may be accomplished via any number of mechanical means, including a press bar outfitted with a scraping knife. The Company has developed a novel process for rendering the BMF char suitable as a soil growth agent. This process comprises several steps including creating the BMF char, expelling detrimental agents within the BMF char, rendering the internal surface area of the BMF char hydrophilic, and adding suitable nutrients and microorganisms to the BMF char to promote plant growth. In addition to acting as a soil enhancing agent, the BMF char can sequester carbon for long periods of time.

Biochar Co-product – Return Carbon to the Soil

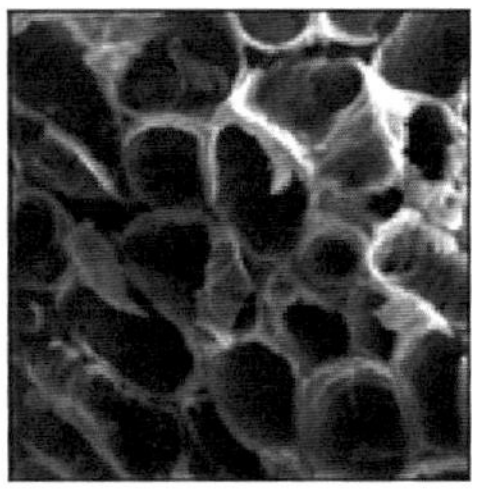

Biochar is a highly porous form of carbon produced by the selective heating of biomass, in the absence of oxygen.

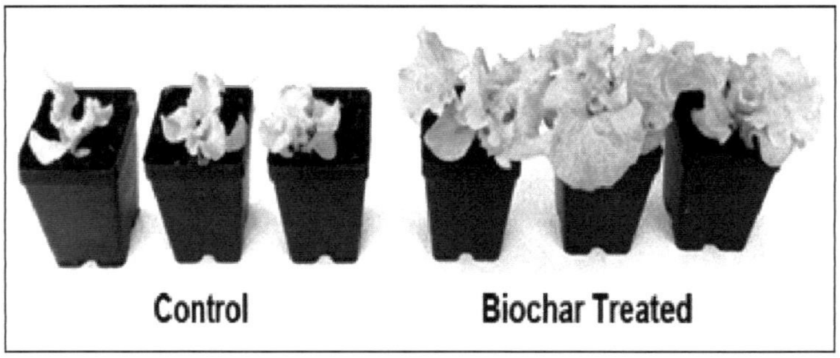

Biochar is very promising as a soil amendment that can retain water and inputs in the soil. Biochar can help maintain a nutrient rich environment and increase productivity of a wide variety of crops.

Drop in Hydrocarbon Fuel Results

Upon completion of catalytic conversion, the devolatilized organic components convert to various liquid fuels such as biodiesel, hydrocarbons, BTX aromatics, jet fuel, light hydrocarbons, gasoline, diesel, methanol, and DME. The organic chemicals can also be useful on their own as co-produced chemicals. By contrast, typical pyrolysis processes do not exhibit a clear volatilization profile such as this. 3rd party testing by independent laboratories shows the compounds consistent with a high octane gasoline as shown in the next graph.

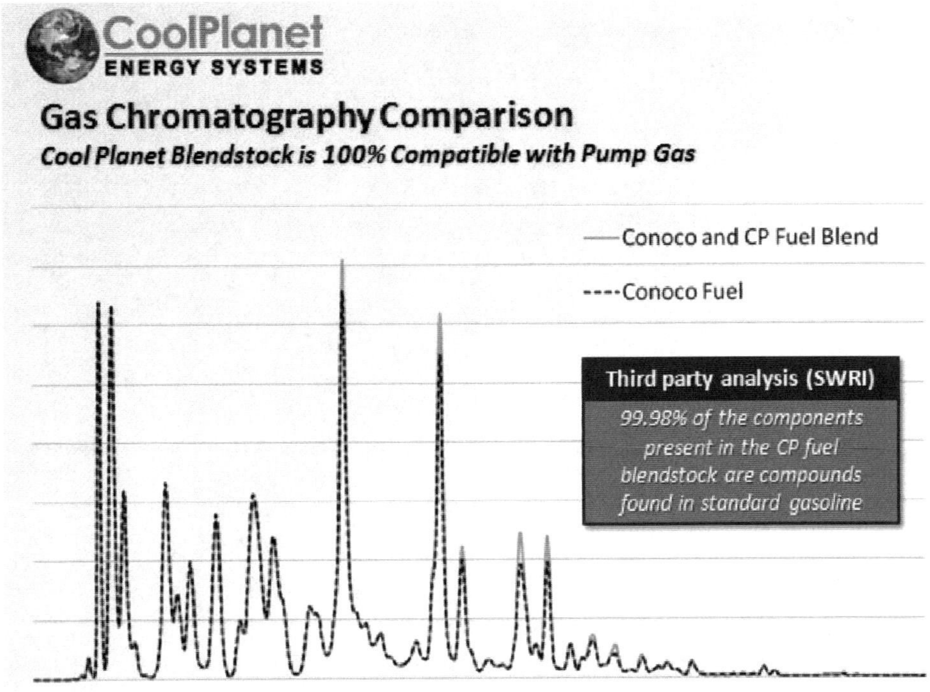

Mass Production of Distributed Units

The Cool Planet refinery equipment is modular, with fractionator and catalyst equipment connected as pairs of processing equipment. This is transported to various biomass locations where the modules are placed within a 30 mile radius of the biomass to be processed. Capex is lower by a factor of 10 compared to existing biomass refineries.

The modular refining units will be manufactured and assembled at a central plant manufacturing facility. Coordinated team will then roll out the modular facilities across the US.

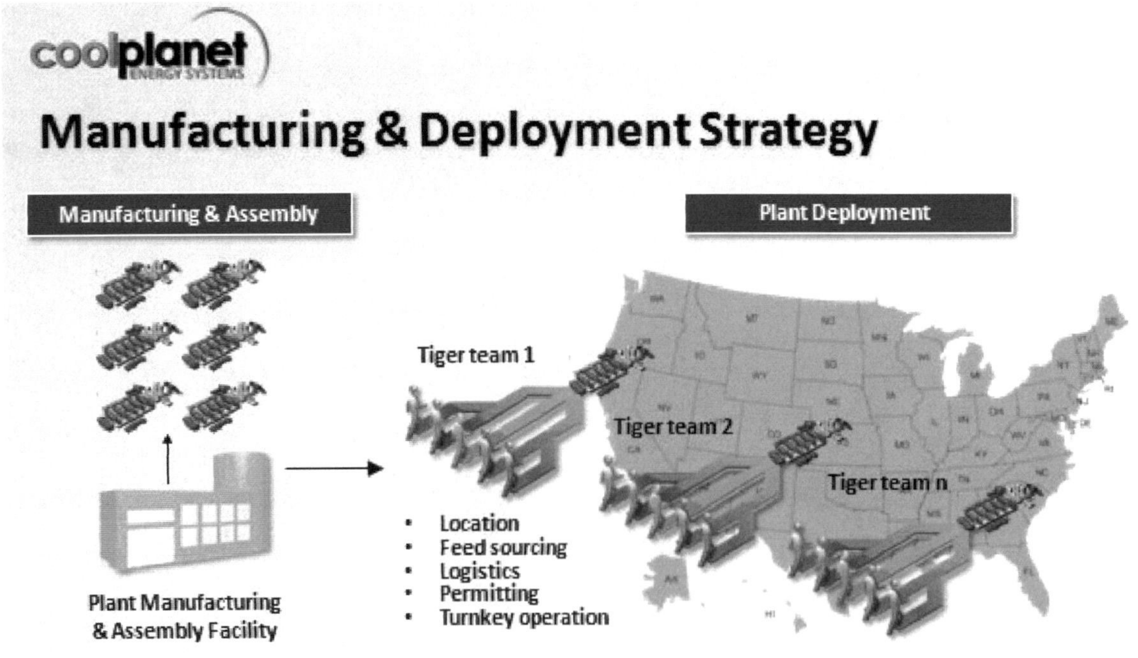

US and Global Roll-out Plan

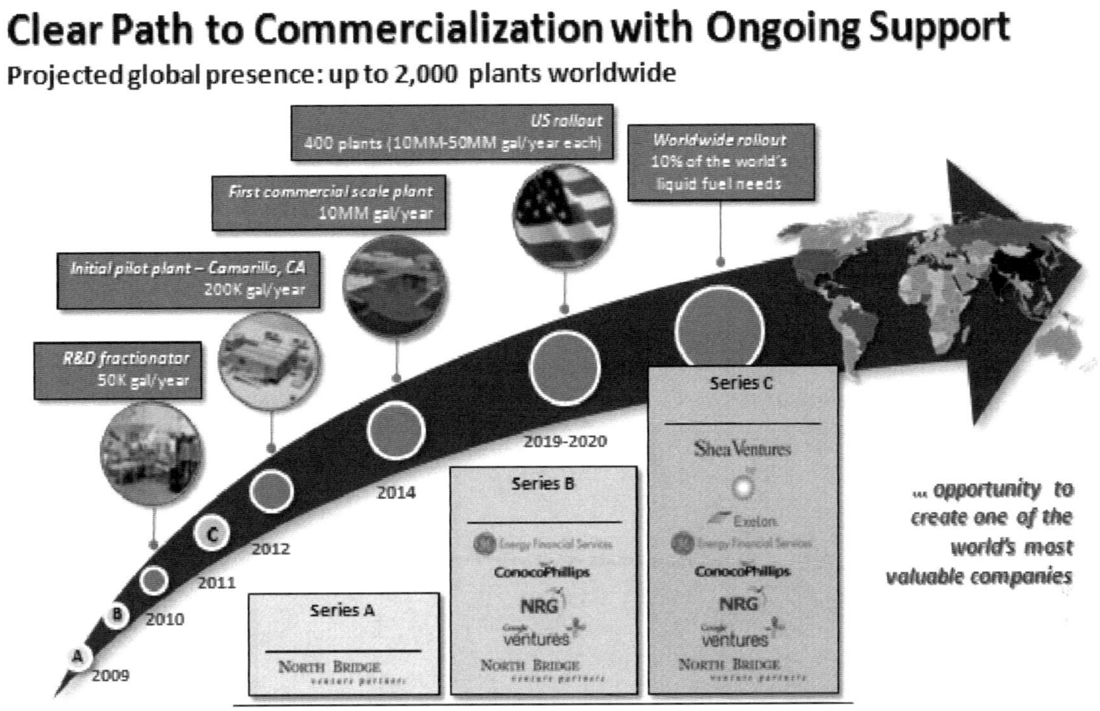

AUTHOR INDEX

AUTHOR INDEX